T0184671

Mathematische Optimierung und Wirtschaftsmathematik | Mathematical Optimization and Economathematics

Reihe herausgegeben von
Ralf Werner, Augsburg, Deutschland
Tobias Harks, Augsburg, Deutschland
Vladimir Shikhman, Chemnitz, Deutschland

In der Reihe werden Arbeiten zu aktuellen Themen der mathematischen Optimierung und der Wirtschaftsmathematik publiziert. Hierbei werden sowohl Themen aus Grundlagen, Theorie und Anwendung der Wirtschafts-, Finanz- und Versicherungsmathematik als auch der Optimierung und des Operations Research behandelt. Die Publikationen sollen insbesondere neue Impulse für weitergehende Forschungsfragen liefern, oder auch zu offenen Problemstellungen aus der Anwendung Lösungsansätze anbieten. Die Reihe leistet damit einen Beitrag der Bündelung der Forschung der Optimierung und der Wirtschaftsmathematik und der sich ergebenden Perspektiven.

Weitere Bände in der Reihe http://www.springer.com/series/15822

Oleg Wilfer

Multi-Composed Programming with Applications to Facility Location

 Springer Spektrum

Oleg Wilfer
Faculty of Mathematics
Chemnitz University of Technology
Chemnitz, Germany

Dissertation, Chemnitz University of Technology, 2017

ISSN 2523-7926   ISSN 2523-7934   (electronic)
Mathematische Optimierung und Wirtschaftsmathematik | Mathematical Optimization and Economathematics
ISBN 978-3-658-30579-6   ISBN 978-3-658-30580-2   (eBook)
https://doi.org/10.1007/978-3-658-30580-2

This Springer Spektrum imprint is published by the registered company Springer Fachmedien Wiesbaden GmbH part of Springer Nature.
The registered company address is: Abraham-Lincoln-Str. 46, 65189 Wiesbaden, Germany

Für meine Eltern

Preface

The goal of this book, based on the doctoral thesis of the author [105], is two-fold. On the one hand, it pursues to provide a contribution to the conjugate duality by proposing a new duality approach for convex multi-composed optimization problems, which can be understood as an umbrella for different meaningful perturbation methods. On the other hand, this book contains investigations on minmax location problems by means of the duality concept introduced in the first part of this work, followed by a numerical approach using epigraphical splitting methods.

After summarizing some elements of the convex analysis as well as introducing key results needed later, we consider an optimization problem with geometric and cone constraints, whose objective function is a composition of $n + 1$ functions. For this problem we propose a conjugate dual problem, where the functions involved in the objective function of the primal problem appear separated. Furthermore, we formulate generalized interior point regularity conditions for strong duality and give necessary and sufficient optimality conditions. As applications of this approach we determine the formulae of the conjugate as well as the biconjugate of the objective function of the primal problem and analyze an optimization problem having as objective function the sum of reciprocals of concave functions.

In the second part of this book we discuss by means of the introduced duality concept four classes of minmax location problems. The first one consists of nonlinear and linear single minmax location problems with geometric constraints, where the maximum of nonlinear or linear functions composed with minimal time functions between pairs of a new and existing points is minimized. The nonlinear location problem is additionally considered with set-up costs. The second class of minmax location problems turns out to be a special case of the first one where the minimal time functions are replaced by gauges.

The third class deals with multifacility location problems as suggested by Drezner (1991), where for each given point the sum of weighted distances to all facilities plus set-up costs is determined and the maximal value of these sums is to be minimized.

As the last and fourth class the classical multifacility location problem with geometrical constraints is considered in a generalized form where the maximum of gauges between pairs of new facilities and the maximum of gauges between pairs of new and existing facilities is minimized.

To each of these location problems associated dual problems will be formulated as well as corresponding duality statements and necessary and sufficient optimality conditions. To illustrate the results of the duality approach and to give a more detailed characterization of the relations between the location problems and their corresponding duals, we give economical as well as geometrical interpretations and consider examples in Euclidean spaces.

This book ends with a numerical approach for solving minmax location problems by epigraphical splitting methods. In this framework, we give formulae for the projections onto the epigraphs of several sums of powers of weighted norms as well as formulae for the projection onto the epigraphs of gauges. Numerical experiments document the usefulness of our approach for the discussed location problems showing that it delivers more exact results in a shorter time and with less costs than other (some quite recent) methods proposed to the same end in the literature.

Oleg Wilfer

Acknowledgements

I am greatly indebted to Professor Dr. Gert Wanka, my doctoral supervisor, for support, patience and assistance on all aspects during my doctoral study. I also want to thank him for countless hints and suggestions for my research.

I would like to express special thanks to Dr. Sorin-Mihai Grad for the many years of unlimited help and support. Thank you, Sorin, for the many helpful advices and valuable discussions concerning this book and beyond.

I am very grateful to Professor Dr. Vladimir Shikhman for the opportunity to transform my PhD thesis into a real book and for his great support. Many thanks, Vladimir, for all the interesting discussions which gave me a new insight into mathematical economics and inspired me for the economic interpretations and applications presented here.

I also wish to express my gratitude to Professor Dr. Radu Ioan Boţ for the opportunity of two fruitful research stays at the Faculty of Mathematics of the University of Vienna as well as the many helpful remarks and comments improving the quality of this book.

My thanks also go to former members of the research group of Professor Dr. Gert Wanka, Robert, Nicole, André, Christopher and Sebastian for the friendly and supportive atmosphere. In this context I also wish to thank my colleagues Thomas and David.

Many thanks go also out to the anonymous reviewers of the articles which build the foundation for this book and their valuable hints and comments and to the referees of my PhD thesis Professor Dr. Andreas Löhne and Professor Dr. Miguel Angel Goberna for hints, remarks and questions that helped me improve the quality and the clarity of the presentation.

Abschließend möchte ich besonders meinen Eltern Olga und Wilfried, meiner Schwester Jeannette sowie meiner Freundin Saba für die Unterstützung, Ermutigung und Liebe danken.

Oleg Wilfer

Contents

List of Symbols and Notations

Spaces and sets

X^*	the topological dual space X^* of X
$\langle x^*, x \rangle$	the value of x^* at x
$w(X^*, X)$	weak* topology on X^* induced by X
$\leqq_K$	the partial ordering induced by the convex cone K
$x \leq_K$	$x \leqq_K y$ and $x \neq y$
0_X	the zero element of X
$+\infty_K$	the greatest element regarding the ordering cone K
$\overline{X}$	the space X to which the element $+\infty_K$ is added
K^*	the dual cone of the cone K
N_S	the normal cone of the set S
$\mathrm{int}(S)$	the interior of the set S
$\mathrm{ri}(S)$	the relative interior of the set S
$\mathrm{cl}(S)$	the closure of the set S
$\mathrm{cone}(S)$	the conic hull of the set S
$\mathrm{core}(S)$	the algebraic interior of the set S
$\mathrm{sqri}(S)$	the strong quasi interior of the set S
$A \times B$	the Cartesian product of two sets
$A + B$	the Minkowski sum of two sets
$|V|$	the cardinality of the index set V
C^0	the polar set of the set C

$\forall$	for all
$\in$	in
$\exists$	there exists (at least one)
$\mathcal{H}$	the Hilbert space $\mathcal{H}$
$\langle \cdot, \cdot \rangle_{\mathcal{H}}$	the inner product in Hilbert space $\mathcal{H}$
$\| \cdot \|_{\mathcal{H}}$	the norm defined by the inner product $\langle \cdot, \cdot \rangle_{\mathcal{H}}$
Argmin f	the set of global minimizers of the function f
$\arg\min_{x \in \mathcal{H}} f(x)$	the unique minimizer of f
P_C	the projection onto the non-empty, closed and convex set C
$\pm\infty$	plus and minus infinite, respectively
$\mathbb{R}$	the set of real numbers
$\overline{\mathbb{R}}$	the extended set of real numbers, $\overline{\mathbb{R}} = \mathbb{R} \cup \{\pm\infty\}$
$\mathbb{R}_+^n$	the non-negative orthant of $\mathbb{R}^n$
$\langle \cdot, \cdot \rangle$	the scalar product in $\mathbb{R}^n$
$\| \cdot \|$	the Euclidean norm in $\mathbb{R}^n$
dom f	the domain of the function f
epi f	the epigraph of the function f
f^*	the conjugate of the function f
f_S^*	the conjugate of the function f regarding the set S
f^{**}	the biconjugate of the function f
∂f	the subdifferential of the function f
δ_A	the indicator function of the set A

Scalar and vector functions

σ_A	the support function of the set A
$\partial f(x)$	the subdifferential of the function f at $x \in X$
$\nabla f(x)$	the gradient of the function f at $x \in X$
$\mathcal{T}^C_{\Omega,f}$	extended perturbed minimal time function
γ_C	the gauge function (a.k.a. Minkowski functional) of the set C
prox_f	the proximity operator of a function f
(z^*F)	the function $\langle z^*, F \rangle$, where F is a vector function and $z^* \in K^*$
$\text{epi}_Q F$	the Q-epigraph of the vector function F
$F \circ G$	the composition of two functions
Id	the identity mapping
0	the zero mapping
$A_{jk},\ B_{ji}$	linear mappings
$v(P^{\cdots})$	the optimal objective value of the optimization problem $(P^{\cdots})$

List of Figures

List of Tables

1 Introduction

The foundation of the conjugate duality theory was laid by Rockafellar in [89] for finite dimensional spaces and latter generalized by Ekeland and Temam in [42] to locally convex spaces. In numerous articles it has been demonstrated that conjugate duality is a powerful instrument to analyze optimization problems and has for that reason a wide range of applications. Over the last 50 years, an important field of applications arises in areas such as vector variational inequalities [1], facility location theory [95], machine learning [16], image restoration [18], portfolio optimization [22] and monotone operator theory [54], to mention only a few of them. In many cases, the objective function of an optimization problem occurring in the mentioned research areas may be written as a composition of two or more functions. This presentation makes not only the derivation of duality assertions easier, but also the handling of optimization problems from the numerical point of view.

But until recently there is no duality approach for the more general situation, namely, where the optimization problem is considered as the minimization of an objective function that is a composition of more than two functions. The advantage of this consideration is that the objective function of a certain optimization problem can be split into a certain number of functions to refine and improve some theoretical and numerical aspects.

Therefore, the goal of this book is to consider an optimization problem with geometric and cone constraints, whose objective function is a composition of finitely many functions and to deliver a detailed duality approach for this type of problems. For short, we call such problems multi-composed optimization problems. In fact, this study is more general than in [14,15,20,23,26,63] and can furthermore be understood as a combination of all kinds of meaningful perturbation methods. To be more precise, we extended the already existing duality schemes to derive a more detailed characterization of the set of optimal solutions and to give a unified framework with a corresponding conjugate dual problem, regularity conditions as well as strong duality statements. As applications we present the formulae of the conjugate and the biconjugate of a multi-composed function, i.e. a function that is a composition of finitely many functions. Moreover, we discuss an optimization problem having as objective function the sum of reciprocals of concave functions.

The results presented in the first part of this book open a new approach to investigating facility location problems. Such kind of optimization problems are known for their numerous applications in areas like computer science, telecommunications, transportation and emergency facilities programming. In the framework of continuous optimization where the distances are measured by gauges, two kinds of location problems are particularly significant. The first one consists of the so-called minsum location problems and has the objective to determine a new point such that the sum of distances between the new and given points is minimal (see [23,26,41,57,68,71,75,78,81]).

© Springer Fachmedien Wiesbaden GmbH, part of Springer Nature 2020
O. Wilfer, *Multi-Composed Programming with Applications to Facility Location*,
Mathematische Optimierung und Wirtschaftsmathematik | Mathematical
Optimization and Economathematics, https://doi.org/10.1007/978-3-658-30580-2_1

The second type contains the so-called minmax location problems, where a new point is sought such that the maximum of distances between the new and given points will be minimized (see [43,45,50,66,70,78,94]). In this book we study more general problems where the distances are replaced by new perturbed minimal time functions and may additionally be composed with a nonlinear function, i.e. we consider besides linear also nonlinear minmax location problems (see [38] and [49]).

Another class of location problems we consider was proposed in 1991 by Drezner in [40] and describes the following emergency scenario. A certain number of emergency calls arise and ask for an ambulance. To each of these demand points an ambulance is sent to load and transport the patient to a hospital. The location of the ambulance-station and the hospital must not be necessary on the same site. This assumption may shorten the response time for the patients, especially for the farthest one, in the situation when for example a hospital is completely overcrowded or short of medical supplies. The aim is now to determine the location of the ambulance-station and the hospital such that the maximum time required before the farthest patient arrives at the hospital will be minimized. In this case the maximum time is naturally defined as the sum of the travel time of the ambulance from the ambulance-station to the patient and the travel time to the hospital plus some set-up costs. Set-up costs like the loading time at the emergency and the unloading time at the hospital of the patient are few examples to cite.

While Drezner suggested a model for the case of the Euclidean norm, Michelot and Plastria [77] work in a higher dimensional space where the distances are measured by a norm. In this book we generalize this location model to the situation where the distances are measured by mixed gauges defined on a Banach space. The goal is then to describe these type of location problems in the framework of the introduced duality concept.

Apart from these classes of location problems, we also consider a more general and complex problem, namely, the so-called multifacility minmax location problem (see [44,77]), which has attracted less attention in the literature compared to the multifacility minsum location problems (see [48,64,69,99]). This may be due to the difficulty in approaching it stressed by the fact that the objective function of the minmax location problem is a composition of a nonlinear outer function defined by a pointwise maximum and an inner vector function given by distances, which are composed with linear operators. In particular, the outer function increases the level of complexity not only from the theoretical but also from the numerical point of view. In other words, the objective of the multifacility minmax location problems is to determine several new points such that either the maximum of distances between pairs of new points or the maximum of distances between new and existing points is minimal.

The last part of this work focuses on solving methods for minmax location problems. In this context we first present formulae for projectors onto the epigraphs of several sums of powers of weighted norms as well as onto the epigraphs of gauges. These formulae allow to combine the epigraphical projection method, developed in [33] for constrained convex optimization problems, with a parallel splitting algorithm (see [3] and [34]) minimizing a finite sum of functions. To show the usefulness of the presented formulae we compare our solving method with the ones presented in [2], [35] and [78] and the computational results prove that our approach delivers optimal solutions to the considered problems faster and cheaper than the other mentioned ones.

Next we give a description of the contents, emphasizing the most important points.

In **Chapter 2** we first collect some elements from the field of convex analysis and present important statements that are used in this book. While in *Section* 2.1 notations and preliminary results are listed on convex sets, *Section* 2.2 is dedicated to convex scalar and vector functions.

After introducing the basics, we consider in **Chapter 3** a multi-composed optimization problem with geometric and cone constraints. We give an equivalent formulation of this problem and use the reformulated optimization problem to construct a corresponding conjugate dual problem to the main problem, followed by a weak duality theorem. One of the advantages of this approach is that the functions involved in the composed objective function of the original problem can be decomposed in the formulation of the conjugate dual problem or, to formulate it more precisely, their conjugates.

Section 3.2 is devoted to generalized interior point regularity conditions guaranteeing strong duality. Moreover, by using the strong duality theorem we formulate some optimality conditions for the original problem and its corresponding conjugate dual problem.

Besides of this approach, we provide in *Section* 3.3 a formula of the conjugate of a multi-composed function. We find also a formula of its biconjugate function and close this section with a theorem which characterizes some topological properties of this function.

In *Section* 3.4, as a further application of our approach, we consider a convex optimization problem having as objective function the sum of reciprocals of concave functions. For this problem we formulate a conjugate dual problem and state a strong duality theorem from which we derive necessary and sufficient optimality conditions. The approach done in this chapter is based on our paper [100].

In **Chapter 4**, which is related to our articles [56, 101, 102] and [104], we analyze four classes of location problems starting with some properties of gauge functions in Section 4.1.1. Moreover, we introduce in *Section* 4.1.2 a new perturbed minimal time function that generalizes the classical minimal time function and state basic properties regarding continuity, convexity and duality.

In *Section* 4.2 we investigate nonlinear single minmax location problems with set-up costs in a Banach space formulated by means of extended perturbed minimal time functions and apply the approach done in Chapter 3 to give necessary and sufficient optimality conditions. Further, in *Subsection* 4.2.1 and *Subsection* 4.3.2 we consider in a more detailed way two linear special cases of the general problem without set-up costs. After presenting associated duality statements, we describe the relation between the optimal solutions of the primal problem and its dual one. In addition, economic interpretations of the primal as well as the dual problem are discussed.

Then, in *Section* 4.3 nonlinear single minmax location problems with gauges are at the
focus of our considerations, which are in fact special cases of the general location problem
considered in Section 4.2 due to the connection between extended perturbed minimal time
functions and gauges. In *Subsection* 4.3.1 we investigate nonlinear single minmax location
problems with set-up costs in a Banach space, while in *Subsection* 4.3.2 the underlying
space is taken to be Hilbert in order to get a deeper insight into the problem. In *Subsection*
4.3.3 as well as in *Subsection* 4.3.4 the location problems will be studied in a Banach space
followed by a characterization to the Euclidean space endowed with a norm. To all of these
location problems necessary and sufficient optimality conditions being delivered together
with geometrical characterizations of the optimal solutions.

In *Section* 4.4 we study extended multifacility location problems introduced by Drezner
in [40]. In *Subsection* 4.4.1 we construct corresponding conjugate dual problems and prove
strong duality from which we derive some optimality conditions. Afterwards, we consider a
special case of these location problems where the weights have a multiplicative structure
like treated by Michelot and Plastria in [77] and describe the relation to their conjugate
dual problems with norms as distance measures. In *Subsection* 4.4.2, we also deal with
location problems without set-up costs. Besides of strong duality assertions and optimality
conditions we give geometrical characterizations of the set of optimal solutions of the
conjugate dual problem as well as illustrating examples.

The analysis of classic multifacility minmax location problems in *Section* 4.5 provides duality statements in the sense of Chapter 3. In concrete terms, this means that we formulate
an associated conjugate dual problem as well as derive necessary and sufficient optimality
conditions in *Subsection* 4.5.1. Further, we introduce another dual problem reducing the
number of dual variables compared to the first formulated dual problem. Continuing in
this vein, we also present statements of strong duality and optimality conditions. As the
most location problems are considered in Euclidean spaces, we particularize in *Subsection*
4.5.2 the latter case in this context and show that we have a full symmetry between the
location problem, its dual problem and the Lagrange dual problem of the dual problem,
which means that the Lagrange dual is identical to the location problem. Finally, we close
this section with an example showing on the one hand how an optimal solution of the
location problem can be recovered from an optimal solution of the associated conjugate dual
problem and on the other hand how we can geometrically interpret an optimal dual solution.

Along with a theoretical consideration, we are interested in **Chapter 5** in a numerical method for solving minmax location problems. For this purpose, we present, after a
brief motivation in *Section* 5.1, formulae of projections onto the epigraphs of several sums
of powers of weighted norms and onto the epigraphs of gauges in *Section* 5.2. Among some
special cases we present also formulae of the projection onto the epigraph of the sum of of
weighted norms as well as for the projection onto a unit ball generated by the weighted sum
of norms and deliver an exact formula of the projection operator onto the epigraph of the
maximum of norms, that are especially useful for the numerical experiments in the last chapter of this book and also for other applications, too. This chapter is based on our paper [103].

Within the **Chapter 6** of this book, we consider in *Section* 6.1 the extended multifacility minmax location problem from Section 4.4 and bring it first into a form of an unconstrained optimization problem where its objective function is a sum of functions. This reformulation allows us then to use the parallel splitting algorithm (see [3, 33, 34]) combined with the formulae from Section 5.2 to solve minmax location problems. In addition, we solve the numerical examples by the method proposed by Cornejo and Michelot in [35], where the sum of powers of weighted norms is split such that the formulae of the projectors onto the epigraphs of the powers of weighted norms are relevant. This splitting scheme makes it necessary to introduce additional variables, which in turn goes at the expense of the numerical performance. It is shown that the parallel splitting algorithm combined with the presented projection formulae that we propose performs very well on these kind of location problems and outperforms the method given in [35].

In *Section* 6.2 we employed a splitting proximal point method to solve minmax location problems with perturbed minimal time functions studied in Subsection 4.2.1 and Subsection 4.3.2 by using the same ideas as in the previous section. More precisely, we present the computational results obtained in MATLAB on concrete examples, successfully comparing these, where possible, with recent similar methods from the literature (see [2, 78, 82]). Moreover, the dual employment of the proximal method turns out to deliver the optimal solution to the considered primal problem faster than the direct usage on the latter. Since our technique successfully solves location optimization problems with large data sets in high dimensions, we envision its future usage on big data problems arising in machine learning. This chapter is based on the articles [56, 103].

2 Notations and preliminary results

This chapter serves as an introduction and aims to make this book as self-contained as possible. We introduce here basic notions from the convex analysis and give important statements on convex sets, convex scalar and vector functions. For readers interested in convex analysis we refer to $[14, 29, 42, 65, 89, 106, 107]$.

2.1 Convex sets

Let X be a Hausdorff locally convex space and X^* its topological dual space endowed with the weak* topology $w(X^*, X)$. For $x \in X$ and $x^* \in X^*$, let $\langle x^*, x \rangle := x^*(x)$ be the value of the linear continuous functional x^* at x.

A set $K \subseteq X$ is called *convex* if it holds $\lambda x + (1 - \lambda)y \in K$ for all $x, y \in K$ and $\lambda \in [0, 1]$ and if K additionally satisfies the condition $\lambda K \subseteq K$ for all $\lambda \geq 0$, then K is said to be a *convex cone* .

Given a set $S \subseteq X$ and $x \in X$, then the *normal cone* to S at x, defined by

$$N_S(\overline{x}) := \{x^* \in X^* : \langle x^*, y - \overline{x} \rangle \leq 0 \ \forall \ y \in S\},$$

is a convex cone.

Consider a convex cone $K \subseteq X$, which induces on X a *partial ordering relation* "$\leqq_K$", defined by $\leqq_K := \{(x, y) \in X \times X : y - x \in K\}$, i.e. for $x, y \in X$ it holds $x \leqq_K y \Leftrightarrow y - x \in K$. Note that we assume that all cones we consider contain the origin, which we denote by 0_X. Further, we attach to X a greatest element with respect to "$\leqq_K$", denoted by $+\infty_K$, which does not belong to X and denote $\overline{X} = X \cup \{+\infty_K\}$. Then it holds $x \leqq_K +\infty_K$ for all $x \in \overline{X}$. We write $x <_K y$ if and only if $x \leqq_K y$ and $x \neq y$. Further, we write $\leqq_{\mathbb{R}_+} =: \leq$ and $\leqq_{\mathbb{R}_+} =: <$.

On $\overline{X}$ we consider the following operations and conventions: $x + (+\infty_K) = (+\infty_K) + x := +\infty_K$ for all $x \in X \cup \{+\infty_K\}$ and $\lambda \cdot (+\infty_K) := +\infty_K$ for all $\lambda \in [0, +\infty]$. Further, $K^* := \{x^* \in X^* : \langle x^*, x \rangle \geq 0 \text{ for all } x \in K\}$ is the *dual cone* of K and we take by convention $\langle x^*, +\infty_K \rangle := +\infty$ for all $x^* \in K^*$. By a slight abuse of notation we denote the extended real space $\overline{\mathbb{R}} = \mathbb{R} \cup \{\pm\infty\}$ and consider on it the following operations and conventions: $\lambda + (+\infty) = (+\infty) + \lambda := +\infty$ for all $\lambda \in [-\infty, +\infty]$, $\lambda + (-\infty) = (-\infty) + \lambda := -\infty$ for all $\lambda \in [-\infty, +\infty)$, $\lambda \cdot (+\infty) := +\infty$ for all $\lambda \in [0, +\infty]$, $\lambda \cdot (+\infty) := -\infty$ for all $\lambda \in [-\infty, 0)$, $\lambda \cdot (-\infty) := -\infty$ for all $\lambda \in (0, +\infty]$, $\lambda \cdot (-\infty) := +\infty$ for all $\lambda \in [-\infty, 0)$ and $0(-\infty) := 0$.

For a set $S \subseteq X$ the *conic hull* is defined by $\text{cone}(S) := \{\lambda x : x \in S, \ \lambda \geq 0\}$. Further, the prefix int we use to denote the *interior* of a set $S \subseteq X$, while the prefixes cl, ri, core

© Springer Fachmedien Wiesbaden GmbH, part of Springer Nature 2020
O. Wilfer, *Multi-Composed Programming with Applications to Facility Location*,
Mathematische Optimierung und Wirtschaftsmathematik I Mathematical
Optimization and Economathematics, https://doi.org/10.1007/978-3-658-30580-2_2

and sqri are used to denote the *closure, relative interior, algebraic interior* and the *strong quasi relative interior*, respectively, where in the case of having a convex set $S \subseteq X$ it holds (see [36])

$$\text{core}(S) = \{x \in S : \text{cone}(S - x) = X\},$$
$$\text{sqri}(S) = \{x \in S : \text{cone}(S - x) \text{ is a closed linear subspace}\}.$$

Note that if $\text{cone}(S - x)$ is a linear subspace, then $x \in S$.

The next statement was given in [8] for the *quasi relative interior*, $\text{qri}(S) = \{x \in S : \text{cl}(\text{cone}(S - x)) \text{ is a linear subspace}\}$, we show the validity for the strong quasi relative interior.

Lemma 2.1. *Let $A \subseteq X$ and $B \subseteq Z$ be non-empty convex subsets. Then, it holds*

$$0_{X \times Z} \in \text{sqri}(A \times B) \Leftrightarrow 0_X \in \text{sqri}(A) \text{ and } 0_Z \in \text{sqri}(B).$$

Proof. First, let us recall that if A and B are convex and $0_X \in A$ and $0_Z \in B$, then

$$\text{cone}(A \times B) = \text{cone}(A) \times \text{cone}(B).$$

Now, let us assume that $0_{X \times Z} \in \text{sqri}(A \times B)$, then $\text{cone}(A \times B)$ is a closed linear subspace of $X \times Z$, which implies that $0_{X \times Z} = (0_X, 0_Z) \in A \times B$. But this means that $\text{cone}(A \times B) = \text{cone}(A) \times \text{cone}(B)$ and hence, $\text{cone}(A)$ and $\text{cone}(B)$ are closed linear subspaces, i.e. $0_X \in \text{sqri}(A)$ and $0_Z \in \text{sqri}(B)$.

On the other hand, let $0_X \in \text{sqri}(A)$ and $0_Z \in \text{sqri}(B)$, then $\text{cone}(A)$ and $\text{cone}(B)$ are closed linear subspaces and so, $0_X \in A$ and $0_Z \in B$. From here follows that $\text{cone}(A \times B) = \text{cone}(A) \times \text{cone}(B)$ and thus, $\text{cone}(A \times B)$ is a closed linear subspace, i.e. $0_{X \times Z} \in \text{sqri}(A \times B)$. $\qquad \square$

2.2 Convex functions

2.2.1 Scalar functions

For a given function $f : X \to \overline{\mathbb{R}}$ we consider its *effective domain* $\text{dom} f := \{x \in X : f(x) < +\infty\}$ and call f *proper* if $\text{dom} f \neq \emptyset$ and $f(x) > -\infty$ for all $x \in X$. The *epigraph* of f is $\text{epi} f = \{(x, r) \in X \times \mathbb{R} : f(x) \leq r\}$. Recall that a function $f : X \to \overline{\mathbb{R}}$ is called *convex* if $f(\lambda x + (1 - \lambda)y) \leq \lambda f(x) + (1 - \lambda)f(y)$ for all $x, y \in X$ and all $\lambda \in [0, 1]$. For a subset $A \subseteq X$, its *indicator function* $\delta_A : X \to \overline{\mathbb{R}}$ is

$$\delta_A(x) := \begin{cases} 0, & \text{if } x \in A, \\ +\infty, & \text{otherwise}, \end{cases} \tag{2.1}$$

and its *support function* $\sigma_A : X^* \to \overline{\mathbb{R}}$ is $\sigma_A(x^*) = \sup_{x \in A} \langle x^*, x \rangle$.

The *conjugate function* of f with respect to the non-empty subset $S \subseteq X$ is defined by

$$f_S^* : X^* \to \overline{\mathbb{R}}, \ f_S^*(x^*) = \sup_{x \in S}\{\langle x^*, x \rangle - f(x)\}. \tag{2.2}$$

In the case $S = X$, f_S^* turns into the classical *Fenchel-Moreau conjugate function* of f denoted by f^*.

A function $f : X \to \overline{\mathbb{R}}$ is called *lower semicontinuous at* $\overline{x} \in X$ if $\liminf_{x \to \overline{x}} f(x) \geq f(\overline{x})$ and when this function is lower semicontinuous at all $x \in X$, then we call it *lower semicontinuous* (l.s.c. for short).

Let $W \subseteq X$ be a non-empty set, then a function $f : X \to \overline{\mathbb{R}}$ is called *K-increasing on* W, if from $x \leq_K y$ follows $f(x) \leq f(y)$ for all $x, y \in W$. When $W = X$, then we call the function f *K-increasing*.

We also use the notion of *subdifferentiability* to formulate optimality conditions. If we take an arbitrary $x \in X$ such that $f(x) \in \mathbb{R}$, then we call the set

$$\partial f(x) := \{x^* \in X^* : f(y) - f(x) \geq \langle x^*, y - x \rangle \ \forall y \in X\} \tag{2.3}$$

the (convex) *subdifferential* of f at x, where the elements are called the *subgradients* of f at x. Moreover, if $\partial f(x) \neq \emptyset$, then we say that f is *subdifferentiable at* x and if $f(x) \notin \mathbb{R}$, then we make the convention that $\partial f(x) := \emptyset$. Note, that the subgradients can be characterized by the conjugate function, especially this means

$$x^* \in \partial f(x) \Leftrightarrow f(x) + f^*(x^*) = \langle x^*, x \rangle \ \forall x \in X, \ x^* \in X^*, \tag{2.4}$$

i.e. the *Young-Fenchel inequality* is fulfilled with equality.

Furthermore, let $\mathcal{H}$ be a real Hilbert space equipped with the inner product $\langle \cdot, \cdot \rangle_{\mathcal{H}}$, where the associated norm $\| \cdot \|_{\mathcal{H}}$ is defined by $\|y\|_{\mathcal{H}} := \sqrt{\langle y, y \rangle_{\mathcal{H}}}$ for all $y \in \mathcal{H}$. If $\mathcal{H} = \mathbb{R}^m$, then $\| \cdot \|_{\mathbb{R}^m}$ is the Euclidean norm associated to the Euclidean inner product on $\mathbb{R}^m$ and we will write for simplicity just $\| \cdot \|$. Some important results in Hilbert spaces can be found in Chapter 5.

2.2.2 Vector functions

Let Z be another Hausdorff locally convex space partially ordered by the convex cone $Q \subseteq Z$ and Z^* its topological dual space endowed with the weak* topology $w(Z^*, Z)$. The *domain* of a vector function $F : X \to \overline{Z} = Z \cup \{+\infty_Q\}$ is dom $F := \{x \in X : F(x) \neq +\infty_Q\}$. F is called *proper* if dom $F \neq \emptyset$.

When $F(\lambda x + (1 - \lambda)y) \leq_Q \lambda F(x) + (1 - \lambda)F(y)$ holds for all $x, y \in X$ and all $\lambda \in [0, 1]$ the function F is said to be *Q-convex*.

The *Q-epigraph* of a vector function F is $\text{epi}_Q F = \{(x, z) \in X \times Z : F(x) \leq_Q z\}$ and when Q is closed we say that F is *Q-epi closed* if $\text{epi}_Q F$ is a closed set.

For a $z^* \in Q^*$ we define the function $(z^* F) : X \to \overline{\mathbb{R}}$ by $(z^* F)(x) := \langle z^*, F(x) \rangle$. Then $\text{dom}(z^* F) = \text{dom } F$. Moreover, it is easy to see that if F is *Q-convex*, then $(z^* F)$ is convex for all $z^* \in Q^*$.

Let us point out that by the operations we defined on a Hausdorff locally convex space attached with a maximal element and on the extended real space, there holds $0f = \delta_{\mathrm{dom}\, f}$ and $(0_{Z^*}F) = \delta_{\mathrm{dom}\, F}$.

The vector function F is called *positively Q-lower semicontinuous at* $x \in X$ if (z^*F) is lower semicontinuous at x for all $z^* \in Q^* \setminus \{0_{X^*}\}$. The function F is called *positively Q-lower semicontinuous* if it is positively Q-lower semicontinuous at every $x \in X$. Note that if F is positively Q-lower semicontinuous, then it is also Q-epi closed, while the inverse statement is not true in general (see: [14, Proposition 2.2.19]). Let us mention that in the case $Z = \mathbb{R}$ and $Q = \mathbb{R}_+$, the notions of Q-epi closedness and positively Q-lower semicontinuity fall into the classical notion of lower semicontinuity.

$F : X \to \overline{Z}$ is called (K, Q)-*increasing on* W, if from $x \leqq_K y$ follows $F(x) \leqq_Q F(y)$ for all $x, y \in W$. When $W = X$, we call this function (K, Q)-*increasing*.

We give now some statements that will be useful later, beginning with one whose proof is straightforward.

Lemma 2.2. *Let V be a Hausdorff locally convex space partially ordered by the convex cone U, $F : X \to \overline{Z}$ be a proper and Q-convex function and $G : Z \to \overline{V}$ be an U-convex and (Q, U)-increasing function on $F(\mathrm{dom}\, F) \subseteq \mathrm{dom}\, G$ with the convention $G(+\infty_Q) = +\infty_U$. Then the function $(G \circ F) : X \to \overline{V}$ is U-convex.*

Lemma 2.3. *(cf. [55]) Let Y be a Hausdorff locally convex space, Q also closed, $h :$ $X \times Y \to \overline{Z}$ and $F : X \to \overline{Z}$ proper vector functions and $G : Y \to Z$ a continuous vector functions, where h is defined by $h(x, y) := F(x) + G(y)$. Then F is Q-epi closed if and only if h is Q-epi closed.*

Proof. "$\Rightarrow$": Let $(x_\alpha, y_\alpha, z_\alpha)_\alpha \subseteq \mathrm{epi}_Q\, h$ be a net such that $(x_\alpha, y_\alpha, z_\alpha) \to (\overline{x}, \overline{y}, \overline{z})$. Then $F(x_\alpha) + G(y_\alpha) \leq z_\alpha$ for any α, followed by $(x_\alpha, z_\alpha - G(y_\alpha))_\alpha \subseteq \mathrm{epi}_Q\, F$ and $(y_\alpha, G(y_\alpha))_\alpha \subseteq \mathrm{epi}_Q\, G$. Because G is continuous and $y_\alpha \to \overline{y}$, it follows that $G(y_\alpha) \to G(\overline{y})$. Then $(x_\alpha, z_\alpha - G(y_\alpha)) \to (\overline{x}, \overline{z} - G(\overline{y})) \in \mathrm{epi}_Q\, F$, because this set is closed. One has then $F(\overline{x}) \leqq_Q \overline{z} - G(\overline{y})$, i.e. $(\overline{x}, \overline{y}, \overline{z}) \in \mathrm{epi}_Q\, h$. As the convergent nets $(x_\alpha)_\alpha$, $(y_\alpha)_\alpha$ and $(z_\alpha)_\alpha$ were arbitrarily chosen, it follows that $\mathrm{epi}_Q\, h$ is closed, i.e. h is Q-epi closed.

"$\Leftarrow$": Let $(x_\alpha, z_\alpha)_\alpha \subseteq \mathrm{epi}_Q\, F$ such that $(x_\alpha, z_\alpha) \to (\overline{x}, \overline{z})$. Take also $(y_\alpha)_\alpha \subseteq Y$ such that $y_\alpha \to \overline{y}$. Because G is continuous, one has $G(y_\alpha) \to G(\overline{y})$. Then $(x_\alpha, y_\alpha, z_\alpha + G(y_\alpha))_\alpha \subseteq \mathrm{epi}_Q\, h$, which is closed, consequently $(\overline{x}, \overline{y}, \overline{z} + G(\overline{y})) \in \mathrm{epi}_Q\, h$, i.e. $F(\overline{x}) + G(\overline{y}) \leqq_Q \overline{z} + G(\overline{y})$. Therefore $F(\overline{x}) \leqq_Q \overline{z}$, i.e. $(\overline{x}, \overline{z}) \in \mathrm{epi}_Q\, F$. As the convergent nets $(x_\alpha)_\alpha$ and $(z_\alpha)_\alpha$ were arbitrarily chosen, it follows that $\mathrm{epi}_Q\, F$ is closed, i.e. F is Q-epi closed. $\square$

Remark 2.1. *Note that a continuous proper vector function $G : Y \to \overline{Z}$, where Y is a Hausdorff locally convex space, has a full domain, thus one can directly take $G : Y \to Z$ in this situation. The question whether the equivalence in Lemma 2.3 remains valid if one considers a proper vector function $G : Y \to \overline{Z}$ that is not necessarily continuous is still open.*

Remark 2.2. *If we set $Y = Z$ and $G(y) = -y$ for all $y \in Y$, then Lemma 2.3 says that F is Q-epi closed if and only if the vector function defined by $(x, y) \in X \times Y \mapsto F(x) - y$ is*

Q-epi closed. For this special case a similar statement can be found in [100, Lemma 2.1], but under the additional hypothesis int $Q \neq \emptyset$.

3 Lagrange duality for multi-composed optimization problems

The goal of this chapter is to consider an optimization problem with geometric and cone constraints, whose objective function is a composition of $n + 1$ functions and to deliver a full duality approach for this type of problems.

By considering such a multi-composed optimization problem there are several ways to formulate a corresponding conjugate dual problem where the composed functions involved in the objective function of the primal problem, or, to be more precise, their conjugates, are separated and to give associated duality statements.

The first method is the direct applying of the perturbation theory (see [14,42,59,60,73,89]). A second approach is presented in this chapter and starts in Section 3.1 by reformulating the primal problem as an optimization problem with set and cone constraints and continues by using the Lagrange duality concept. The question is now, which of these two methods is more suitable? It is shown in Section 3.2 that the second method asks for weaker hypotheses on the involved functions of the primal problem for guaranteeing strong duality. As applications, we present the formulae of the conjugate and the biconjugate of a multi-composed function in Section 3.3, i.e. a function that is a composition of $n + 1$ functions. Moreover, we discuss in the Section 3.4 an optimization problem having as objective function the sum of reciprocals of concave functions. This chapter is based on the paper [100].

3.1 The multi-composed optimization problem and its conjugate dual

As already mentioned, our aim is to formulate a conjugate dual problem to an optimization problem with geometric and cone constraints having as objective function the composition of $n + 1$ functions. In other words, we consider the following problem

$$(P^C) \quad \inf_{x \in \mathcal{A}} (f \circ F^1 \circ ... \circ F^n)(x),$$
$$\mathcal{A} = \{x \in S : g(x) \in -Q\},$$

where Z is a Hausdorff locally convex space partially ordered by the convex cone $Q \subseteq Z$ and X_i is a Hausdorff locally convex space partially ordered by the convex cone $K_i \subseteq X_i$, $i = 0, ..., n - 1$. Moreover,

- S is a non-empty subset of the Hausdorff locally convex space X_n,

© Springer Fachmedien Wiesbaden GmbH, part of Springer Nature 2020
O. Wilfer, *Multi-Composed Programming with Applications to Facility Location*,
Mathematische Optimierung und Wirtschaftsmathematik | Mathematical
Optimization and Economathematics, https://doi.org/10.1007/978-3-658-30580-2_3

- $f : X_0 \to \overline{\mathbb{R}}$ is proper and K_0-increasing on $F^1(\mathrm{dom}\, F^1) + K_0 \subseteq \mathrm{dom}\, f$,

- $F^i : X_i \to \overline{X}_{i-1} = X_{i-1} \cup \{+\infty_{K_{i-1}}\}$ is proper and (K_i, K_{i-1})-increasing on $F^{i+1}(\mathrm{dom}\, F^{i+1}) + K_i \subseteq \mathrm{dom}\, F^i$ for $i = 1, ..., n-2$,

- $F^{n-1} : X_{n-1} \to \overline{X}_{n-2} = X_{n-2} \cup \{+\infty_{K_{n-2}}\}$ is proper and (K_{n-1}, K_{n-2})-increasing on $F^n(\mathrm{dom}\, F^n \cap \mathcal{A}) + K_{n-1} \subseteq \mathrm{dom}\, F^{n-1}$,

- $F^n : X_n \to \overline{X}_{n-1} = X_{n-1} \cup \{+\infty_{K_{n-1}}\}$ is a proper function and

- $g : X_n \to \overline{Z}$ is a proper function fulfilling

$$S \cap g^{-1}(-Q) \cap ((F^n)^{-1} \circ ... \circ (F^1)^{-1})(\mathrm{dom}\, f) \neq \emptyset. \tag{3.1}$$

Additionally, we make the convention that $f(+\infty_{K_0}) = +\infty$ and $F^i(+\infty_{K_i}) = +\infty_{K_{i-1}}$, i.e. $f : \overline{X}_0 \to \overline{\mathbb{R}}$ and $F^i : \overline{X}_i \to \overline{X}_{i-1}$, $i = 1, ..., n-1$.

Remark 3.1. *For the rest of this paper it is preferable to make the following arrangement. In the situation when $n = 1$ we set $\{1, ..., n-1\} = \{1, ..., n-2\} = \emptyset$ and when $n = 2$, $\{1, ..., n-2\} = \emptyset$. In particular, this means for the case $n = 1$ that $F^1 : X_1 \to \overline{X}_0$ is a proper function and for the case $n = 2$ that $F^1 : \overline{X}_1 \to \overline{X}_0$ is a proper and (K_1, K_0)-increasing function on $F^2(\mathrm{dom}\, F^2 \cap \mathcal{A}) + K_1 \subseteq \mathrm{dom}\, F^1$ and $F^2 : X_2 \to \overline{X}_1$ a proper function.*

Let us now consider the following problem

$$(\widetilde{P}^C) \qquad \inf_{(y^0, ..., y^n) \in \widetilde{\mathcal{A}}} \widetilde{f}(y^0, ..., y^n),$$

where

$$\widetilde{\mathcal{A}} = \{(y^0, ..., y^{n-1}, y^n) \in X_0 \times ... \times X_{n-1} \times S :$$
$$g(y^n) \in -Q,\ h^i(y^i, y^{i-1}) \in -K_{i-1},\ i = 1, ..., n\}.$$

The functions $\widetilde{f} : X_0 \times ... \times X_n \to \overline{\mathbb{R}}$ and $h^i : X_i \times X_{i-1} \to \overline{X}_{i-1}$ are defined as

$$\widetilde{f}(y^0, ..., y^n) = f(y^0) \text{ and } h^i(y^i, y^{i-1}) = F^i(y^i) - y^{i-1} \text{ for } i = 1, ..., n.$$

Lemma 3.1. *Let $(y^0, ..., y^n)$ be feasible to $(\widetilde{P}^C)$, then it holds $f((F^1 \circ ... \circ F^n)(y^n)) \leq f(y^0)$.*

Proof. Let $(y^0, ..., y^n)$ be feasible to $(\widetilde{P}^C)$, then we have

$$F^n(y^n) \leq_{K_{n-1}} y^{n-1}, ..., F^1(y^1) \leq_{K_0} y^0.$$

Moreover, since F^{n-1} is (K_{n-1}, K_{n-2})-increasing on $F^n(\mathrm{dom}\, F^n \cap \mathcal{A}) + K_{n-1}$ and F^i is (K_i, K_{i-1})-increasing on $F^{i+1}(\mathrm{dom}\, F^{i+1}) + K_i$ for $i = 1, ..., n-2$, it follows $(F^{n-1} \circ F^n)(y^n) \leq_{K_{n-2}} F^{n-1}(y^{n-1}) \leq_{K_{n-2}} y^{n-2}$ and so on $(F^1 \circ ... \circ F^n)(y^n) \leq_{K_0} F^1(y^1) \leq_{K_0} y^0$. Since f is K_0-increasing on $F^1(\mathrm{dom}\, F^1) + K_0$ we get the desired inequality $f((F^1 \circ ... \circ F^n)(y^n)) \leq f(y^0)$. $\qquad\square$

Remark 3.2. *If F^n is an affine function, then it can be useful to set $K_{n-1} = \{0_{X_{n-1}}\}$, because in this case F^{n-1} does not need to be monotone to ensure the inequality of the previous lemma.*

If we denote by $v(P^C)$ and $v(\widetilde{P}^C)$ the optimal objective values of the problems (P^C) and $(\widetilde{P}^C)$, respectively, then the following relation between the optimal objective values is always true.

Theorem 3.1. *It holds $v(P^C) = v(\widetilde{P}^C)$.*

Proof. Let x be a feasible element to (P^C) and set $y^n = x$, $y^{n-1} = F^n(y^n)$, $y^{n-2} = F^{n-1}(y^{n-1})$, ..., $y^0 = F^1(y^1)$. If there exists an $i \in \{2, ..., n\}$ such that $F^i(y^i) \notin \mathrm{dom}\, F^{i-1}$ or $F^1(y^1) \notin \mathrm{dom}\, f$ or there exists an $i \in \{1, ..., n\}$ such that $F^i(y^i) = +\infty_{K_{i-1}}$, then it obviously holds $f((F^1 \circ ... \circ F^n)(y^n)) = +\infty \geq v(\widetilde{P}^C)$. Otherwise it holds $F^i(y^i) - y^{i-1} = 0 \in -K_{i-1}$ for $i = 1, ..., n$. Moreover, by the feasibility of y^n it holds $g(y^n) \in -Q$, which implies the feasibility of $(y^0, ..., y^n)$ to the problem $(\widetilde{P}^C)$ and $f((F^1 \circ ... \circ F^n)(y^n)) = f(y^0) = \widetilde{f}(y^0, ..., y^n) \geq v(\widetilde{P}^C)$. Hence it holds $f((F^1 \circ ... \circ F^n)(y^n)) \geq v(\widetilde{P}^C)$ for all y^n feasible to (P^C), which means that $v(P^C) \geq v(\widetilde{P}^C)$.

Let now $(y^0, ..., y^n)$ be feasible to $(\widetilde{P}^C)$. If $y^0 \notin \mathrm{dom}\, f$, then obviously we have $v(P^C) \leq f((F^1 \circ ... \circ F^n)(y^n)) \leq f(y^0) = \widetilde{f}(y^0, ..., y^n) = +\infty$. On the other hand, since $(y^0, ..., y^n)$ is feasible to $(\widetilde{P}^C)$ it holds $h^i(y^i, y^{i-1}) \in -K_{i-1}$ for $i = 1, ..., n$ (i.e. $F^i(y^i) - y^{i-1} \in -K_{i-1}$ for $i = 1, ..., n$) and $g(y^n) \in -Q$. By Lemma 3.1 we have $v(P^C) \leq f((F^1 \circ ... \circ F^n)(y^n)) \leq f(y^0) = \widetilde{f}(y^0, ..., y^n)$ and by taking the infimum over $(y^0, ..., y^n)$ on the right-hand side we get $v(P^C) \leq v(\widetilde{P}^C)$.

Summarizing, we get the desired result $v(P^C) = v(\widetilde{P}^C)$. $\qquad\square$

Remark 3.3. *The assumption that f is K_0-increasing on $F^1(\mathrm{dom}\, F^1) + K_0 \subseteq \mathrm{dom}\, f$ was made to allow functions which are not necessarily monotone on their whole effective domain. But in some situations the inclusion $F^1(\mathrm{dom}\, F^1) + K_0 \subseteq \mathrm{dom}\, f$ may not be fulfilled. As an example consider the convex optimization problem (P^G) in Section 3.4.*

To overcome these circumstances one can alternatively assume that f is K_0-increasing on $\mathrm{dom}\, f$ and $F^1(\mathrm{dom}\, F^1) \subseteq \mathrm{dom}\, f$. For the functions $F^1, ..., F^{n-1}$ one can formulate in the same way alternative assumptions. To be more precise, we can alternatively ask that F^i is (K_i, K_{i-1})-increasing on $\mathrm{dom}\, F^i$ and $F^{i+1}(\mathrm{dom}\, F^{i+1}) \subseteq \mathrm{dom}\, F^i$, $i = 1, ..., n - 2$, and F^{n-1} is (K_{n-1}, K_{n-2})-increasing on $\mathrm{dom}\, F^{n-1}$ and $F^n(\mathrm{dom}\, F^n \cap \mathcal{A}) \subseteq \mathrm{dom}\, F^{n-1}$. One can observe that under these alternative assumptions Lemma 3.1 and especially Theorem 3.1 still hold.

As we have seen by Theorem 3.1, the problem (P^C) can be associated to the problem $(\widetilde{P}^C)$. In the next step we want to determine the corresponding conjugate dual problems to the problems (P^C) and $(\widetilde{P}^C)$.

As we take a careful look at the optimization problem $(\widetilde{P}^C)$, we can see that this problem can be rewritten in the form

$$(\widetilde{P}^C) \quad \inf_{\substack{\widetilde{y} \in \widetilde{S}, \\ \widetilde{h}(\widetilde{y}) \in -\widetilde{K}}} \widetilde{f}(\widetilde{y}), \tag{3.2}$$

where $\widetilde{y} := (y^0, ..., y^n) \in \widetilde{X} := X_0 \times ... \times X_n$, $\widetilde{Z} := X_0 \times ... \times X_{n-1} \times Z$ ordered by $\widetilde{K} := K_0 \times ... \times K_{n-1} \times Q$, $\widetilde{S} := X_0 \times ... \times X_{n-1} \times S$ and $\widetilde{h} : \widetilde{X} \to \overline{\widetilde{Z}} = \widetilde{Z} \cup \{+\infty_{\widetilde{K}}\}$ is defined as

$$\widetilde{h}(\widetilde{y}) := \begin{cases} (h^1(y^1, y^0), ..., h^n(y^n, y^{n-1}), g(y^n)), & \text{if } (y^i, y^{i-1}) \in \operatorname{dom} h^i, \ i = 1, ..., n, \\ & y^n \in \operatorname{dom} g, \\ +\infty_{\widetilde{K}}, & \text{otherwise.} \end{cases}$$

Note that by the definition of h^i we have

$$\operatorname{dom} h^i = \operatorname{dom} F^i \times X_{i-1}, \ i = 1, ..., n,$$

which yields

$$\operatorname{dom} \widetilde{h} = X_0 \times \operatorname{dom} F^1 \times ... \times (\operatorname{dom} F^n \cap \operatorname{dom} g). \tag{3.3}$$

At this point, let us additionally remark that the assumption from the beginning,

$$S \cap g^{-1}(-Q) \cap ((F^n)^{-1} \circ ... \circ (F^1)^{-1})(\operatorname{dom} f) \neq \emptyset, \tag{3.4}$$

implies also that $\operatorname{dom} \widetilde{f} \cap \widetilde{S} \cap \widetilde{h}^{-1}(-\widetilde{K}) \neq \emptyset$, but the inverse is not true. This means

$$S \cap g^{-1}(-Q) \cap ((F^n)^{-1} \circ ... \circ (F^1)^{-1})(\operatorname{dom} f) \neq \emptyset$$
$$\Leftrightarrow \exists (y^0, y^1, ..., y^{n-1}, y^n) \in \operatorname{dom} f \times X_1 \times ... \times X_{n-1} \times S \text{ such that}$$
$$F^1(y^1) - y^0 = 0 \in -K_0, ..., F^n(y^n) - y^{n-1} = 0 \in -K_{n-1} \text{ and } g(y^n) \in -Q$$
$$\Rightarrow \exists \widetilde{y} \in \widetilde{S} \cap \operatorname{dom} \widetilde{f} \text{ such that } \widetilde{h}(\widetilde{y}) \in -\widetilde{K}$$
$$\Leftrightarrow \operatorname{dom} \widetilde{f} \cap \widetilde{S} \cap \widetilde{h}^{-1}(-\widetilde{K}) \neq \emptyset.$$

The corresponding Lagrange dual problem $(\widetilde{D}^C)$ with $\widetilde{z}^* := (z^{0*}, ..., z^{(n-1)*}, z^{n*}) \in \widetilde{K}^* := K_0^* \times ... \times K_{n-1}^* \times Q^*$ as the dual variable to the problem $(\widetilde{P}^C)$ is

$$(\widetilde{D}^C) \quad \sup_{\widetilde{z}^* \in \widetilde{K}^*} \inf_{\widetilde{y} \in \widetilde{S}} \{\widetilde{f}(\widetilde{y}) + \langle \widetilde{z}^*, \widetilde{h}(\widetilde{y}) \rangle\},$$

which can equivalently be written as

$$(\widetilde{D}^C) \quad \sup_{\substack{z^{n*} \in Q^*, \ z^{i*} \in K_i^* \\ i=0,...,n-1}} \inf_{\substack{y^n \in S, \ y^i \in X_i \\ i=0,...,n-1}} \left\{ \widetilde{f}(y^0, ..., y^n) + \sum_{i=1}^n \langle z^{(i-1)*}, h^i(y^i, y^{i-1}) \rangle + \langle z^{n*}, g(y^n) \rangle \right\}.$$

Through the definitions we made above for $\tilde{f}$ and h^i and since we set $x = y^n$, we can deduce the conjugate dual problem (D^C) to problem (P^C)

$$(D^C) \quad \sup_{\substack{z^{n*} \in Q^*, \ z^{i*} \in K_i^* \\ i=0,...,n-1}} \inf_{\substack{x \in S, \ y^i \in X_i \\ i=0,...,n-1}} \left\{ f(y^0) + \langle z^{(n-1)*}, F^n(x) - y^{n-1} \rangle + \langle z^{n*}, g(x) \rangle \right.$$

$$\left. + \sum_{i=1}^{n-1} \langle z^{(i-1)*}, F^i(y^i) - y^{i-1} \rangle \right\}$$

$$= \sup_{\substack{z^{n*} \in Q^*, \ z^{i*} \in K_i^*, \\ i=0,...,n-1}} \left\{ \inf_{x \in S} \{ \langle z^{(n-1)*}, F^n(x) \rangle + \langle z^{n*}, g(x) \rangle \} - \sup_{y^0 \in X_0} \{ \langle z^{0*}, y^0 \rangle - f(y^0) \} \right.$$

$$\left. - \sum_{i=1}^{n-1} \sup_{y^i \in X_i} \{ \langle z^{i*}, y^i \rangle - \langle z^{(i-1)*}, F^i(y^i) \rangle \} \right\}.$$

Hence, the conjugate dual problem (D^C) to problem (P^C) has the following form

$$(D^C) \quad \sup_{\substack{z^{n*} \in Q^*, \ z^{i*} \in K_i^*, \\ i=0,...,n-1}} \left\{ \inf_{x \in S} \{ \langle z^{(n-1)*}, F^n(x) \rangle + \langle z^{n*}, g(x) \rangle \} - f^*(z^{0*}) \right.$$

$$\left. - \sum_{i=1}^{n-1} (z^{(i-1)*} F^i)^*(z^{i*}) \right\}. \tag{3.5}$$

The optimal objective values of the problems $(\widetilde{D}^C)$ and (D^C) are of course equal, i.e. $v(\widetilde{D}^C) = v(D^C)$. The next result arises from the definition of the dual problem and is always fulfilled.

Theorem 3.2 (weak duality). *Between the primal problem (P^C) and its conjugate dual problem weak duality always holds, i.e. $v(P^C) \geq v(D^C)$.*

Proof. By Theorem 3.1.1 in [14] it holds $v(\widetilde{P}^C) \geq v(\widetilde{D}^C)$. Moreover, by Theorem 3.1 and since $v(\widetilde{D}^C) = v(D^C)$ we have $v(P^C) = v(\widetilde{P}^C) \geq v(\widetilde{D}^C) = v(D^C)$. □

Remark 3.4. *Let Z_i be a locally convex Hausdorff space partially ordered by the non-empty convex cone Q_i, $i = 0, ..., n - 1$. Then the introduced concept covers also optimization problems of the form*

$$(P^{CC}) \quad \inf_{x \in \mathcal{L}} \varphi(x),$$

with

$$\mathcal{L} := \{ x \in S : (G^1 \circ ... \circ G^n)(x) \in -Q_0 \},$$

where $\varphi : X_n \to \overline{\mathbb{R}}$ is proper, $G^i : \overline{Z}_i \to \overline{Z}_{i-1}$ is proper and (Q_i, Q_{i-1})-increasing on $G^{i+1}(\text{dom}\, G^{i+1}) + Q_i \subseteq \text{dom}\, G^i$, $i = 1, ..., n-1$, and $G^n : X_n \to \overline{Z}_{n-1}$ is proper. The problem (P^{CC}) can equivalently be rewritten as

$$(P^{CC}) \quad \inf_{x \in X_n} \{\varphi(x) + \delta_S(x) + (\delta_{-Q_0} \circ G^1 \circ ... \circ G^n)(x)\}$$

and by setting $X_0 := \mathbb{R} \times Z_0$, $K_0 := \mathbb{R}_+ \times Q_0$, $X_i := \mathbb{R} \times Z_i$, $K_i := \mathbb{R}_+ \times Q_i$, $i = 1, ..., n-1$ and by defining the following functions

- $f : \overline{X}_0 \times \mathbb{R}$, $f(y^0) := y_1^0 + \delta_{-Q_0}(y_2^0)$ with $y^0 = (y_1^0, y_2^0) \in X_0$,

- $F^i : \overline{X}_i \to \overline{X}_{i-1}$, $F^i(y_1^i, y_2^i) := (y_1^i, G^i(y_2^i))$, $i = 1, ..., n-1$ with $y^i = (y_1^i, y_2^i) \in X_i$,

- $F^n : X_n \to \overline{X}_{n-1}$, $F^n(x) := (\varphi(x) + \delta_S(x), G^n(x))$,

the problem (P^{CC}) turns into a special case of the problem (P^C)

$$(P^{CC}) \quad \inf_{x \in X_n} (f \circ F^1 \circ ... \circ F^n)(x)$$

with $\mathcal{A} \equiv X_n$.

3.2 Regularity conditions, strong duality and optimality conditions

In this section we want to characterize strong duality through the so-called generalized interior point regularity conditions. Besides we provide some optimality conditions for the primal problem and its corresponding conjugate dual problem. For this purpose we additionally assume for the rest of this chapter that $S \subseteq X_n$ is a convex set, f is a convex function, F^i is a K_{i-1}-convex function for $i = 1, ..., n$ and g is a Q-convex function. Hence, as can be easily seen, $(f \circ F^1 \circ ... \circ F^n)$ is a convex function and (P^C) is a convex optimization problem. Moreover, the problem $(\widetilde{P}^C)$ is also convex.

Remark 3.5. *Let us point out that for the convexity of $(f \circ F^1 \circ ... \circ F^n)$ we ask that the function f be convex and K_0-increasing on $F^1(\text{dom}\, F^1) + K_0$ and the functions F^i be K_{i-1}-convex and fulfill also the property of monotonicity for $i = 1, ..., n-1$, while the function F^n needs just be K_{n-1}-convex (see Theorem 2.2). It turns out, especially in the context of location problems, that in the case when F^n is an affine function, then the condition of monotonicity of F^{n-1} is too restrictive. But fortunately, this circumstance can be bypassed by the fact that the composition of an affine function and a function, which fulfills the property of convexity, fulfills also the property of convexity. That is to say, that one can omit the monotonicity of F^{n-1} by setting $K_{n-1} = \{0_{X_{n-1}}\}$ to preserve the property of convexity of $(f \circ F^1 \circ ... \circ F^n)$. Let us pay also attention to Remark 3.2, i.e. one can choose $K_{n-1} = \{0_{X_{n-1}}\}$.*

To derive regularity conditions which secure strong duality for the pair (P^C)-(D^C), we first consider regularity conditions for strong duality between the problems $(\tilde{P}^C)$ and $(\tilde{D}^C)$, which were presented in [14]. The first one is the well-known *Slater constraint qualification*

$$(\widetilde{RC}_1^C) \quad \Big| \quad \exists \tilde{y}' \in \operatorname{dom} \tilde{f} \cap \tilde{S} \text{ such that } \tilde{h}(\tilde{y}') \in -\operatorname{int} \tilde{K}.$$

Using the definitions of $\tilde{f}$ and $\tilde{h}$ as well as $\tilde{S}$ and $\tilde{K}$ we get

$$\operatorname{dom} \tilde{f} \cap \tilde{S} = (\operatorname{dom} f \times X_1 \times ... \times X_n) \cap (X_0 \times X_1 \times ... \times X_{n-1} \times S)$$
$$= \operatorname{dom} f \times X_1 \times ... \times X_{n-1} \times S \qquad (3.6)$$

and

$$\operatorname{int} \tilde{K} = \operatorname{int}(K_0 \times ... \times K_{n-1} \times Q) = \operatorname{int} K_0 \times ... \times \operatorname{int} K_{n-1} \times \operatorname{int} Q.$$

Therefore the condition $(\widetilde{RC}_1^C)$ can in the context of the primal-dual pair (P^C)-(D^C) be rewritten as follows

$$(RC_1^C) \quad \Big| \quad \exists (y^{0'}, y^{1'}, ..., y^{(n-1)'}, y^{n'}) \in \operatorname{dom} f \times X_1 \times ... \times X_{n-1} \times S \text{ such that}$$
$$F^i(y^{i'}) - y^{(i-1)'} \in -\operatorname{int} K_{i-1}, \; i = 1, ..., n, \text{ and } g(y^{n'}) \in -\operatorname{int} Q.$$

The condition (RC_1^C) can also equivalently be formulated as

$$(RC_{1'}^C) \quad \Big| \quad \exists x' \in S \text{ such that } g(x') \in -\operatorname{int} Q \text{ and } F^n(x') \in (F^{n-1})^{-1}((F^{n-2})^{-1}(...$$
$$(F^1)^{-1}(\operatorname{dom} f - \operatorname{int} K_0) - \operatorname{int} K_1...) - \operatorname{int} K_{n-2}) - \operatorname{int} K_{n-1}.$$

This can be seen as follows: The assumption that there exists $x' \in S$ such that

$$F^n(x') \in (F^{n-1})^{-1}((F^{n-2})^{-1}(...(F^1)^{-1}(\operatorname{dom} f - \operatorname{int} K_0) - \operatorname{int} K_1...) - \operatorname{int} K_{n-2}) - \operatorname{int} K_{n-1}$$

implies that there exists $(y^{0'}, ..., y^{(n-1)'}) \in X_0 \times ... \times X_{n-1}$ such that

$$y^{(n-1)'} \in (F^{n-1})^{-1}((F^{n-2})^{-1}(...(F^1)^{-1}(\operatorname{dom} f - \operatorname{int} K_0) - \operatorname{int} K_1...) - \operatorname{int} K_{n-2})$$
$$y^{(n-2)'} \in (F^{n-2})^{-1}((F^{n-3})^{-1}(...(F^1)^{-1}(\operatorname{dom} f - \operatorname{int} K_0) - \operatorname{int} K_1...) - \operatorname{int} K_{n-3})$$
$$\vdots$$
$$y^{1'} \in (F^1)^{-1}(\operatorname{dom} f - \operatorname{int} K_0)$$
$$y^{0'} \in \operatorname{dom} f.$$

Therefore, by setting $x' = y^{n'}$ the elements $(y^{0'}, ..., y^{n'}) \in \operatorname{dom} f \times X_1 \times ... \times X_{n-1} \times S$ fulfill $F^n(y^{n'}) - y^{(n-1)'} \in -\operatorname{int} K_{n-1}, ..., F^1(y^{1'}) - y^{0'} \in -\operatorname{int} K_0$ and from here we can now affirm that the condition (RC_1^C) is fulfilled.

On the other hand, if there exists $(y^{0'}, ..., y^{n'}) \in \operatorname{dom} f \times X_1 \times ... \times X_{n-1} \times S$ such that $g(y^{n'}) \in -\operatorname{int} Q$ and $F^i(y^{i'}) - y^{(i-1)'} \in -\operatorname{int} K_{i-1}$ for $i = 1, ..., n$, then we set $y^{n'} = x'$ and get

$$F^n(x') - y^{(n-1)'} \in -\operatorname{int} K_{n-1} \Rightarrow F^n(x') \in y^{(n-1)'} - \operatorname{int} K_{n-1}. \qquad (3.7)$$

Further, we have

$$F^{n-1}(y^{(n-1)'}) - y^{(n-2)'} \in -\operatorname{int} K_{n-2} \Rightarrow F^{n-1}(y^{(n-1)'}) \in y^{(n-2)'} - \operatorname{int} K_{n-2}$$
$$\Rightarrow y^{(n-1)'} \in (F^{n-1})^{-1}(y^{(n-2)'} - \operatorname{int} K_{n-2}). \tag{3.8}$$

From (3.7) and (3.8) follows

$$F^n(x') \in (F^{n-1})^{-1}(y^{(n-2)'} - \operatorname{int} K_{n-2}) - \operatorname{int} K_{n-1}. \tag{3.9}$$

Since

$$F^{n-2}(y^{(n-2)'}) - y^{(n-3)'} \in -\operatorname{int} K_{n-3} \Rightarrow F^{n-2}(y^{(n-2)'}) \in y^{(n-3)'} - \operatorname{int} K_{n-3}$$
$$\Rightarrow y^{(n-2)'} \in (F^{n-2})^{-1}(y^{(n-3)'} - \operatorname{int} K_{n-3})$$

we get for (3.9)

$$F^n(x') \in (F^{n-1})^{-1}((F^{n-2})^{-1}(y^{(n-3)'} - \operatorname{int} K_{n-3}) - \operatorname{int} K_{n-2}) - \operatorname{int} K_{n-1}.$$

If we continue in this manner until $y^{0'} \in \operatorname{dom} f$ we get finally

$$F^n(x') \in (F^{n-1})^{-1}((F^{n-2})^{-1}(...(F^1)^{-1}(\operatorname{dom} f - \operatorname{int} K_0) - \operatorname{int} K_1...) - \operatorname{int} K_{n-2}) - \operatorname{int} K_{n-1}.$$

This means that $(RC_{1'}^C)$ is equivalent to (RC_1^C). Additionally, we consider a class of regularity conditions which assume that the underlying spaces are Fréchet spaces:

$$(\widetilde{RC_2^C}) \quad \left| \begin{array}{l} \widetilde{X} \text{ and } \widetilde{Z} \text{ are Fréchet spaces, } \widetilde{S} \text{ is closed, } \widetilde{f} \text{ is lower semicontinuous,} \\ \widetilde{h} \text{ is } \widetilde{K}\text{-epi closed and } 0_{\widetilde{Z}} \in \operatorname{sqri}(\widetilde{h}(\operatorname{dom} \widetilde{f} \cap \widetilde{S} \cap \operatorname{dom} \widetilde{h}) + \widetilde{K}). \end{array} \right.$$

If we exchange sqri for core or int we get stronger versions of this regularity condition:

$$(\widetilde{RC_{2'}^C}) \quad \left| \begin{array}{l} \widetilde{X} \text{ and } \widetilde{Z} \text{ are Fréchet spaces, } \widetilde{S} \text{ is closed, } \widetilde{f} \text{ is lower semicontinuous,} \\ \widetilde{h} \text{ is } \widetilde{K}\text{-epi closed and } 0_{\widetilde{Z}} \in \operatorname{core}(\widetilde{h}(\operatorname{dom} \widetilde{f} \cap \widetilde{S} \cap \operatorname{dom} \widetilde{h}) + \widetilde{K}), \end{array} \right.$$

$$(\widetilde{RC_{2''}^C}) \quad \left| \begin{array}{l} \widetilde{X} \text{ and } \widetilde{Z} \text{ are Fréchet spaces, } \widetilde{S} \text{ is closed, } \widetilde{f} \text{ is lower semicontinuous,} \\ \widetilde{h} \text{ is } \widetilde{K}\text{-epi closed and } 0_{\widetilde{Z}} \in \operatorname{int}(\widetilde{h}(\operatorname{dom} \widetilde{f} \cap \widetilde{S} \cap \operatorname{dom} \widetilde{h}) + \widetilde{K}), \end{array} \right.$$

where the last two conditions are equivalent (see [14]). If we work in finite dimensional spaces the regularity condition $(\widetilde{RC_2^C})$ can be written in the following way (see [14])

$$(\widetilde{RC_3^C}) \quad \left| \begin{array}{l} \dim(\operatorname{lin}(\widetilde{h}(\operatorname{dom} \widetilde{f} \cap \widetilde{S} \cap \operatorname{dom} \widetilde{h}) + \widetilde{K})) < +\infty \text{ and} \\ 0_{\widetilde{Z}} \in \operatorname{ri}(\widetilde{h}(\operatorname{dom} \widetilde{f} \cap \widetilde{S} \cap \operatorname{dom} \widetilde{h}) + \widetilde{K}). \end{array} \right.$$

To derive corresponding regularity conditions for the primal-dual pair (P^C)-(D^C) formulated with the involved functions we first consider the formulae (3.3) and (3.6), which imply that

$$\widetilde{h}(\operatorname{dom} \widetilde{f} \cap \widetilde{S} \cap \operatorname{dom} \widetilde{h})$$
$$= \widetilde{h}(\operatorname{dom} f \times \operatorname{dom} F^1 \times \ldots \times \operatorname{dom} F^{n-1} \times (\operatorname{dom} F^n \cap \operatorname{dom} g \cap S))$$
$$= h^1(\operatorname{dom} F^1 \times \operatorname{dom} f) \times h^2(\operatorname{dom} F^2 \times \operatorname{dom} F^1) \times \ldots \times$$
$$\quad h^{n-1}(\operatorname{dom} F^{n-1} \times \operatorname{dom} F^{n-2}) \times$$
$$\quad h^n((\operatorname{dom} F^n \cap \operatorname{dom} g \cap S) \times \operatorname{dom} F^{n-1}) \times g(\operatorname{dom} F^n \cap \operatorname{dom} g \cap S)$$
$$= (F^1(\operatorname{dom} F^1) - \operatorname{dom} f) \times (F^2(\operatorname{dom} F^2) - \operatorname{dom} F^1) \times \ldots \times$$
$$\quad (F^{n-1}(\operatorname{dom} F^{n-1}) - \operatorname{dom} F^{n-2}) \times$$
$$\quad (F^n(\operatorname{dom} F^n \cap \operatorname{dom} g \cap S) - \operatorname{dom} F^{n-1}) \times g(\operatorname{dom} F^n \cap \operatorname{dom} g \cap S)$$

and from here we get by Lemma 2.1 that

$$0_{\widetilde{Z}} \in \operatorname{sqri}\Big((F^1(\operatorname{dom} F^1) - \operatorname{dom} f + K_0) \times \ldots$$
$$\times (F^{n-1}(\operatorname{dom} F^{n-1}) - \operatorname{dom} F^{n-2} + K_{n-2})$$
$$\times (F^n(\operatorname{dom} F^n \cap \operatorname{dom} g \cap S) - \operatorname{dom} F^{n-1} + K_{n-1})$$
$$\times (g(\operatorname{dom} F^n \cap \operatorname{dom} g \cap S) + Q)\Big)$$

is equivalent to

$$0_{X_0} \in \operatorname{sqri}(F^1(\operatorname{dom} F^1) - \operatorname{dom} f + K_0),$$
$$0_{X_i} \in \operatorname{sqri}(F^i(\operatorname{dom} F^i) - \operatorname{dom} F^{i-1} + K_{i-1}), \ i = 2, \ldots, n-1,$$
$$0_{X_n} \in \operatorname{sqri}(F^n(\operatorname{dom} F^n \cap \operatorname{dom} g \cap S) - \operatorname{dom} F^{n-1} + K_{n-1}) \text{ and}$$
$$0_Z \in \operatorname{sqri}(g(\operatorname{dom} F^n \cap \operatorname{dom} g \cap S) + Q).$$

Now, let

$$\varrho : X_0 \times \ldots \times X_n \times X_0 \times \ldots \times X_{n-1} \times Z \to X_0^2 \times \ldots \times X_{n-1}^2 \times X_n \times Z \quad (3.10)$$

be defined by

$$\varrho(y^0, \ldots, y^n, v^0, \ldots, v^n) := (y^0, v^0, \ldots, y^n, v^n). \quad (3.11)$$

Further, let us define the functions

$$\varrho_{X_i}^n : X_i \times X_{i-1} \times X_{i-1} \to X_{i-1} \times X_{i-1} \times X_i \quad (3.12)$$

by

$$\varrho_{X_i}^n(y^i, y^{i-1}, v^{i-1}) := (y^{i-1}, v^{i-1}, y^i), \ i = 1, \ldots, n. \quad (3.13)$$

Obviously, the defined functions are homeomorphisms and map open sets into open sets and closed sets into closed sets. More precisely, this means that $\varrho(\text{epi}_{\widetilde{K}}\,\widetilde{h})$ is closed if and only if $\text{epi}_{\widetilde{K}}\,\widetilde{h}$ is a closed set and $\varrho^n_{X_i}(\text{epi}_{K_{i-1}}\,h^i)$ is closed if and only if $\text{epi}_{K_{i-1}}\,h^i$ is a closed set, $i = 1, ..., n$. Furthermore, we have

$$
\begin{aligned}
\text{epi}_{\widetilde{K}}\,\widetilde{h} \;=\; &\{(y^0, ..., y^n, v^0, ..., v^n) \in X_0 \times ... \times X_n \times X_0 \times ... \times X_{n-1} \times Z : \\
&\quad (y^1, y^0, v^0) \in \text{epi}_{K_0}\,h^1, \\
&\quad\;\; \vdots \\
&\quad (y^n, y^{n-1}, v^{n-1}) \in \text{epi}_{K_{n-1}}\,h^n, \\
&\quad (y^n, v^n) \in \text{epi}_Q\,g\} \\
=\; &\{(y^0, ..., y^n, v^0, ..., v^n) \in X_0 \times ... \times X_n \times X_0 \times ... \times X_{n-1} \times Z : \\
&\quad (y^0, v^0, y^1) \in \varrho^n_{X_1}(\text{epi}_{K_0}\,h^1), \\
&\quad\;\; \vdots \\
&\quad (y^{n-1}, v^{n-1}, y^n) \in \varrho^n_{X_n}(\text{epi}_{K_{n-1}}\,h^n), \\
&\quad (y^n, v^n) \in \text{epi}_Q\,g\} \\
=\; &\Big\{(y^0, ..., y^n, v^0, ..., v^n) \in X_0 \times ... \times X_n \times X_0 \times ... \times X_{n-1} \times Z : \\
&\quad (y^0, v^0, y^1, v^1, y^2, v^2, ..., y^{n-1}, v^{n-1}, y^n, v^n) \in \\
&\quad \big(\varrho^n_{X_1}(\text{epi}_{K_0}\,h^1)\big) \times X_1 \times X_2^2 \times ... \times X_{n-1}^2 \times X_n \times Z, \\
&\quad\;\; \vdots \\
&\quad (y^0, v^0, y^1, v^1, ..., y^{n-2}, v^{n-2}, y^{n-1}, v^{n-1}, y^n, v^n) \in \\
&\quad X_0^2 \times X_1^2 \times ... \times X_{n-2}^2 \times \big(\varrho^n_{X_n}(\text{epi}_{K_{n-1}}\,h^n)\big) \times Z, \\
&\quad (y^0, v^0, ..., y^{n-1}, v^{n-1}, y^n, v^n) \in X_0^2 \times ... \times X_{n-1}^2 \times \text{epi}_Q\,g\Big\} \\
=\; &\Big\{(y^0, ..., y^n, v^0, ..., v^n) \in X_0 \times ... \times X_n \times X_0 \times ... \times X_{n-1} \times Z : \\
&\quad (y^0, v^0, ..., y^{i-2}, v^{i-2}, y^{i-1}, v^{i-1}, y^i, v^i, y^{i+1}, v^{i+1}, ..., y^n, v^n) \in \\
&\quad X_0^2 \times ... \times X_{i-2}^2 \times \big(\varrho^n_{X_i}(\text{epi}_{K_{i-1}}\,h^i)\big) \times X_i \times X_{i+1}^2 \times ... \times X_n \times Z, \\
&\quad i = 1, ..., n, \; (y^0, v^0, ..., y^{n-1}, v^{n-1}, y^n, v^n) \in X_0^2 \times ... \times X_{n-1}^2 \times \text{epi}_Q\,g\Big\},
\end{aligned}
$$

so we can write

$$
\varrho(\text{epi}_{\widetilde{K}}\,\widetilde{h}) = \left(\bigcap_{i=1}^n \big(X_0^2 \times ... \times X_{i-2}^2 \times \big(\varrho^n_{X_i}(\text{epi}_{K_{i-1}}\,h^i)\big) \times X_i \times X_{i+1}^2 \times ... \times X_n \times Z\big)\right)
$$
$$
\bigcap \big(X_0^2 \times ... \times X_{n-1}^2 \times \text{epi}_Q\,g\big)
$$

and get as a consequence that $\mathrm{epi}_{\widetilde{K}}\,\widetilde{h}$ is closed if $\mathrm{epi}_{K_{i-1}}\,h^i$, $i = 1, ..., n$, and $\mathrm{epi}_Q\,g$ are closed sets.

Vice versa, let $\mathrm{epi}_{\widetilde{K}}\,\widetilde{h}$ be closed, i.e. $\varrho(\mathrm{epi}_{\widetilde{K}}\,\widetilde{h})$ is closed, and

$$(y_\alpha^1, y_\alpha^0, v_\alpha^0)_\alpha \subseteq \mathrm{epi}_{K_0}\,h^1, ..., (y_\alpha^n, y_\alpha^{n-1}, v_\alpha^{n-1})_\alpha \subseteq \mathrm{epi}_{K_{n-1}}\,h^n \text{ and } (y_\alpha^n, v_\alpha^n)_\alpha \subseteq \mathrm{epi}_Q\,g,$$

i.e. $(y_\alpha^0, ..., y_\alpha^n, v_\alpha^0, ..., v_\alpha^n)_\alpha \subseteq \mathrm{epi}_{\widetilde{K}}\,\widetilde{h}$. As $\mathrm{epi}_{\widetilde{K}}\,\widetilde{h}$ is closed, we have that

$$(y_\alpha^0, ..., y_\alpha^n, v_\alpha^0, ..., v_\alpha^n)_\alpha \to (y^0, ..., y^n, v^0, ..., v^n) \in \mathrm{epi}_{\widetilde{K}}\,\widetilde{h}, \tag{3.14}$$

but this means that

$$(y^1, y^0, v^0) \in \mathrm{epi}_{K_0}\,h^1, ..., (y^n, y^{n-1}, v^{n-1}) \in \mathrm{epi}_{K_{n-1}}\,h^n \text{ and } (y^n, v^n) \in \mathrm{epi}_Q\,g,$$

which implies the closedness of $\mathrm{epi}_{K_0}\,h^1, ..., \mathrm{epi}_{K_{n-1}}\,h^n$ and $\mathrm{epi}_Q\,g$.

Besides, we know by Lemma 2.3 that for a non-empty closed convex cone K_{i-1} it holds that $\mathrm{epi}_{K_{i-1}}\,h^i$ is closed if and only if $\mathrm{epi}_{K_{i-1}}\,F^i$ is closed, $i = 1, ..., n$. Bringing now the last facts together implies that for non-empty closed convex cones K_{i-1}, $i = 1, ..., n$, it holds that $\mathrm{epi}_{\widetilde{K}}\,\widetilde{h}$ is closed if and only if $\mathrm{epi}_Q\,g$ and $\mathrm{epi}_{K_{i-1}}\,F^i$ are closed sets, $i = 1, ..., n$.

Moreover, since $\widetilde{S}$ is closed if and only if S is closed and $\widetilde{f}$ is lower semicontinuous if and only if f is lower semicontinuous (follows from the fact that $\mathrm{epi}\,f$ is closed if and only if $\mathrm{epi}\,\widetilde{f}$ is closed), we get the following regularity condition for the primal-dual pair (P^C)-(D^C) (note that if X_i is a Fréchet space, $i = 0, ..., n$, then $\widetilde{X} = X_0 \times ... \times X_n$ is a Fréchet space, too)

(RC_2^C) | $X_0, ..., X_n$ and Z are Fréchet spaces, f is l.s.c., S is closed, g is Q-epi, closed, K_{i-1} is closed, F^i is K_{i-1}-epi closed, $i = 1, ..., n$,
$0_{X_0} \in \mathrm{sqri}(F^1(\mathrm{dom}\,F^1) - \mathrm{dom}\,f + K_0)$,
$0_{X_{i-1}} \in \mathrm{sqri}(F^i(\mathrm{dom}\,F^i) - \mathrm{dom}\,F^{i-1} + K_{i-1}), i = 2, ..., n - 1$,
$0_{X_{n-1}} \in \mathrm{sqri}(F^n(\mathrm{dom}\,F^n \cap \mathrm{dom}\,g \cap S) - \mathrm{dom}\,F^{n-1} + K_{n-1})$ and
$0_Z \in \mathrm{sqri}(g(\mathrm{dom}\,F^n \cap \mathrm{dom}\,g \cap S) + Q)$.

In the same way we get equivalent formulations of the regularity conditions $(RC_{2'}^C)$ and $(RC_{2''}^C)$ using core and int, respectively, instead of sqri. The same holds also for the condition (RC_3^C).

As we have seen, the condition (RC_i^C) is equivalent to $\widetilde{(RC_i^C)}$, $i \in \{1, 2, 2', 2'', 3\}$. Moreover, since on the one hand Theorem 3.1 is always fulfilled and on the other hand the optimal objective values between $(\widetilde{D}^C)$ and (D^C) are equal, it holds the following theorem (see Theorem 3.2.9 and 3.2.10 in [14]).

Theorem 3.3 (strong duality). *If one of the conditions (RC_i^C), $i \in \{1, 1', 2, 2', 2'', 3\}$, is fulfilled, then between (P^C) and (D^C) strong duality holds, i.e. $v(P^C) = v(D^C)$ and the conjugate dual problem has an optimal solution.*

Remark 3.6. *If for some $i \in \{1, ..., n\}$ the function F^i is positively K_{i-1}-lower semicontinuous, then we can omit asking that F^i is K_{i-1}-epi closed in the regularity conditions (RC_i^C), $i \in \{2, 2', 2''\}$, because the positive K_{i-1}-lower semicontinuity of F^i implies the positive K_{i-1}-lower semicontinuity of h^i, which then implies the K_{i-1}-epi closedness of h^i.*

Remark 3.7. *Besides the used regularity conditions there are also the so-called closedness type conditions guaranteeing strong duality. Such regularity conditions were studied in different contexts, like strong duality, subdifferential calculus etc. (see [25]). These types of regularity conditions were also studied in [47] and [46], to cite only few of them.*

We have also extensively studied closedness type conditions in the context of multi-composed optimization problems with the focus on stable strong duality and ϵ-optimality conditions in our article [55]. As applications we considered problems from fractional programming and entropy optimization (see also [5, 6, 13, 30, 37, 51, 93]).

We come now to the point where we can give necessary and sufficient optimality conditions for the primal-dual pair $v(P^C)$-$v(D^C)$.

Theorem 3.4 (optimality conditions). *(a) Suppose that one of the regularity conditions (RC_i^C), $i \in \{1, 1', 2, 2', 2'', 3\}$, is fulfilled and let $\overline{x} \in \mathcal{A}$ be an optimal solution to the problem (P^C). Then there exists $(\overline{z}^{0*}, ..., \overline{z}^{(n-1)*}, \overline{z}^{n*}) \in K_0^* \times ... \times K_{n-1}^* \times Q^*$, an optimal solution to (D^C), such that*

(i) $f((F^1 \circ ... \circ F^n)(\overline{x})) + f^*(\overline{z}^{0*}) = \langle \overline{z}^{0*}, (F^1 \circ ... \circ F^n)(\overline{x}) \rangle$,

(ii) $(\overline{z}^{(i-1)*}F^i)((F^{i+1} \circ ... \circ F^n)(\overline{x})) + (\overline{z}^{(i-1)*}F^i)^*(\overline{z}^{i*}) = \langle \overline{z}^{i*}, (F^{i+1} \circ ... \circ F^n)(\overline{x}) \rangle$, $i = 1, ..., n-1$,

(iii) $(\overline{z}^{(n-1)*}F^n)(\overline{x}) + (\overline{z}^{n*}g)(\overline{x}) + ((\overline{z}^{(n-1)*}F^n) + (\overline{z}^{n*}g))_S^*(0_{X_n^*}) = 0$,

(iv) $\langle \overline{z}^{n*}, g(\overline{x}) \rangle = 0$.

(b) If there exists $\overline{x} \in \mathcal{A}$ such that for some $(\overline{z}^{0*}, ..., \overline{z}^{(n-1)*}, \overline{z}^{n*}) \in K_0^* \times ... \times K_{n-1}^* \times Q^*$ the conditions (i)-(iv) are fulfilled, then $\overline{x}$ is an optimal solution to (P^C), $(\overline{z}^{0*}, ..., \overline{z}^{n*})$ is an optimal solution to (D^C) and $v(P^C) = v(D^C)$.

Proof. First, we consider part (a). By Theorem 3.3, strong duality holds for the primal-dual pair (P^C)-(D^C), which means that there exists $(\overline{z}^{0*}, ..., \overline{z}^{(n-1)*}, \overline{z}^{n*}) \in K_0^* \times ... \times K_{n-1}^* \times Q^*$, an optimal solution to (D^C), such that the following equality holds

$$(f \circ F^1 \circ ... \circ F^n)(\overline{x})$$

$$= \inf_{x \in S}\{\langle \overline{z}^{(n-1)*}, F^n(x) \rangle + \langle \overline{z}^{n*}, g(x) \rangle\} - f^*(\overline{z}^{0*}) - \sum_{i=1}^{n-1}(\overline{z}^{(i-1)*}F^i)^*(\overline{z}^{i*}).$$

Furthermore, since by definition it holds

$$\sum_{i=1}^{n}(\overline{z}^{(i-1)*}F^{i})((F^{i+1}\circ...\circ F^{n})(\overline{x}))$$

$$= \langle \overline{z}^{0*},(F^{1}\circ...\circ F^{n})(\overline{x})\rangle + \sum_{i=1}^{n-1}\langle \overline{z}^{i*},(F^{i+1}\circ...\circ F^{n})(\overline{x})\rangle,$$

the assertions (i)-(iv) can be deduced immediately by the following consideration

$$(f\circ F^{1}\circ...\circ F^{n})(\overline{x}) + f^{*}(\overline{z}^{0*}) + \sum_{i=1}^{n-1}(\overline{z}^{(i-1)*}F^{i})^{*}(\overline{z}^{i*})$$

$$+ ((\overline{z}^{(n-1)*}F^{n}) + (\overline{z}^{n*}g))_{S}^{*}(0_{X_{n}^{*}}) = 0$$

$$\Leftrightarrow (f\circ F^{1}\circ...\circ F^{n})(\overline{x}) + f^{*}(\overline{z}^{0*}) + \sum_{i=1}^{n-1}(\overline{z}^{(i-1)*}F^{i})^{*}(\overline{z}^{i*})$$

$$+ ((\overline{z}^{(n-1)*}F^{n}) + (\overline{z}^{n*}g))_{S}^{*}(0_{X_{n}^{*}}) + (\overline{z}^{n*}g)(\overline{x}) - \langle \overline{z}^{n*},g(\overline{x})\rangle$$

$$+ \sum_{i=1}^{n}(\overline{z}^{(i-1)*}F^{i})((F^{i+1}\circ...\circ F^{n})(\overline{x}))$$

$$- \langle \overline{z}^{0*},(F^{1}\circ...\circ F^{n})(\overline{x})\rangle - \sum_{i=1}^{n-1}\langle \overline{z}^{i*},(F^{i+1}\circ...\circ F^{n})(\overline{x})\rangle = 0$$

$$\Leftrightarrow [(f\circ F^{1}\circ...\circ F^{n})(\overline{x}) + f^{*}(\overline{z}^{0*}) - \langle \overline{z}^{0*},(F^{1}\circ...\circ F^{n})(\overline{x})\rangle]+$$

$$\sum_{i=1}^{n-1}[(\overline{z}^{(i-1)*}F^{i})((F^{i+1}\circ...\circ F^{n})(\overline{x})) + (\overline{z}^{(i-1)*}F^{i})^{*}(\overline{z}^{i*}) - \langle \overline{z}^{i*},(F^{i+1}\circ...\circ F^{n})(\overline{x})\rangle]$$

$$+ [(\overline{z}^{(n-1)*}F^{n})(\overline{x}) + (\overline{z}^{n*}g)(\overline{x}) + ((\overline{z}^{(n-1)*}F^{n}) + (\overline{z}^{n*}g))_{S}^{*}(0_{X_{n}^{*}})]$$

$$+ [-\langle \overline{z}^{n*},g(\overline{x})\rangle] = 0.$$

By the Young-Fenchel inequality and the constraints of the primal and dual problem, all the terms within the brackets are non-negative and consequently must be equal to zero. Concerning the proof of part (b) we observe that all considerations and calculations within the proof of part (a) can be done in the reverse direction. $\qquad\square$

Remark 3.8. *The conditions (i)-(iv) can equivalently be expressed as*

(i) $\overline{z}^{0*} \in \partial f((F^{1}\circ...\circ F^{n})(\overline{x})),$

(ii) $\overline{z}^{i*} \in \partial(\overline{z}^{(i-1)*}F^{i})((F^{i+1}\circ...\circ F^{n})(\overline{x})),\ i=1,...,n-1,$

(iii) $0_{X_{n}^{*}} \in \partial((\overline{z}^{(n-1)*}F^{n}) + (\overline{z}^{n*}g) + \delta_{S})(\overline{x}),$

(iv) $\langle \overline{z}^{n*},g(\overline{x})\rangle = 0.$

Remark 3.9. *In fact, we can write every optimization problem in the form of a multi-composed problem by splitting the objective function into $n + 1$ composed functions. Allowing n to be greater than one, as considered until now in the literature, is very fruitful for dealing with a big variety of optimization problems because many such problems cannot be described with an objective function composed by only two functions (case $n = 1$, cf. the optimization problem presented in Section 3.4). The problems and results in several papers (cf. [14, 15, 20, 23, 24, 26, 63]) turn out to be special cases of our approach for the situation when only two functions are composed in the objective function, i.e. $n = 1$. For this reason our analysis sums up the approaches and the results in the mentioned papers into one unified theory. Similarly to the difficulties of the usual perturbation theory, where every perturbation is accompanied by additional conditions, like monotonicity and convexity, needed for strong duality, we derive situations in our approach, where for each additional splitting function involved in the multi-composed objective function, also additional properties must be fulfilled to ensure strong duality.*

But viewed from the other side, one can derive a conjugate dual problem, where the associated objective function is easier to handle due to the fact that the conjugate functions of the functions involved in the objective function of the primal problem are split. Furthermore, one can derive by the corresponding optimality conditions listed above a more detailed characterization of the set of optimal solutions of the dual and the primal problems as well as their relationship to each other. Such information can be used on the one hand to give a geometrical interpretation of the dual problem and on the other hand to improve the approach of optimization problems from the numerical point of view. In Chapter 4 we consider in this context minmax location problems as an application of the presented theory.

3.3 The conjugate function of a multi-composed function

Before we continue with our further approach we want to calculate the conjugate of the function $(f \circ F^1 \circ \ldots \circ F^n)$, or, to be more precise, we determine to the function

$$\gamma(x) = (f \circ F^1 \circ \ldots \circ F^n)(x), \ x \in X_n, \tag{3.15}$$

its conjugate function

$$\gamma^*(x^*) = \sup_{x \in X_n} \{\langle x^*, x \rangle - (f \circ F^1 \circ \ldots \circ F^n)(x)\}, \ x^* \in X_n^*. \tag{3.16}$$

With this in mind, we consider for fixed $x^* \in X_n^*$ the problem

$$(P^K) \quad \inf_{x \in X_n} \{(f \circ F^1 \circ \ldots \circ F^n)(x) - \langle x^*, x \rangle\} \tag{3.17}$$

and the equivalent primal problem

$$(\widetilde{P}^K) \quad \inf_{\substack{(y^0, \ldots, y^n) \in X_0 \times \ldots \times X_n, \\ F^i(y^i) - y^{i-1} \in -K_{i-1}, \ i=1,\ldots,n}} \{\widetilde{f}(y^0, y^1, \ldots, y^n) - \langle x^*, y^n \rangle\}.$$

In the same way like in the proof of Theorem 3.1 one can show that it holds $v(P^K) = v(\widetilde{P}^K)$ (where $v(P^K)$ and $v(\widetilde{P}^K)$ denote the optimal objective values of the problems (P^K) and $(\widetilde{P}^K)$, respectively). The corresponding Lagrange dual problem to problem $(\widetilde{P}^K)$ looks like

$$(\widetilde{D}^K) \quad \sup_{\substack{z^{i*} \in K_i^*, \\ i=0,\ldots,n-1}} \inf_{\substack{y^i \in X_i, \\ i=0,\ldots,n}} \left\{ \widetilde{f}(y^0, y^1, ..., y^n) + \sum_{i=1}^{n} \langle z^{(i-1)*}, F^i(y^i) - y^{i-1} \rangle - \langle x^*, y^n \rangle \right\}$$

$$= \sup_{\substack{z^{i*} \in K_i^*, \\ i=0,\ldots,n-1}} \left\{ - \sup_{y^0 \in X_0} \{ \langle z^{0*}, y^0 \rangle - f(y^0) \} - \right.$$

$$\left. \sup_{\substack{y^n \in S, \; y^i \in X_i, \\ i=1,\ldots,n-1}} \left\{ \sum_{i=1}^{n-1} \langle z^{i*}, y^i \rangle + \langle x^*, y^n \rangle - \sum_{i=1}^{n} \langle z^{(i-1)*}, F^i(y^i) \rangle \right\} \right\}$$

$$= \sup_{\substack{z^{i*} \in K_i^*, \\ i=0,\ldots,n-1}} \left\{ - f^*(z^{0*}) - (z^{(n-1)*} F^n)^*(x^*) - \sum_{i=1}^{n-1} (z^{(i-1)*} F^i)^*(z^{i*}) \right\}.$$

Hence, we define the conjugate dual problem corresponding to the primal problem (P^K) as

$$(D^K) \quad \sup_{\substack{z^{i*} \in K_i^*, \\ i=0,\ldots,n-1}} \left\{ - f^*(z^{0*}) - (z^{(n-1)*} F^n)^*(x^*) - \sum_{i=1}^{n-1} (z^{(i-1)*} F^i)^*(z^{i*}) \right\}.$$

Let us notice that for all $x^* \in X_n^*$ one has $\operatorname{dom} \widetilde{f} = \operatorname{dom}(\widetilde{f} + \langle x^*, \cdot \rangle)$. To guarantee strong duality between the problem (P^K) and its conjugate dual problem (D^K), we use the regularity conditions we introduced above. Therefore, we set $Z = X$ ordered by the trivial cone $Q = X$ and define the function $g : X \to X$ by $g(x) := x$ such that g is Q-epi closed and

$$0_X \in \operatorname{sqri}(g(X) + Q) = \operatorname{sqri}(X + Q) = X.$$

Hence, we get for the pair (P^K)-(D^K) the following regularity conditions. The first one looks like

$$(RC_1^K) \quad \left| \begin{array}{l} \exists (y^{0'}, y^{1'}, ..., y^{n'}) \in \operatorname{dom} f \times X_1 \times ... \times X_n \text{ such that} \\ F^i(y^{i'}) - y^{(i-1)'} \in -\operatorname{int} K_{i-1}, \; i = 1, ..., n \end{array} \right.$$

and can also be written as

$$(RC_{1'}^K) \quad \left| \begin{array}{l} \exists x' \in X_n \text{ such that } F^n(x') \in (F^{n-1})^{-1}((F^{n-2})^{-1}(... \\ (F^1)^{-1}(\operatorname{dom} f - \operatorname{int} K_0) - \operatorname{int} K_{n-1}...) - \operatorname{int} K_{n-2}) - \operatorname{int} K_{n-1}. \end{array} \right.$$

For the interior point regularity condition we get

$$(RC_2^K) \quad \left| \begin{array}{l} X_0, ..., X_n \text{ are Fréchet spaces, } f \text{ is l.s.c.,} \\ K_{i-1} \text{ is closed, } F^i \text{ is } K_{i-1}\text{-epi closed, } i = 1, ..., n, \\ 0_{X_0} \in \operatorname{sqri}(F^1(\operatorname{dom} F^1) - \operatorname{dom} f + K_0) \text{ and} \\ 0_{X_{i-1}} \in \operatorname{sqri}(F^i(\operatorname{dom} F^i) - \operatorname{dom} F^{i-1} + K_{i-1}), i = 2, ..., n. \end{array} \right.$$

In the same way we get representations for (RC_i^K), $i = 2', 2'', 3$.
By Theorem 3.3 we can state the following one:

Theorem 3.5 (strong duality). *Let $f : \overline{X}_0 \to \overline{\mathbb{R}}$ be proper, convex and K_0-increasing on $F^1(\text{dom } F^1) + K_0$, $F^i : \overline{X}_i \to \overline{X}_{i-1}$, be proper, K_{i-1}-convex and (K_i, K_{i-1})-increasing on $F^{i+1}(\text{dom } F^{i+1}) + K_i$, $i = 1, ..., n-1$ and $F^n : X_n \to \overline{X}_{n-1}$ be proper and K_{n-1}-convex. If one of the conditions (RC_i^K), $i \in \{1, 1', 2, 2', 2'', 3\}$, is fulfilled, then between (P^K) and (D^K) strong duality holds, i.e. $v(P^K) = v(D^K)$ and the conjugate dual problem has an optimal solution.*

Furthermore, it holds the following theorem.

Theorem 3.6. *Let $f : \overline{X}_0 \to \overline{\mathbb{R}}$ be proper, convex and K_0-increasing on $F^1(\text{dom } F^1) + K_0$, $F^i : \overline{X}_i \to \overline{X}_{i-1}$, be proper, K_{i-1}-convex and (K_i, K_{i-1})-increasing on $F^{i+1}(\text{dom } F^{i+1}) + K_i$, $i = 1, ..., n-1$ and $F^n : X_n \to \overline{X}_{n-1}$ be proper and K_{n-1}-convex. If one of the regularity conditions (RC_i^K), $i \in \{1, 1', 2, 2', 2'', 3\}$, is fulfilled, then the conjugate function of γ is given by*

$$\gamma^*(x^*) = \min_{\substack{z^{i*} \in K_i^*, \\ i=0,...,n-1}} \left\{ f^*(z^{0*}) + (z^{(n-1)*} F^n)^*(x^*) + \sum_{i=1}^{n-1} (z^{(i-1)*} F^i)^*(z^{i*}) \right\} \tag{3.18}$$

for all $x^* \in X_n^*$.

Proof. By using Theorem 3.5 it follows that

$$\gamma^*(x^*) = \sup_{x \in X} \{ \langle x^*, x \rangle - (f \circ F^1 \circ ... \circ F^n)(x) \}$$

$$= \min_{\substack{y^{i*} \in K_i^*, \\ i=0,...,n-1}} \left\{ f^*(y^{0*}) + (y^{(n-1)*} F^n)^*(x^*) + \sum_{i=1}^{n-1} (y^{(i-1)*} F^i)^*(y^{i*}) \right\} \forall x^* \in X_n^*.$$

$\square$

Remark 3.10. *The advantage of the introduced concept is that a "complicated" function γ can be split into $n + 1$ "simple" functions such that the calculation of the conjugate can be simplified by calculating just the conjugates of the $n + 1$ "simple" functions.*

Example 3.1. *Let us consider the following generalized signomial function $\gamma : \mathbb{R}^n \times \mathbb{R}^n \to \mathbb{R}$ defined by*

$$\gamma(x, y) = \begin{cases} \max \left\{ \frac{1}{x_1^{p_1} y_1^{q_1}}, ..., \frac{1}{x_n^{p_n} y_n^{q_n}} \right\}, & \text{if } (x, y) \in \text{int } \mathbb{R}_+^n \times \text{int } \mathbb{R}_+^n \\ +\infty, & \text{otherwise}, \end{cases}$$

with $p_i, q_i \geq 0$ for all $i = 1, ..., n$, and $x = (x_1, ..., x_n)^T \in \mathbb{R}^n$, $y = (y_1, ..., y_n)^T \in \mathbb{R}^n$. Then, we split the function γ into the functions

- $f : \overline{\mathbb{R}}^n \to \overline{\mathbb{R}}$ defined by

$$f(y^0) := \begin{cases} \max\{y_1^0, ..., y_n^0\}, & \text{if } y^0 = (y_1^0, ..., y_n^0)^T \in \mathbb{R}_+^n, \\ +\infty, & \text{otherwise,} \end{cases} \quad (3.19)$$

- $F^1 : \overline{\mathbb{R}}^n \to \overline{\mathbb{R}}^n$, defined by

$$F^1(y^1) := \begin{cases} (e^{y_1^1}, ..., e^{y_n^1})^T, & \text{if } y^1 = (y_1^1, ..., y_n^1)^T \in \mathbb{R}^n \\ +\infty_{\mathbb{R}_+^n}, & \text{otherwise,} \end{cases} \quad (3.20)$$

and

- $F^2 : \mathbb{R}^n \times \mathbb{R}^n \to \overline{\mathbb{R}^n}$, defined by

$$F^2(x,y) := \begin{cases} (-p_1 \ln x_1 - q_1 \ln y_1, ..., -p_n \ln x_n - q_n \ln y_n)^T, & \text{if } x, y \in \text{int } \mathbb{R}_+^n, \\ +\infty_{\mathbb{R}_+^n}, & \text{otherwise,} \end{cases} \quad (3.21)$$

such that γ is writeable as

$$\gamma(x,y) = (f \circ F^1 \circ F^2)(x,y) \quad (3.22)$$

and set $K_0 = K_1 = \mathbb{R}_+^n$. Without much effort one can observe that f is proper, convex and $\mathbb{R}_+^n$-increasing on $F^1(\text{dom } F^1) + \mathbb{R}_+^n = \text{int } \mathbb{R}_+^n + \mathbb{R}_+^n = \text{int } \mathbb{R}_+^n \subseteq \mathbb{R}_+^n$, F^1 is proper, $\mathbb{R}_+^n$-convex and $(\mathbb{R}_+^n, \mathbb{R}_+^n)$-increasing on $F^2(\text{dom } F^2) + \mathbb{R}_+^n = \mathbb{R}^n$ and F^2 is proper and $\mathbb{R}_+^n$-convex. Moreover, it is easy to verify that the regularity condition $(RC_{1'}^K)$ looks in this special case like

$$(RC_{1'}^{K_e}) \quad | \quad \exists (x', y') \in \mathbb{R}^n \times \mathbb{R}^n \text{ such that } -p_i \ln x_i' - q_i \ln y_i' \in \mathbb{R}, \ i = 1, ..., n,$$

which, of course, is always fulfilled. Thus, we can apply the formula (3.18) of Theorem 3.6 for the determination of the conjugate function of γ:

$$\gamma^*(x^*, y^*) = \min_{z^{0*}, z^{1*} \in \mathbb{R}_+^n} \left\{ f^*(z^{0*}) + (z^{0*}F^1)^*(z^{1*}) + (z^{1*}F^2)^*(x^*, y^*) \right\} \ \forall (x^*, y^*) \in \mathbb{R}^n \times \mathbb{R}^n.$$

$$(3.23)$$

Now, we have to calculate the conjugate functions involved in the formula (3.23). We have for $z^{0*} = (z_1^{0*}, ..., z_n^{0*})^T \in \mathbb{R}_+^n$:

$$\begin{aligned} f^*(z^{0*}) &= \sup_{(y_1^0, ..., y_n^0)^T \in \mathbb{R}^n} \left\{ \sum_{i=1}^n z_i^{0*} y_i^0 - f(y^0) \right\} \\ &= \sup_{(y_1^0, ..., y_n^0)^T \in \mathbb{R}_+^n} \left\{ \sum_{i=1}^n z_i^{0*} y_i^0 - \max\{y_1^0, ..., y_n^0\} \right\} \\ &= \sup_{(y_1^0, ..., y_n^0)^T \in \mathbb{R}_+^n} \left\{ \sum_{i=1}^n z_i^{0*} y_i^0 - \inf_{\substack{t \in \mathbb{R}_+, \ y_i^0 \leq t, \\ i=1,...,n}} t \right\} = \sup_{\substack{y_i^0 \in \mathbb{R}_+, \ t \in \mathbb{R}_+, \ y_i^0 \leq t, \\ i=1,...,n}} \left\{ \sum_{i=1}^n z_i^{0*} y_i^0 - t \right\}. \end{aligned}$$

As one may see, f^* can be expressed as a supremum of a linear function and thus, by elementary calculations, we have that

$$f^*(z^{0*}) = \begin{cases} 0, & if \ \sum_{i=1}^{n} z_i^{0*} \leq 1, \ (z_1^{0*}, ..., z_n^{0*})^T \in \mathbb{R}_+^n, \\ +\infty, & otherwise. \end{cases} \tag{3.24}$$

From (3.23) and (3.24) follows for the conjugate function of γ

$$\gamma^*(x^*, y^*) = \min_{\substack{z_i^{0*}, \ z_i^{1*} \in \mathbb{R}_+, \ i=1,...,n, \\ \sum_{i=1}^{n} z_i^{0*} \leq 1}} \left\{ (z^{0*} F^1)^*(z^{1*}) + (z^{1*} F^2)^*(x^*, y^*) \right\}. \tag{3.25}$$

Furthermore, we have for $z_i^{0*} \geq 0$, $i = 1, ..., n$,

$$\begin{aligned} (z^{0*} F^1)^*(z^{1*}) &= \sup_{y_i^1 \in \mathbb{R}, \ i=1,...,n} \left\{ \sum_{i=1}^{n} z_i^{1*} y_i^1 - \sum_{i=1}^{n} z_i^{0*} e^{y_i^1} \right\} \\ &= \sum_{i=1}^{n} \sup_{y_i^1 \in \mathbb{R}} \{ z_i^{1*} y_i^1 - z_i^{0*} e^{y_i^1} \} \end{aligned}$$

with (see [14] or also [29])

$$\sup_{y_i^1 \in \mathbb{R}} \{ z_i^{1*} y_i^1 - z_i^{0*} e^{y_i^1} \} = \begin{cases} z_i^{1*} \left(\ln \frac{z_i^{1*}}{z_i^{0*}} - 1 \right), & if \ z_i^{0*}, \ z_i^{1*} > 0, \\ 0, & if \ z_i^{1*} = 0, \ z_i^{0*} \geq 0, \\ +\infty, & otherwise, \end{cases} \tag{3.26}$$

for $i = 1, ..., n$ and for $z_i^{1*} \geq 0$, $i = 1, ..., n$, it holds

$$\begin{aligned} (z^{1*} F^2)^*(x^*, y^*) &= \sup_{x_i, y_i > 0, \ i=1,...,n} \left\{ \sum_{i=1}^{n} x_i^* x_i + \sum_{i=1}^{n} y_i^* y_i + \sum_{i=1}^{n} z_i^{1*} p_i \ln x_i + \sum_{i=1}^{n} z_i^{1*} q_i \ln y_i \right\} \\ &= \sum_{i=1}^{n} \left(\sup_{x_i > 0} \{ x_i^* x_i + z_i^{1*} p_i \ln x_i \} + \sup_{y_i > 0} \{ y_i^* y_i + z_i^{1*} q_i \ln y_i \} \right) \end{aligned}$$

for all $x^* = (x_1^*, ..., x_n^*)^T$, $y^* = (y_1^*, ..., y_n^*)^T \in \mathbb{R}^n$, where (see [29])

$$\sup_{x_i > 0} \{ x_i^* x_i + z_i^{1*} p_i \ln x_i \} = \begin{cases} -z_i^{1*} p_i \left(1 + \ln \left(-\frac{x_i^*}{z_i^{1*} p_i} \right) \right), & if \ x_i^* < 0, \ z_i^{1*}, \ p_i > 0, \\ 0, & if \ x_i^* \leq 0 \ and \ z_i^{1*} = 0 \ or \ x_i^* \leq 0 \ and \ p_i = 0, \\ +\infty, & otherwise, \end{cases} \tag{3.27}$$

and likewise

$$\sup_{y_i > 0} \{ y_i^* y_i + z_i^{1*} q_i \ln y_i \} = \begin{cases} -z_i^{1*} q_i \left(1 + \ln \left(-\frac{y_i^*}{z_i^{1*} q_i} \right) \right), & if \ y_i^* < 0, \ z_i^{1*}, \ q_i > 0, \\ 0, & if \ y_i^* \leq 0 \ and \ z_i^{1*} = 0 \ or \ y_i^* \leq 0 \ and \ q_i = 0, \\ +\infty, & otherwise, \end{cases} \tag{3.28}$$

for $i = 1, ..., n$. Finally, we define the function $\xi : \mathbb{R} \to \{0, 1\}$ by

$$\xi(x) = \begin{cases} 1, & \text{if } x > 0, \\ 0, & \text{otherwise}, \end{cases} \tag{3.29}$$

which leads, by using (3.25), (3.26), (3.27), (3.28) and (3.29), to the following formula of the conjugate function of γ

$$\gamma^*(x^*, y^*) = \min_{\substack{\sum_{n=1}^{n} z_i^{0*} \leq 1, \ z_i^{0*} \geq 0, \\ z_i^{1*} \geq 0, \ i=1,...,n}} \left\{ \sum_{i=1}^{n} z_i^{1*}[(\ln z_i^{1*} - \ln z_i^{0*} - 1)\, \xi(z_i^{0*}) \right.$$

$$\left. - p_i \left(1 + \ln x_i^* - \ln z_i^{1*} p_i\right) - q_i \left(1 + \ln y_i^* - \ln z_i^{1*} q_i\right)] \right\}$$

for all $x_i^*, \ y_i^* \geq 0$, $i = 1, ..., n$, with the convention $0 \ln 0 = 0$.

Next, we give an alternative representation for γ. But, first pay attention to the following function

$$\beta(x^*) := \inf_{\substack{z^{i*} \in X_i^*, \\ i=0,...,n-1}} \left\{ f^*(z^{0*}) + (z^{(n-1)*} F^n)^*(x^*) + \sum_{i=1}^{n-1} (z^{(i-1)*} F^i)^*(z^{i*}) \right\} \ \forall x^* \in X_n^*.$$

If $f : X_0 \to \overline{\mathbb{R}}$ is a K_0-increasing function on $\{F^1(\text{dom } F^1) + K_0\} - K_0$, it follows by [14, Proposition 2.3.11] that

$$f^*(z^{0*}) = +\infty \ \forall z^{0*} \notin K_0^*, \ \text{i.e. } \text{dom } f^* \subseteq K_0^*,$$

and thus it holds

$$\beta(x^*) = \inf_{\substack{z^{0*} \in K_0^*, \ z^{i*} \in X_i^*, \\ i=1,...,n-1}} \left\{ f^*(z^{0*}) + (z^{(n-1)*} F^n)^*(x^*) + \sum_{i=1}^{n-1} (z^{(i-1)*} F^i)^*(z^{i*}) \right\}$$

for all $x^* \in X_n^*$. Moreover, if $F^1 : X_1 \to \overline{X}_0$ is (K_1, K_0)-increasing on $\{F^2(\text{dom } F^2) + K_1\} - K_1$, then $(z^{0*} F^1) : X_1 \to \overline{\mathbb{R}}$ is K_1-increasing on $\{F^2(\text{dom } F^2) + K_1\} - K_1$ for $z^{0*} \in K_0^*$. By using again [14, Proposition 2.3.11] one gets for $z^{0*} \in K_0^*$

$$(z^{0*} F^1)^*(z^{1*}) = +\infty \ \forall z^{1*} \notin K_1^*, \ \text{i.e. } \text{dom}(z^{0*} F^1) \subseteq K_1^*$$

and we can write

$$\beta(x^*) = \inf_{\substack{z^{0*} \in K_0^*, \ z^{1*} \in K_1^*, \\ z^{i*} \in X_i^*, \ i=2,...,n-1}} \left\{ f^*(z^{0*}) + (z^{(n-1)*} F^n)^*(x^*) + \sum_{i=1}^{n-1} (z^{(i-1)*} F^i)^*(z^{i*}) \right\} \ \forall x^* \in X_n^*.$$

If we proceed in this way, it follows that

$$(z^{(i-1)*}F^i)^*(z^{i*}) = +\infty \ \forall z^{i*} \notin K_i^*, \text{ i.e. } \operatorname{dom}(z^{(i-1)*}F^i)^* \subseteq K_i^*, \ i = 2, ..., n-1,$$

and therefore, it holds

$$\beta(x^*) = \inf_{\substack{z^{i*} \in K_i^*, \\ i=0,...,n-1}} \left\{ f^*(z^{0*}) + (z^{(n-1)*}F^n)^*(x^*) + \sum_{i=1}^{n-1}(z^{(i-1)*}F^i)^*(z^{i*}) \right\}$$

for all $x^* \in X_n^*$. For the conjugate function of β one has

$$
\begin{aligned}
\beta^*(x) &= \sup_{x^* \in X_n^*} \left\{ \langle x^*, x \rangle - \beta(x^*) \right\} \\
&= \sup_{x^* \in X_n^*} \left\{ \langle x^*, x \rangle - \inf_{\substack{z^{i*} \in X_i^*, \\ i=0,...,n-1}} \left\{ f^*(z^{0*}) + (z^{(n-1)*}F^n)^*(x^*) + \sum_{i=1}^{n-1}(z^{(i-1)*}F^i)^*(z^{i*}) \right\} \right\} \\
&= \sup_{\substack{x^* \in X_n^*, \ z^{i*} \in X_i^*, \\ i=0,...,n-1}} \left\{ \langle x^*, x \rangle - f^*(z^{0*}) - (z^{(n-1)*}F^n)^*(x^*) - \sum_{i=1}^{n-1}(z^{(i-1)*}F^i)^*(z^{i*}) \right\} \\
&= \sup_{\substack{z^{i*} \in X_i^*, \\ i=0,...,n-1}} \left\{ \sup_{x^* \in X_n^*} \{ \langle x^*, x \rangle - (z^{(n-1)*}F^n)^*(x^*) \} - f^*(z^{0*}) - \sum_{i=1}^{n-1}(z^{(i-1)*}F^i)^*(z^{i*}) \right\} \\
&= \sup_{\substack{z^{i*} \in X_i^*, \\ i=0,...,n-1}} \left\{ (z^{(n-1)*}F^n)^{**}(x) \} - f^*(z^{0*}) - \sum_{i=1}^{n-1}(z^{(i-1)*}F^i)^*(z^{i*}) \right\} \quad (3.30)
\end{aligned}
$$

for all $x \in X_n$. Since F^n is proper and K_{n-1}-convex and if we ask that F^n is also positively K_{n-1}-lower semicontinuous, (3.30) can by using the Fenchel-Moreau Theorem be written as

$$\beta^*(x) = \sup_{\substack{z^{i*} \in X_i^*, \\ i=0,...,n-1}} \left\{ (z^{(n-1)*}F^n)(x) - f^*(z^{0*}) - \sum_{i=1}^{n-1}(z^{(i-1)*}F^i)^*(z^{i*}) \right\} \quad (3.31)$$

for all $x \in X_n$. If we additionally ask that the function F^i is positively K_{i-1}-lower semicontinuous, $i = 1, ..., n-1$, and if we assume that f is lower semicontinuous, then one gets for (3.31) by using again the Fenchel-Moreau Theorem

$$
\begin{aligned}
\beta^*(x) = \sup_{\substack{z^{i*} \in X_i^*, \\ i=0,...,n-2}} \Big\{ &\sup_{z^{(n-1)*} \in X_{n-1}^*} \{ \langle z^{(n-1)*}, F^n(x) \rangle - (z^{(n-2)*}F^{(n-1)})^*(z^{(n-1)*}) \} - \\
&f^*(z^{0*}) - \sum_{i=1}^{n-2}(z^{(i-1)*}F^i)^*(z^{i*}) \Big\}
\end{aligned}
$$

$$= \sup_{\substack{z^{i*} \in X_i^*, \\ i=0,\dots,n-2}} \left\{ (z^{(n-2)*}F^{n-1})^{**}(F^n(x)) - f^*(z^{0*}) - \sum_{i=1}^{n-2}(z^{(i-1)*}F^i)^*(z^{i*}) \right\}$$

$$= \sup_{\substack{z^{i*} \in X_i^*, \\ i=0,\dots,n-2}} \left\{ (z^{(n-2)*}F^{n-1})(F^n(x)) - f^*(z^{0*}) - \sum_{i=1}^{n-2}(z^{(i-1)*}F^i)^*(z^{i*}) \right\}$$

$$= \sup_{\substack{z^{i*} \in X_i^*, \\ i=0,\dots,n-3}} \left\{ \sup_{z^{(n-2)*} \in X_{n-2}^*} \left\{ \langle z^{(n-2)*}, F^{n-1}(F^n(x)) \rangle - (z^{(n-3)*}F^{(n-2)})^*(z^{(n-2)*}) \right\} \right.$$
$$\left. - f^*(z^{0*}) - \sum_{i=1}^{n-3}(z^{(i-1)*}F^i)^*(z^{i*}) \right\}$$

$$= \sup_{\substack{z^{i*} \in X_i^*, \\ i=0,\dots,n-3}} \left\{ (z^{(n-3)*}F^{n-2})(F^{n-1}(F^n(x))) - f^*(z^{0*}) - \sum_{i=1}^{n-3}(z^{(i-1)*}F^i)^*(z^{i*}) \right\}$$

$$\vdots$$

$$= \sup_{z^{0*} \in X_0^*} \left\{ \langle z^{0*}, (F^1 \circ \dots \circ F^n)(x) \rangle - f^*(z^{0*}) \right\} = f^{**}((F^1 \circ \dots \circ F^n)(x))$$

$$= (f \circ F^1 \circ \dots \circ F^n)(x) = \gamma(x) \ \forall x \in X_n.$$

Since the weak duality always holds, i.e. $v(P^K) \geq v(D^K)$, we have $\gamma^*(x^*) \leq \beta(x^*)$ for all $x^* \in X_n^*$. Moreover, it holds $\gamma(x) \geq \gamma^{**}(x)$ for all $x \in X_n$ and from here it follows that $\gamma(x) \geq \gamma^{**}(x) \geq \beta^*(x) = \gamma(x)$, $x \in X_n$, i.e. $\gamma(x) = \gamma^{**}(x)$ for all $x \in X_n$. The latter means that γ is proper, convex and lower semicontinuous. Summarizing, we get the following theorem:

Theorem 3.7. *Let $f : \overline{X}_0 \to \overline{\mathbb{R}}$ be a proper, convex, K_0-increasing on $\{F^1(\text{dom } F^1) + K_0\} - K_0$ and lower semicontinuous function, $F^i : \overline{X}_i \to \overline{X}_{i-1}$ be a proper, K_{i-1}-convex, (K_i, K_{i-1})-increasing on $\{F^{i+1}(\text{dom } F^{i+1}) + K_i\} - K_i$ and positively K_{i-1}-lower semicontinuous function, $i = 1, \dots, n-1$, and $F^n : X_n \to \overline{X}_{n-1}$ be a proper, K_{n-1}-convex and positively K_{n-1}-lower semicontinuous function. Then the function $\gamma = f \circ F^1 \circ \dots \circ F^n : X_n \to \overline{\mathbb{R}}$ is proper, convex and lower semicontinuous and can alternatively be written as*

$$(f \circ F^1 \circ \dots \circ F^n)(x) = \sup_{\substack{z^{i*} \in X_i^*, \\ i=0,\dots,n-1}} \left\{ (z^{(n-1)*}F^n)(x) - f^*(z^{0*}) - \sum_{i=1}^{n-1}(z^{(i-1)*}F^i)^*(z^{i*}) \right\}$$

for all $x \in X_n$.

Remark 3.11. *Besides the introduced duality concept there is a second way to construct a corresponding conjugate dual problem to (P^C) and to formulate associated duality statements, where the conjugates of the functions involved in the objective function of the original problem are split.*

This dual approach is characterized by the direct applying of the perturbation theory by defining an associated perturbation function of the following form

$$\Phi(x, y^0, ..., y^{n+1}) := \begin{cases} f(F^1(...F^{n-1}(F^n(x + y^n) + y^{n-1})...) + y^0), & \text{if } g(x) \in y^{n+1} - Q, \\ +\infty, & \text{otherwise,} \end{cases}$$

where $(y^0, ..., y^n, y^{n+1}) \in X_0 \times ... \times X_n \times Z$ *are the dual variables.*

If we use this method in the context of the generalized interior point regularity conditions, then we have to impose for strong duality that the perturbation function Φ *is lower semicontinuous (see [14]). But this means, as shown in Theorem 3.7, that we have to ensure that the functions* F^i *are all positively* K_{i-1}-*lower semicontinuous, respectively. In contrast, to employ the proposed method in this chapter, we only need to secure that each of these functions is* K_{i-1}-*epi closed, respectively. It is well known that if a function* F^i *is positively* K_{i-1}-*lower semicontinuous, then it is also* K_{i-1}-*epi closed, while the inverse statement is not true in general (see Proposition 2.2.19 and Example 2.2.6. in [14]). In this sense the method introduced in this book asks for weaker hypothesis on the involved functions for guaranteeing strong duality.*

Finally, let us turn to the question why we did not apply the Fenchel-Lagrange duality theory to the reformulated primal problem $(\widetilde{P}^C)$ *with set and cone constraints. The reason is that even though that the functions* F^n *and* g *can not be split directly, one derives more complicated and stronger regularity conditions compared to the ones proposed in this work.*

3.4 An optimization problem having as objective function the sum of reciprocals of concave functions

Let E_i be a non-empty convex subset of X, $i = 1, ..., n$, where X is a locally convex Hausdorff space partially ordered by the closed and convex cone K. Then, we consider a convex optimization problem having as objective function the sum of reciprocals of concave functions $h_i : E_i \to \mathbb{R}$ with strict positive values, $i = 1, .., n$, and geometric and cone constraints, i.e., the optimization problem that we discuss in this section (cf. the definitions from Section 3.1) is given by

$$(P^G) \qquad \inf_{\substack{x \in S, \\ g(x) \in -Q}} \left\{ \sum_{i=1}^n \frac{1}{h_i(x)} \right\}.$$

Optimization problems of this type arise, for instance, in the study of power functions by setting $h_i : \mathbb{R}_+ \to \mathbb{R}$, $h_i(x) = c_i x^{p_i}$ with $c_i p_i(p_i - 1) \leq 0$, $i = 1, ..., n$, (see [92]) and have a wide range of applications in economics, engineering and finance.

To apply the results from the previous section to (P^G), i.e. to characterize strong duality and to derive optimality conditions, we assume that the function $-h_i$ is K-increasing on E_i, $i = 1, ..., n$, and set $X_0 = \mathbb{R}^n$, $K_0 = \mathbb{R}^n_+$, $X_1 = X^n$, $K_1 = K^n$ and $X_2 = X$. Additionally, we define the following functions

- $f : \overline{\mathbb{R}}^n \to \overline{\mathbb{R}}$,

$$f(y^0) = \begin{cases} -\sum_{i=1}^{n} \frac{1}{y_i^0}, & \text{if } y_i^0 < 0, \ i = 1, ..., n, \\ +\infty, & \text{otherwise}, \end{cases}$$

- $F^1 : X^n \to \overline{\mathbb{R}}^n$,

$$F^1(y^1) = \begin{cases} (-h_1(y_1^1), ..., -h_n(y_n^1))^T, & \text{if } y_i^1 \in E_i, \ i = 1, ..., n, \\ +\infty_{\mathbb{R}_+^n}, & \text{otherwise} \end{cases}$$

and

- $F^2 : X \to X^n$, $F^2(x) := (x, ..., x) \in X^n$

and we assume that $F^2(S \cap \operatorname{dom} g) \subseteq E_1 \times ... \times E_n$ (cf. Remark 3.3). From here, it follows that the problem (P^G) can equivalently be written as

$$(P^G) \qquad \inf_{\substack{x \in S, \\ g(x) \in -Q}} \left\{ (f \circ F^1 \circ F^2)(x) \right\}$$

and by using the formula from Section 3.1 its corresponding conjugate dual problem (D^G) turns into

$$(D^G) \qquad \sup_{\substack{z^{0*} \in \mathbb{R}_+^n, \ z^{1*} \in (K^*)^n, \\ z^{2*} \in Q^*}} \left\{ \inf_{x \in S} \left\{ \left\langle \sum_{i=1}^{n} z_i^{1*}, x \right\rangle + \langle z^{2*}, g(x) \rangle \right\} - f^*(z^{0*}) - (z^{0*} F^1)^*(z^{1*}) \right\}.$$

Furthermore, one has (see [15], [58] or [62]):

$$f^*(z^{0*}) = \sum_{i=1}^{n} \sup_{y_i < 0} \left\{ z_i^{0*} y_i^0 + \frac{1}{y_i^0} \right\} = -2 \sum_{i=1}^{n} \sqrt{z_i^{0*}}$$

for all $z_i^{0*} \geq 0$, $i = 1, ..., n$, and since

$$(z^{0*} F^1)^*(z^{1*}) = \sum_{i=1}^{n} \sup_{y_i^1 \in E_i} \left\{ \langle z_i^{1*}, y_i^1 \rangle + z_i^{0*} h_i(y_i^1) \right\} = \sum_{i=1}^{n} (-z_i^{0*} h_i)_{E_i}^*(z_i^{1*})$$

holds, one gets for the conjugate dual problem

$$(D^G) \qquad \sup_{\substack{z^{0*} \in \mathbb{R}_+^n, \ z^{1*} \in (K^*)^n, \\ z^{2*} \in -Q^*}} \left\{ -(z^{2*} g)_S^* \left(-\sum_{i=1}^{n} z_i^{1*} \right) + \sum_{i=1}^{n} \left(2\sqrt{z_i^{0*}} - (-z_i^{0*} h_i)_{E_i}^*(z_i^{1*}) \right) \right\}.$$

It is easy to observe that f is proper, $\mathbb{R}_+^n$-increasing on $\operatorname{dom} f = -\operatorname{int}(\mathbb{R}_+^n)$, convex and lower semicontinuous, F^1 is proper, $(K^n, \mathbb{R}_+^n)$-increasing on $\operatorname{dom} F^1 = E_1 \times ... \times E_n$ and $\mathbb{R}_+^n$-convex and that $F^1(\operatorname{dom} F^1) \subseteq \operatorname{int}(-\mathbb{R}_+^n) = \operatorname{dom} f$ (in this context pay attention on Remark 3.3). For that reason we can now attach the regularity condition (RC_1^C), specialized for the optimization problem (P^G),

(RC_1^G) $\quad \mid\quad \exists (y^{0'}, y^{1'}, y^{2'}) \in (-\infty, 0)^n \times X^n \times S$ such that $h_i(y_i^{1'}) + y_i^{0'} > 0,$
$\qquad\qquad y_i^{2'} - y_i^{1'} \in -\operatorname{int} K,\ i = 1, ..., n,$ and $g(y^{2'}) \in -\operatorname{int} Q.$

As h_i is a concave function with strict positive values on E_i, there exist $y_i^{0*} < 0$ and $y_i^{1*} \in E_i$ such that $h_i(y_i^{1'}) + y_i^{0'} > 0,\ i = 1, ..., n,$ and hence (RC_1^G) reduces to

(RC_1^G) $\quad \mid\quad \exists (y^{1'}, y^{2'}) \in X^n \times S$ such that $y^{2'} - y_i^{1'} \in -\operatorname{int} K,\ i = 1, ..., n,$
$\qquad\qquad$ and $g(y^{2'}) \in -\operatorname{int} Q.$

or, equivalently, in the light of $(RC_{1'}^G)$,

$(RC_{1'}^G)$ $\quad \mid\quad \exists x' \in S$ such that $x' \in E_i - \operatorname{int} K,\ i = 1, ..., n,$ and $g(x') \in -\operatorname{int} Q.$

The generalized interior point regularity conditions (RC_2^C), specialized for (P^G), looks like

(RC_2^G) $\quad \mid\quad X$ and Z are Fréchet spaces, S is closed, g is Q-epi closed,
$\qquad\qquad -h_i$ is lower semicontinuous, $0_X \in \operatorname{sqri}(\operatorname{dom} g \cap S - E_i + K),$
$\qquad\qquad i = 1, ..., n,$ and $0_Z \in \operatorname{sqri}(g(\operatorname{dom} g \cap S) + Q).$

In the same way one can formulate a specialized regularity condition (RC_i^G) in respect to the condition (RC_i^C) for $i \in \{2', 2'', 3\}$.

Remark 3.12. *Recall, that in respect to Remarks 3.2 and 3.5 the function F^1 does not need to be monotone, because F^2 is a linear function. In this case we set, as mentioned in Remark 3.2, $K_1 = \{0_{X^n}\} = \{0_X\}^n$. But pay attention to the circumstance that the regularity conditions (RC_1^G) and $(RC_{1'}^G)$ are no more applicable in this framework, as $\operatorname{int}\{0_X\} = \emptyset$.*

By Theorems 3.3 and 3.4 the strong duality statement and the optimality conditions follows immediately.

Theorem 3.8 (strong duality). *If one of the conditions (RC_i^G), $i \in \{1, 1', 2, 2', 2'', 3\}$, is fulfilled, then between (P^G) and (D^G) strong duality holds, i.e. $v(P^G) = v(D^G)$ and the conjugate dual problem has an optimal solution.*

Theorem 3.9 (optimality conditions). *(a) Suppose that one of the regularity conditions (RC_i^G), $i \in \{1, 1', 2, 2', 2'', 3\}$, is fulfilled and let $\overline{x} \in S$ be an optimal solution to the problem (P^G). Then there exists $(\overline{z}^{0*}, \overline{z}^{1*}, \overline{z}^{2*}) \in \mathbb{R}_+^n \times (K^*)^n \times Q^*$, an optimal solution to (D^G), such that*

(i) $\displaystyle\sum_{i=1}^n \frac{1}{h_i(\overline{x})} - 2 \sum_{i=1}^n \sqrt{\overline{z}_i^{0*}} = -\sum_{i=1}^n \overline{z}_i^{0*} h_i(\overline{x}),$

(ii) $\displaystyle\sum_{i=1}^n (-\overline{z}_i^{0*} h_i)_{E_i}^* (\overline{z}_i^{1*}) - \sum_{i=1}^n \overline{z}_i^{0*} h_i(\overline{x}) = \left\langle \sum_{i=1}^n \overline{z}_i^{1*}, \overline{x} \right\rangle,$

(iii) $\displaystyle\langle \overline{z}^{2*}, g(\overline{x})\rangle + (\overline{z}^{2*} g)_S^* \left(-\sum_{i=1}^n \overline{z}_i^{1*}\right) = \left\langle -\sum_{i=1}^n \overline{z}_i^{1*}, \overline{x} \right\rangle,$

(iv) $\langle \overline{z}^{2*}, g(\overline{x}) \rangle = 0.$

(b) If there exists $\overline{x} \in S$ such that for some $(\overline{z}^{0*}, \overline{z}^{1*}, \overline{z}^{2*}) \in \mathbb{R}_+^n \times (K^*)^n \times Q^*$ the conditions (i)-(iv) are fulfilled, then $\overline{x}$ is an optimal solution to (P^G), $(\overline{z}^{0*}, \overline{z}^{1*}, \overline{z}^{2*})$ is an optimal solution to (D^G) and $v(P^G) = v(D^G)$.

Remark 3.13. *In view of the Young-Fenchel inequality, we can refine the conditions (i) and (ii) of Theorem 3.9 like follows*

(i) $\overline{z}_i^{0*} h_i(\overline{x}) = 2\sqrt{\overline{z}_i^{0*}} - \frac{1}{h_i(\overline{x})}$, $i = 1, ..., n,$

(ii) $(-\overline{z}_i^{0*} h_i)_{E_i}^*(\overline{z}_i^{1*}) - \overline{z}_i^{0*} h_i(\overline{x}) = \langle \overline{z}_i^{1*}, \overline{x} \rangle$, $i = 1, ..., n.$

In the end of this section we give, for completeness, alternative representations of the optimality conditions presented in Theorem 3.9 and refined in the previous remark.

Remark 3.14. *In accordance with Remarks 3.8 and 3.13 the optimality conditions (i)-(iv) of Theorem 3.9 can equivalently be rewritten as*

(i) $\overline{z}_i^{0*} \in \partial \left(-\frac{1}{\cdot} \right) (-h_i(\overline{x}))$, $i = 1, ..., n,$

(ii) $\overline{z}_i^{1*} \in \partial(-\overline{z}_i^{0*} h_i)(\overline{x})$, $i = 1, ..., n,$

(iii) $-\sum_{i=1}^{n} \overline{z}_i^{1*} \in \partial((\overline{z}^{2*} g) + \delta_S)(\overline{x}),$

(iv) $\langle \overline{z}^{2*}, g(\overline{x}) \rangle = 0.$

Remark 3.15. *One may see that the function F^2 has been introduced in order to split the functions h_i, $i = 1, ..., n$, and g or, more precisely, to decompose their conjugate functions in the formulation of the dual problem (D^G). As a further advantage one gets a detailed characterization of the set of optimality conditions presented in Theorem 3.9, Remark 3.13 and Remark 3.14. Other duality schemes may be employed for approaching this kind of optimization problems, too, however, the separation of the conjugates of the involved functions in the corresponding dual problems may fail to happen. This also underlines the benefit of the introduced multi-composed duality concept.*

4 Duality results for minmax location problems

In the recent years, location problems attracted a considerable attention in the scientific community and a large number of papers studying minsum and minmax location problems have been published (see [23, 26, 38–41, 43, 45, 49, 50, 57, 66, 68–70, 77, 81, 94, 94, 95]). This is due to the fact that location problems cover many practical situations occurring for example in geometry, physics, economics, urban area models, computer science, telecommunication and also in health management and emergency facilities location programming.

In this chapter, which is mainly based on our articles [56, 101, 102] and [104], minmax location problems form the focal point of our approach. As suggested, for instance, in [2, 87], solving general minmax location problems as considered in this book could prove to be useful in dealing with some classes of constrained optimization, too, like the ones that appear in machine learning.

In particular, we investigate nonlinear single minmax location problems that are generalizations of the classical *Sylvester problem* in location theory - not to be confused with Sylvester's line problem, and of the celebrated Apollonius problem, formulated by means of an extended perturbed minimal time function via the duality theory developed in the previous chapter, which allows us to formulate more detailed dual problems as well as associated duality statements as in the mentioned papers. This approach is necessary in order to be able to numerically solve such problems and their corresponding dual problems by means of a proximal method, which will be presented in Chapter 6.

In addition to the single minmax location problems defined by extended perturbed minimal time functions, we study single minmax location problems defined by gauges, extended multifacility and classical multifacility minmax location problems and to each of them we consider different settings to specialize the associated duality results. While most of the theoretical results are provided in the general framework of Banach spaces, in the more restrictive setting of Hilbert spaces we were also able to provide characterizations of the optimal solutions of the considered problems by means of the dual optimal solutions.

In order to introduce the general single nonlinear minmax location problems we propose in the next section a new perturbed minimal time function that generalizes the classical minimal time function, introduced over four decades ago and recently reconsidered by Mordukhovich and Nam in a series of papers (see, for instance, [79, 80, 84, 85]) and the book [78], and several of its recent extensions (cf. [76, 84, 85, 98, 108]). The motivation to investigate problems formulated by the minimal time functions comes from both theoretical and practical reasons, as location type problems arise, as already mentioned, in various

© Springer Fachmedien Wiesbaden GmbH, part of Springer Nature 2020
O. Wilfer, *Multi-Composed Programming with Applications to Facility Location*,
Mathematische Optimierung und Wirtschaftsmathematik | Mathematical
Optimization and Economathematics, https://doi.org/10.1007/978-3-658-30580-2_4

areas of research and real life, applications from these fields being mentioned in our paper as possible interpretations of our results.

But first, some properties of gauges will be listed in the next section. Gauge functions are a generalization of norms and can be understood as infimal distances to sets. The use of these functions allows to consider more general location models, especially, in situations when asymmetric distance measures are of interest.

4.1 Some properties of gauges and perturbed minimal time functions

Let us start this section by proving the following statements that we also shall use in the sequel.

Lemma 4.1. Let $a_i \in \mathbb{R}_+$ be a given point and $h_i : \mathbb{R} \to \overline{\mathbb{R}}$ with $h_i(x) \in \mathbb{R}_+$, if $x \in \mathbb{R}_+$, and $h_i(x) = +\infty$, otherwise, be a proper, lower semicontinuous and convex function, $i = 1, ..., n$. Then the conjugate of the function $g : \mathbb{R}^n \to \overline{\mathbb{R}}$ defined by

$$g(x_1, ..., x_n) := \begin{cases} \max\{h_1(x_1) + a_1, ..., h_n(x_n) + a_n\}, & \text{if } x_i \in \mathbb{R}_+, \ i = 1, ..., n, \\ +\infty, & \text{otherwise,} \end{cases}$$

is given by $g^* : \mathbb{R}^n \to \mathbb{R}$,

$$g^*(x_1^*, ..., x_n^*) = \min_{\substack{\sum_{i=1}^{n} z_i^{0*} \le 1, \ z_i^{0*} \ge 0, \\ i=1,...,n}} \left\{ \sum_{i=1}^{n} [(z_i^{0*} h_i)^*(x_i^*) - z_i^{0*} a_i] \right\}.$$

Proof. We set $X_0 = X_1 = \mathbb{R}^n$ and $K_0 = \mathbb{R}_+^n$. Further, we define the function $f : \overline{\mathbb{R}^n} \to \overline{\mathbb{R}}$ by

$$f(y_1^0, ..., y_n^0) := \begin{cases} \max\{y_1^0 + a_1, ..., y_n^0 + a_n\}, & \text{if } y_i^0 \in \mathbb{R}_+, \ i = 1, ..., n, \\ +\infty, & \text{otherwise,} \end{cases}$$

and the function $F^1 : \mathbb{R}^n \to \overline{\mathbb{R}^n}$ by

$$F^1(x_1, ..., x_n) := \begin{cases} (h_1(x_1), ..., h_n(x_n))^T, & \text{if } x_i \in \mathbb{R}_+, \ i = 1, ..., n, \\ +\infty_{\mathbb{R}_+^n}, & \text{otherwise.} \end{cases}$$

Hence, the function g can be written as

$$g(x_1, ..., x_n) = (f \circ F^1)(x_1, ..., x_n).$$

It can easy be verified that the function f is proper, convex, lower semicontinuous and $\mathbb{R}_+^n$-increasing on $F^1(\text{dom } F^1) + K_0 \subseteq \mathbb{R}_+^n$ (as f is the pointwise supremum of proper, convex and lower semicontinuous functions) and the function F^1 is proper, $\mathbb{R}_+^n$-epi closed and $\mathbb{R}_+^n$-convex.

Therefore, it follows by Theorem 3.6 (note also that $0_{\mathbb{R}^n} \in \operatorname{sqri}(F^1(\operatorname{dom} F^1) - \operatorname{dom} f + K_0) = \operatorname{sqri}(F^1(\operatorname{dom} F^1) - \mathbb{R}_+^n + \mathbb{R}_+^n) = \mathbb{R}^n$) that

$$g^*(x_1^*, ..., x_n^*) = \min_{\substack{y_i^{0*} \in \mathbb{R}_+, \\ i=1,...,n}} \{f^*(y_1^{0*}, ..., y_n^{0*}) + ((y_1^{0*}, ..., y_n^{0*})^T F^1)^*(x_1^*, ..., x_n^*)\}.$$

For the conjugate of the function f we have

$$
\begin{aligned}
f^*(y^{0*}) &= \sup_{y_i^0 \in \mathbb{R},\ i=1,...,n} \left\{ \sum_{i=1}^n y_i^{0*} y_i^0 - f(y_1^0, ..., y_n^0) \right\} \\
&= \sup_{y_i^0 \in \mathbb{R}_+,\ i=1,...,n} \left\{ \sum_{i=1}^n y_i^{0*} y_i^0 - \max_{1 \le i \le n} \{y_i^0 + a_i\} \right\} \\
&= \sup_{y_i^0 \in \mathbb{R}_+,\ i=1,...,n} \left\{ \sum_{i=1}^n y_i^{0*} y_i^0 - \min_{\substack{t \in \mathbb{R}_+,\, y_i^0 + a_i \le t, \\ i=1,...,n}} t \right\} \\
&= \sup_{\substack{t \in \mathbb{R}_+,\ y_i^0 \in \mathbb{R}_+, \\ y_i^0 + a_i \le t,\ i=1,...,n}} \left\{ \sum_{i=1}^n y_i^{0*} y_i^0 - t \right\}.
\end{aligned}
\tag{4.1}
$$

Now, let us consider for any $y^{0*} \in \mathbb{R}_+^n$ the following primal optimization problem

$$(P^{max}) \qquad \inf_{\substack{t \in \mathbb{R}_+,\ y_i^0 \in \mathbb{R}_+, \\ y_i^0 + a_i \le t,\ i=1,...,n}} \left\{ t - \sum_{i=1}^n y_i^{0*} y_i^0 \right\}. \tag{4.2}$$

and its corresponding Lagrange dual problem

$$
\begin{aligned}
(D^{max}) \quad & \sup_{\lambda_i \ge 0,\ i=1,...,n} \inf_{\substack{t \in \mathbb{R}_+,\ y_i^0 \in \mathbb{R}_+, \\ i=1,...,n}} \left\{ t - \sum_{i=1}^n y_i^{0*} y_i^0 + \sum_{i=1}^n \lambda_i (y_i^0 + a_i - t) \right\} \\
= & \sup_{\substack{\lambda_i \ge 0, \\ i=1,...,n}} \left\{ - \sup_{t \in \mathbb{R}_+} \left\{ \left(\sum_{i=1}^n \lambda_i - 1 \right) t \right\} \right. \\
& \left. - \sup_{\substack{y_i^0 \in \mathbb{R}_+, \\ i=1,...,n}} \left\{ \sum_{i=1}^n (y_i^{0*} - \lambda_i) y_i^0 \right\} + \sum_{i=1}^n \lambda_i a_i \right\} = \sup_{\substack{\sum_{i=1}^n \lambda_i \le 1,\ \lambda_i \ge 0, \\ y_i^{0*} \le \lambda_i,\ i=1,...,n}} \left\{ \sum_{i=1}^n \lambda_i a_i \right\}.
\end{aligned}
$$

As the Slater constraint qualification is fulfilled, it holds $v(P^{max}) = v(D^{max})$ and the dual has an optimal solution, thus one gets for the conjugate function of f

$$f^*(y^{0*}) = \min_{\substack{\sum_{i=1}^n \lambda_i \le 1,\ \lambda_i \ge 0, \\ y_i^{0*} \le \lambda_i,\ i=1,...,n}} \left\{ - \sum_{i=1}^n \lambda_i a_i \right\}. \tag{4.3}$$

Furthermore, one has

$$((y_1^{0*}, ..., y_n^{0*})^T F^1)^*(x_1^*, ..., x_n^*)$$

$$= \sup_{x_i \in \mathbb{R}, \ i=1,...,n} \left\{ \sum_{i=1}^n x_i^* x_i - (y_1^{0*}, ..., y_n^{0*})^T F^1(x_1, ..., x_n) \right\}$$

$$= \sup_{x_i \in \mathbb{R}_+, \ i=1,...,n} \left\{ \sum_{i=1}^n x_i^* x_i - \sum_{i=1}^n y_i^{0*} h_i(x_i) \right\}$$

$$= \sum_{i=1}^n \sup_{x_i \in \mathbb{R}_+} \{ x_i^* x_i - y_i^{0*} h_i(x_i) \} = \sum_{i=1}^n (y_i^{0*} h_i)^*(x_i^*), \tag{4.4}$$

and so, the conjugate function of g turns into

$$g^*(x_1^*, ..., x_n^*) = \min_{\substack{y_i^{0*} \geq 0, \\ i=1,...,n}} \left\{ \min_{\substack{\sum_{i=1}^n \lambda_i \leq 1, \ \lambda_i \geq 0, \\ y_i^{0*} \leq \lambda_i, \ i=1,...,n}} \left\{ -\sum_{i=1}^n \lambda_i a_i \right\} + \sum_{i=1}^n (y_i^{0*} h_i)^*(x_i^*) \right\}$$

$$= \min_{\substack{\sum_{i=1}^n \lambda_i \leq 1, \ \lambda_i \geq 0, \\ 0 \leq y_i^{0*} \leq \lambda_i, \ i=1,...,n}} \left\{ \sum_{i=1}^n [(y_i^{0*} h_i)^*(x_i^*) - \lambda_i a_i] \right\}. \tag{4.5}$$

We fix $x_i^* \in \mathbb{R}^n$, $i = 1, ..., n$, and emphasize that the problem

$$(P^g) \qquad \min_{\substack{\sum_{i=1}^n \lambda_i \leq 1, \ \lambda_i \geq 0, \\ 0 \leq y_i^{0*} \leq \lambda_i, \ i=1,...,n}} \left\{ \sum_{i=1}^n [(y_i^{0*} h_i)^*(x_i^*) - \lambda_i a_i] \right\} \tag{4.6}$$

is equivalent to

$$(\widetilde{P}^g) \qquad \min_{\substack{\sum_{i=1}^n z_i^{0*} \leq 1, \ z_i^{0*} \geq 0, \\ i=1,...,n}} \left\{ \sum_{i=1}^n [(z_i^{0*} h_i)^*(x_i^*) - z_i^{0*} a_i] \right\} \tag{4.7}$$

in the sense that $v(P^g) = v(\widetilde{P}^g)$ (where $v(P^g)$ and $v(\widetilde{P}^g)$ denote the optimal objective values of the problems (P^g) and $(\widetilde{P}^g)$, respectively).

To see this, take first a feasible element $(\lambda_1, ..., \lambda_n, y_1^{0*}, ..., y_n^{0*}) \in \mathbb{R}_+^n \times \mathbb{R}_+^n$ of the problem (P^g) and set $z_i^{0*} = \lambda_i$, $i = 1, ..., n$, then it follows from $\sum_{i=1}^n \lambda_i \leq 1$, $\lambda_i, y_i^{0*} \geq 0$, $y_i^{0*} \leq \lambda_i$, $i = 1, ..., n$, that $\sum_{i=1}^n z_i^{0*} \leq 1$, $z_i^{0*} \geq 0$, $i = 1, ..., n$, i.e. $(z_1^{0*}, ..., z_n^{0*})$ is feasible to the problem $(\widetilde{P}^g)$. From $y_i^{0*} \leq z_i^{0*}$, we have that $y_i^{0*} h_i(x_i) \leq z_i^{0*} h_i(x_i)$ and by [14, Proposition 2.3.2.(c)] follows that $(y_i^{0*} h_i)^*(x_i^*) \geq (z_i^{0*} h_i)^*(x_i^*)$. Hence it holds

$$\sum_{i=1}^n [(y_i^{0*} h_i)^*(x_i^*) - \lambda_i a_i] \geq \sum_{i=1}^n [(z_i^{0*} h_i)^*(x_i^*) - z_i^{0*} a_i] \geq v(\widetilde{P}^g) \tag{4.8}$$

for all $(\lambda_1, ..., \lambda_n, y_1^{0*}, ..., y_n^{0*})$ feasible to (P^g), i.e. $v(P^g) \geq v(\widetilde{P}^g)$.

Now, take a feasible element $(z_1^{0*}, ..., z_n^{0*}) \in \mathbb{R}_+^n$ of the problem $(\widetilde{P}^g)$ and set $y_i^{0*} = \lambda_i = z_i^{0*}$ for all $i = 1, ..., n$, then we have from $\sum_{i=1}^n z_i^{0*} \leq 1$, $z_i^{0*} \geq 0$, $i = 1, ..., n$, that $\sum_{i=1}^n \lambda_i \leq 1$, $\lambda_i, y_i^{0*} \geq 0$, $y_i^{0*} = \lambda_i$, $i = 1, ..., n$, which means that $(\lambda_1, ..., \lambda_n, y_1^{0*}, ..., y_n^{0*})$ is a feasible element of (P^g) and it holds

$$\sum_{i=1}^n [(z_i^{0*}h_i)^*(x_i^*) - z_i^{0*}a_i] = \sum_{i=1}^n [(y_i^{0*}h_i)^*(x_i^*) - \lambda_i a_i] \geq v(P^g) \tag{4.9}$$

for all $(z_1^{0*}, ..., z_n^{0*})$ feasible to $v(\widetilde{P}^g)$, which implies $v(P^g) \leq v(\widetilde{P}^g)$. Finally, it follows that $v(P^g) = v(\widetilde{P}^g)$ and thus, the conjugate function of g is given by

$$g^*(x_1^*, ..., x_n^*) = \min_{\substack{\sum_{i=1}^n z_i^{0*} \leq 1, \; z_i^{0*} \geq 0, \\ i=1,...,n}} \left\{ \sum_{i=1}^n [(z_i^{0*}h_i)^*(x_i^*) - z_i^{0*}a_i] \right\} \tag{4.10}$$

and takes only finite values. □

Lemma 4.2. *Let $a_i \in \mathbb{R}_+$ be a given point and $h_i : \mathbb{R} \to \overline{\mathbb{R}}$ with $h_i(x) \in \mathbb{R}_+$, if $x \in \mathbb{R}_+$, and $h_i(x) = +\infty$, otherwise, be a proper, lower semicontinuous and convex function, $i = 1, ..., n$. Then the function $g : \mathbb{R}^n \to \overline{\mathbb{R}}$,*

$$g(x_1, ..., x_n) = \begin{cases} \max\{h_1(x_1) + a_1, ..., h_n(x_n) + a_n\}, & \text{if } x_i \in \mathbb{R}_+, \; i = 1, ..., n, \\ +\infty, & \text{otherwise,} \end{cases}$$

can equivalently be expressed as

$$g(x_1, ..., x_n) = \max_{\substack{\sum_{i=1}^n z_i^{0*} \leq 1, \; z_i^{0*} \geq 0, \\ i=1,...,n}} \left\{ \sum_{i=1}^n z_i^{0*}[h_i(x_i) + a_i] \right\} \; \forall x_i \in \mathbb{R}, \; i = 1, ..., n.$$

Proof. By Lemma 4.1 and the definition of the conjugate function we have for the biconjugate function of g

$$g^{**}(x_1, ..., x_n) = \sup_{\substack{x_i^* \in \mathbb{R}, \\ i=1,...,n}} \left\{ \sum_{i=1}^n x_i^* x_i - \min_{\substack{\sum_{i=1}^n z_i^{0*} \leq 1, \\ z_i^{0*} \geq 0, i=1,...,n}} \left\{ \sum_{i=1}^n [(z_i^{0*}h_i)^*(x_i^*) - z_i^{0*}a_i] \right\} \right\}$$

$$= \sup_{\substack{x_i^* \in \mathbb{R}, \; z_i^{0*} \geq 0, \\ i=1,...,n, \; \sum_{i=1}^n z_i^{0*} \leq 1}} \left\{ \sum_{i=1}^n x_i^* x_i - \sum_{i=1}^n [(z_i^{0*}h_i)^*(x_i^*) - z_i^{0*}a_i] \right\}$$

$$\tag{4.11}$$

$$= \sup_{\substack{z_i^{0*} \geq 0,\ i=1,...,n, \\ \sum_{i=1}^{n} z_i^{0*} \leq 1}} \left\{ \sum_{i=1}^{n} \left[\sup_{x_i^* \in \mathbb{R}} \{ x_i^* x_i - (z_i^{0*} h_i)^*(x_i^*) \} + z_i^{0*} a_i \right] \right\}$$

$$= \sup_{\substack{z_i^{0*} \geq 0,\ i=1,...,n, \\ \sum_{i=1}^{n} z_i^{0*} \leq 1}} \left\{ \sum_{i=1}^{n} [(z_i^{0*} h_i)^{**}(x_i) + z_i^{0*} a_i] \right\} \ \forall x_i \in \mathbb{R}, \ i = 1, ..., n.$$

$$(4.12)$$

As h_i, $i = 1, ..., n$, are proper, convex and lower semicontinuous functions it follows by the Fenchel-Moreau Theorem that

$$g^{**}(x_1, ..., x_n) = \sup_{\substack{z_i^{0*} \geq 0,\ i=1,...,n, \\ \sum_{i=1}^{n} z_i^{0*} \leq 1}} \left\{ \sum_{i=1}^{n} [z_i^{0*} h_i(x_i) + z_i^{0*} a_i] \right\} \tag{4.13}$$

for all $x_i \in \mathbb{R}$, $i = 1, ..., n$, and moreover, as g is also a proper, convex and lower semicontinuous function it follows by using again the Fenchel-Moreau Theorem that $g = g^{**}$, i.e.

$$g(x_1, ..., x_n) = \max_{\substack{z_i^{0*} \geq 0,\ i=1,...,n, \\ \sum_{i=1}^{n} z_i^{0*} \leq 1}} \left\{ \sum_{i=1}^{n} [z_i^{0*} h_i(x_i) + z_i^{0*} a_i] \right\} \tag{4.14}$$

for all $x_i \in \mathbb{R}$, $i = 1, ..., n$. $\square$

Remark 4.1. *Note that the statement in Lemma 4.2 can also be proved in a simpler way, as the maximum of finitely many real numbers is the maximum over the convex hull of finitely many real numbers.*

Remark 4.2. *If we consider the situation when the given points a_i, $i = 1, ..., n$, are arbitrary, i.e. $a_i \in \mathbb{R}$, then it can easily be verified that the conjugate function of f in (4.1) looks like*

$$f^*(y^{0*}) = \sup_{\substack{y_i^0 \in \mathbb{R}_+,\ t \in \mathbb{R}, \\ y_i^0 + a_i \leq t,\ i=1,...,n}} \left\{ \sum_{i=1}^{n} y_i^{0*} y_i^0 - t \right\} \tag{4.15}$$

(notice that here $t \in \mathbb{R}$ instead of $t \in \mathbb{R}_+$).
If we now construct to the conjugate function in (4.15) a primal problem in the sense of (P^{max}) in (4.2), then the corresponding Lagrange dual problem (D^{max}) has the form

$$(D^{max}) \qquad \sup_{\substack{\sum_{i=1}^{n} \lambda_i = 1,\ \lambda_i \geq 0, \\ y_i^{0*} \leq \lambda_i,\ i=1,...,n}} \left\{ \sum_{i=1}^{n} \lambda_i a_i \right\}.$$

Analogously to the calculations done above in (4.3) - (4.14) one derives for the conjugate function of g,

$$g^*(x_1^*, ..., x_n^*) = \min_{\substack{\sum_{i=1}^{n} z_i^{0*} = 1, \; z_i^{0*} \geq 0, \\ i=1,...,n}} \left\{ \sum_{i=1}^{n} [(z_i^{0*} h_i)^*(x_i^*) - z_i^{0*} a_i] \right\},$$

while its biconjugate is then given by

$$g^{**}(x_1, ..., x_n) = g(x_1, ..., x_n) = \max_{\substack{\sum_{i=1}^{n} z_i^{0*} = 1, \; z_i^{0*} \geq 0, \\ i=1,...,n}} \left\{ \sum_{i=1}^{n} z_i^{0*} [h_i(x_i) + a_i] \right\}$$

for all $x_i \in \mathbb{R}$, $i = 1, ..., n$.

4.1.1 Gauges

In the following, let X be a Hausdorff locally convex space partially ordered by the convex cone $K \subseteq X$ and X^* its topological dual space endowed with the weak* topology $w(X^*, X)$. Further, let Y_i be another Hausdorff locally convex space partially ordered by the convex cone $Q_i \subseteq Y_i$ and Y_i^* its topological dual space endowed with the weak* topology $w(Y_i^*, Y_i)$. In order to introduce perturbed minimal time functions, one needs first to define a *gauge*. In the literature one can find different functions called gauges, see, for instance, [53] or [89, Section 15]. In the following we call *gauge function* (known in the literature also as the *Minkowski functional*) of a set $C \subseteq X$ the function $\gamma_C : X \to \overline{\mathbb{R}}$, defined by

$$\gamma_C(x) := \inf\{\lambda > 0 : x \in \lambda C\}.$$

Note that the gauge function can also take the value $+\infty$ if there does not exists an element $\lambda > 0$ such that $x \in \lambda C$, as by definition it holds $\inf \emptyset = +\infty$. From the definition it follows that $\text{dom}\, \gamma_C = \text{cone}\, C$ if $0 \in C$ and $\text{dom}\, \gamma_C = \text{cone}\, C \setminus \{0\}$ if $0 \notin C$. As $\gamma_\emptyset \equiv +\infty$, we consider further the set C to be nonempty.

It is well-known that for the gauge function the generalized Cauchy-Schwarz inequality is fulfilled. The proof of this inequality is rare to find and for this reason we give it here.

Lemma 4.3. *It holds $\gamma_C(x)\gamma_{C^0}(x^*) \geq \langle x^*, x \rangle$ for all $x \in X$, $x^* \in X^*$.*

Proof. Let $x^* \in X^*$ and $x \in X$. If $\langle x^*, x \rangle \leq 0$, then there is nothing to prove, as the gauge and the dual gauge are nonnegative functions. Let $\langle x^*, x \rangle > 0$. If $\gamma_C(x) > 0$ then we have that

$$\gamma_{C^0}(x^*) \geq \sup_{\lambda > 0, \; \gamma_C(x) \leq \frac{1}{\lambda}} \lambda \langle x^*, x \rangle = \langle x^*, x \rangle \sup_{0 < \lambda \leq \frac{1}{\gamma_C(x)}} \lambda = \frac{1}{\gamma_C(x)} \langle x^*, x \rangle,$$

i.e. $\gamma_C(x)\gamma_{C^0}(x^*) \geq \langle x^*, x \rangle$.

Otherwise, if $\gamma_C(x) = 0$, then one has that

$$\gamma_{C^0}(x) \geq \sup_{\lambda > 0, \ 0 = \gamma_C(x) \leq \frac{1}{\lambda}} \lambda \langle x^*, x \rangle = +\infty,$$

i.e. $\gamma_C(x)\gamma_{C^0}(x^*) = 0(+\infty) \geq \langle x^*, x \rangle$.   □

When in the literature the question of continuity of the gauge function arises, then it is often assumed that $0_X \in \operatorname{int} C$ (see [3, 14, 29, 52, 65, 106, 107]). We start with a statement where this assumption is weakened to $0_X \in C$.

Theorem 4.1. Let $C \subseteq X$ be a convex and closed set with $0_X \in C$, then the gauge function γ_C is proper, convex and lower semicontinuous.

Proof. Let us define the function $g : X^* \to \overline{\mathbb{R}}$ by

$$g(x^*) := \begin{cases} 0, & \text{if } \sigma_C(x^*) \leq 1, \\ +\infty, & \text{otherwise.} \end{cases}$$

It is obvious that g is proper, convex and lower semicontinuous. For the corresponding conjugate function of g one has

$$g^*(x) = \sup_{x^* \in X^*} \{ \langle x^*, x \rangle - g(x^*) \} = \sup_{\substack{x^* \in X^*, \\ \sigma_C(x^*) \leq 1}} \langle x^*, x \rangle.$$

There is $g^*(x) = \sup_{x^* \in X^*} \{ \langle x^*, x \rangle - g(x^*) \} \geq \langle 0_{X^*}, x \rangle - g(0_{X^*}) = 0$ since $g(0_{X^*}) = 0$ for all $x \in X$, and $g^*(0_X) = \sup_{x^* \in X^*} \{ -g(x^*) \} = 0$, i.e. g^* is proper. At this point it is important to say that from $0_X \in C$ follows that $\gamma_C(0_X) = 0$, i.e. $g^*(0_X) = \gamma_C(0_X)$.

Let us now assume that $x \neq 0_X$ and consider for fixed $x \in X$ the following convex optimization problem

$$(P^\gamma) \qquad \inf_{\substack{x^* \in X^*, \\ \sigma_C(x^*) \leq 1}} \langle -x^*, x \rangle.$$

As $\sigma_C(0_{X^*}) = 0 < 1$, the Slater condition is fulfilled and hence, it holds strong duality between the problem (P^γ) and its corresponding Lagrange dual problem

$$(D_L^\gamma) \qquad \sup_{\lambda \geq 0} \inf_{x^* \in X^*} \{ \langle -x^*, x \rangle + \lambda(\sigma_C(x^*) - 1) \}.$$

Therefore, the conjugate function of g can be represented for $x \neq 0_X$ as

$$\begin{aligned} g^*(x) &= \sup_{\substack{x^* \in X^*, \\ \sigma_C(x^*) \leq 1}} \langle x^*, x \rangle = -\max_{\lambda \geq 0} \inf_{x^* \in X^*} \{ \langle -x^*, x \rangle + \lambda(\sigma_C(x^*) - 1) \} \\ &= \min_{\lambda \geq 0} \left\{ \lambda + \sup_{x^* \in X^*} \{ \langle x^*, x \rangle - \lambda \sigma_C(x^*) \} \right\}. \end{aligned} \tag{4.16}$$

For $\lambda = 0$ we verify two conceivable cases.
(a) If $\sigma_C(x^*) < +\infty$, then $0 \cdot \sigma_C(x^*) = 0$ and therefore,

$$\sup_{x^* \in X^*} \{\langle x^*, x \rangle - 0 \cdot \sigma_C(x^*)\} = \sup_{x^* \in X^*} \langle x^*, x \rangle = \begin{cases} 0, & \text{if } x = 0_X, \\ +\infty, & \text{if } x \neq 0_X. \end{cases}$$

As by assumption $x \neq 0_X$, we have $\sup_{x^* \in X^*} \langle x^*, x \rangle = +\infty$, but this has no effect on the minimum in 4.16.
(b) If $\sigma_C(x^*) = +\infty$, then one has by convention that $\lambda \cdot \sigma_C(x^*) = 0 \cdot (+\infty) = +\infty$ and hence,

$$\langle x^*, x \rangle - \lambda \sigma_C(x^*) = \langle x^*, x \rangle - \infty = -\infty,$$

which has no effect on $\sup_{x^* \in X^*} \{\langle x^*, x \rangle - \lambda \sigma_C(x^*)\}$, since σ_C is proper.
Hence, as the cases (a) and (b) are not relevant for g^*, we can omit the situation when $\lambda = 0$ and can write

$$g^*(x) = \inf_{\lambda > 0} \left\{ \lambda + \lambda \sup_{x^* \in X^*} \left\{ \left\langle x^*, \frac{1}{\lambda} x \right\rangle - \sigma_C(x^*) \right\} \right\}.$$

Moreover, as C is a non-empty, closed and convex subset of X, the conjugate of the support function σ_C is the indicator function δ_C, i.e.

$$g^*(x) = \inf_{\lambda > 0} \left\{ \lambda + \lambda \delta_C \left(\frac{1}{\lambda} x \right) \right\} = \inf_{\lambda > 0, \, \frac{1}{\lambda} x \in C} \lambda = \inf\{\lambda > 0 : x \in \lambda C\}.$$

Taking the situations where $x = 0_X$ and $x \neq 0_X$ together implies that $g^*(x) = \gamma_C(x)$ for all $x \in X$. Hence, γ_C is the conjugate function of g and by the definition of the conjugate function it follows that γ_C is convex and lower semicontinuous. This completes the proof. $\square$

Lemma 4.4. *Let $C \subseteq X$ be a convex and closed set with $0_X \in C$, then the conjugate of the gauge function γ_C is given by*

$$\gamma_C^*(x^*) := \begin{cases} 0, & \text{if } \sigma_C(x^*) \leq 1, \\ +\infty, & \text{otherwise.} \end{cases}$$

Proof. In the proof of Theorem 4.1 we have shown that γ_C is the conjugate function of g, i.e. $\gamma_C = g^*$, and as g is proper, convex and lower semicontinuous we have $g = g^{**}$. As g^{**} is also the conjugate function of γ_C, it holds $\gamma_C^* = g$. $\square$

Remark 4.3. *(see [52, 65]) Let C be convex and $0_X \in \text{int } C$, then the gauge function γ_C is not only convex but also sublinear and the following properties holds*

$$\gamma_C(x) \geq 0 \; \forall x \in X,$$
$$\gamma_C(0_X) = 0,$$
$$\gamma_C(\mu x) = \mu \gamma_C(x) \; \forall \mu \geq 0, \; \forall x \in X,$$
$$\gamma_C(x_1 + x_2) \leq \gamma_C(x_1) + \gamma_C(x_2) \; \forall x_1, \, x_2 \in X.$$

Moreover, γ_C is well-defined, which means that $\mathrm{dom}\,\gamma_C = X$, as well as continuous and

$$\mathrm{int}\, C = \{x \in X : \gamma_C(x) < 1\}, \ \mathrm{cl}(C) = \{x \in X : \gamma_C(x) \le 1\}$$

(see [52]).

Remark 4.4. *Let $C_i \subseteq Y_i$ be a closed and convex set with $0_{Y_i} \in \mathrm{int}\, C_i$ and $\gamma_{C_i} : Y_i \to \mathbb{R}$ be a gauge function of the set C_i, $i = 1, ..., n$. Then γ_{C_i} is continuous, $i = 1, ..., n$, and moreover, it is an easy exercise to check that the function $\gamma_C : Y_1 \times ... \times Y_n \to \mathbb{R}$ defined by $\gamma_C(x_1, ..., x_n) := \sum_{i=1}^n \gamma_{C_i}(x_i)$, is a gauge function fulfilling the properties listed in Remark 4.3. Especially, it holds that γ_C is continuous such that $C := \{(x_1, ..., x_n) \in Y_1 \times ... \times Y_n : \gamma_C(x_1, ..., x_n) \le 1\}$.*

Definition 4.1. *Let $C \subseteq X$. The polar set of C is defined by*

$$C^0 := \left\{ x^* \in X^* : \sup_{x \in C} \langle x^*, x \rangle \le 1 \right\} = \{x^* \in X^* : \sigma_C(x^*) \le 1\}$$

and by means of the polar set the dual gauge is defined by

$$\gamma_{C^0}(x^*) := \sup_{x \in C} \langle x^*, x \rangle = \sigma_C(x^*).$$

Remark 4.5. *Note that C^0 is a convex and closed set containing the origin and by the definition of the dual gauge follows that the conjugate function of γ_C can equivalently be expressed by*

$$\gamma_C^*(x^*) := \begin{cases} 0, & \text{if } \gamma_{C^0}(x^*) \le 1, \\ +\infty, & \text{otherwise,} \end{cases} = \begin{cases} 0, & \text{if } x^* \in C^0, \\ +\infty, & \text{otherwise,} \end{cases} = \delta_{C^0}(x^*) \ \forall x^* \in X^*.$$

Furthermore, if C is a convex cone, then $C^0 = \{x^* \in X^* : \sigma_C(x^*) \le 0\}$, i.e. $-C^0$ is the dual cone of C.

Lemma 4.5. *Let $\gamma_{C_i} : Y_i \to \mathbb{R}$ be a gauge of the closed and convex set $C_i \subseteq Y_i$ with $0_{Y_i} \in \mathrm{int}\, C_i$, $i = 1, ..., n$. If the gauge $\gamma_C : Y_1 \times ... \times Y_n \to \mathbb{R}$ is defined by*

$$\gamma_C(x) := \sum_{i=1}^n \gamma_{C_i}(x_i), \ x = (x_1, ..., x_n) \in Y_1 \times ... \times Y_n,$$

then its associated dual gauge $\gamma_{C^0} : Y_1^* \times ... \times Y_n^* \to \mathbb{R}$ is given by

$$\gamma_{C^0}(x^*) = \max_{1 \le i \le n} \left\{ \gamma_{C_i^0}(x_i^*) \right\}, \ x^* = (x_1^*, ..., x_n^*) \in Y_1^* \times ... \times Y_n^*. \tag{4.17}$$

Proof. As C_i is closed, convex and $0_{Y_i} \in \mathrm{int}\, C_i$, the gauge γ_{C_i} is continuous, convex and well-defined, $i = 1, ..., n$, and thus, the gauge γ_C is also continuous, convex and well-defined. In the following, let $\widetilde{X}^* = Y_1^* \times ... \times Y_n^*$ be the topological dual space of $\widetilde{X} := Y_1 \times ... \times Y_n$

where for $x = (x_1, ..., x_n) \in \widetilde{X}$ and $x^* = (x_1^*, ..., x_n^*) \in \widetilde{X}^*$ we define $\langle x^*, x \rangle := \sum_{i=1}^{n} \langle x_i^*, x_i \rangle$. Hence, for the associated dual gauge of γ_C holds

$$\gamma_{C^0}(x^*) = \sup_{x \in C} \langle x^*, x \rangle.$$

Now, we fix $x^* \in \widetilde{X}^*$ and consider the problem

$$(P^{\gamma^0}) \quad \inf_{x \in C} \langle -x^*, x \rangle = \inf_{x \in \widetilde{X}, \, \gamma_C(x) \leq 1} \langle -x^*, x \rangle,$$

where its associated Lagrange dual problem is

$$(D_L^{\gamma^0}) \quad \sup_{\lambda \geq 0} \inf_{x \in \widetilde{X}} \{ \langle -x^*, x \rangle + \lambda(\gamma_C(x) - 1) \} = \sup_{\lambda \geq 0} \left\{ -\lambda + \inf_{x \in \widetilde{X}} \{ \langle -x^*, x \rangle + \lambda \gamma_C(x) \} \right\}$$

$$= \sup_{\lambda \geq 0} \left\{ -\lambda - \sup_{x \in \widetilde{X}} \{ \langle x^*, x \rangle - \lambda \gamma_C(x) \} \right\} = \sup_{\lambda \geq 0} \left\{ -\lambda - (\lambda \gamma_C)^*(x^*) \right\}. \tag{4.18}$$

For $\lambda > 0$ it holds (see Lemma 4.4 and Remark 4.5)

$$(\lambda \gamma_C)^*(x^*) = \sup_{x \in \widetilde{X}} \{ \langle x^*, x \rangle - \lambda \gamma_C(x) \} = \sup_{\substack{x_i \in Y_i, \\ i=1,...,n}} \left\{ \sum_{i=1}^{n} \langle x_i^*, x_i \rangle - \lambda \sum_{i=1}^{n} \gamma_{C_i}(x_i) \right\}$$

$$= \sum_{i=1}^{n} \sup_{x_i \in Y_i} \{ \langle x_i^*, x_i \rangle - \lambda \gamma_{C_i}(x_i) \}$$

$$= \sum_{i=1}^{n} \lambda \gamma_{C_i}^* \left(\frac{1}{\lambda} x_i^* \right) = \begin{cases} 0, & \text{if } \sigma_{C_i}(x_i^*) \leq \lambda \; \forall i = 1, ..., n, \\ +\infty, & \text{otherwise} \end{cases}$$

$$= \begin{cases} 0, & \text{if } \gamma_{C_i^0}(x_i^*) \leq \lambda \; \forall i = 1, ..., n, \\ +\infty, & \text{otherwise} \end{cases} \tag{4.19}$$

and for $\lambda = 0$ we have

$$(0 \cdot \gamma_C)^*(x^*) = \sup_{x \in \widetilde{X}} \{ \langle x^*, x \rangle \} = \begin{cases} 0, & \text{if } x_i^* = 0_{Y_i^*} \; \forall i = 1, ..., n, \\ +\infty, & \text{otherwise}. \end{cases} \tag{4.20}$$

As $\gamma_{C_i^0}(0_{Y_i^*}) = \sup_{x_i \in C_i} \langle 0_{Y_i^*}, x_i \rangle = 0$, one gets by (4.19) and (4.20) for the Lagrange dual problem $(D_L^{\gamma^0})$ that

$$(D_L^{\gamma^0}) \quad \sup_{\lambda \geq 0} \{ -\lambda - (\lambda \gamma_C)^*(x^*) \} = \sup_{\lambda \geq 0} \left\{ -\lambda : \gamma_{C_i^0}(x_i^*) \leq \lambda \; \forall i = 1, ..., n \right\}$$

and since for the primal-dual pair (P^{γ^0})-$(D_L^{\gamma^0})$ the Slater constraint qualification is fulfilled, it holds strong duality. From the last statement we derive an alternative formula for the dual gauge γ_{C^0},

$$\gamma_{C^0}(x^*) = \sup_{x \in C} \langle x^*, x \rangle = \min_{\lambda \geq 0} \left\{ \lambda : \gamma_{C_i^0}(x_i^*) \leq \lambda \; \forall i = 1, ..., n \right\} = \max_{1 \leq i \leq n} \left\{ \gamma_{C_i^0}(x_i^*) \right\}.$$

$\square$

Now, it is natural to ask, whether the dual gauge of $\max_{1 \le i \le n}\{\gamma_{C_i}(\cdot)\}$ is $\sum_{i=1}^{n}\gamma_{C_i^0}(\cdot)$. The next lemma gives a positive answer.

Lemma 4.6. *Let* $\gamma_{C_i} : Y_i \to \mathbb{R}$ *be a gauge of the closed and convex set* $C_i \subseteq Y_i$ *with* $0_{Y_i} \in \operatorname{int} C_i$, $i = 1, ..., n$. *If the gauge* $\gamma_C : Y_1 \times ... \times Y_n \to \mathbb{R}$ *is defined by*

$$\gamma_C(x) := \max_{1 \le i \le n}\{\gamma_{C_i}(x_i)\}, \quad x = (x_1, ..., x_n) \in Y_1 \times ... \times Y_n,$$

then its associated dual gauge $\gamma_{C^0} : Y_1^* \times ... \times Y_n^* \to \mathbb{R}$ *is given by*

$$\gamma_{C^0}(x^*) = \sum_{i=1}^{n} \gamma_{C_i^0}(x_i^*), \quad x^* = (x_1^*, ..., x_n^*) \in Y_1^* \times ... \times Y_n^*. \tag{4.21}$$

Proof. The main ideas here are similar to the ones in the proof of Lemma 4.5. As γ_C is the pointwise maximum of n continuous, convex and well-defined gauges, it is clear that γ_C is continuous, convex and well-defined and for the corresponding dual gauge of γ_C holds $\gamma_{C^0}(x^*) = \sup_{x \in C}\{\langle x^*, x \rangle\}$.

For fixed $x^* := (x_1^*, ..., x_n^*) \in \widetilde{X}^* = Y_1^* \times ... \times Y_n^*$ we consider the problem

$$(\widetilde{P}^{\gamma^0}) \quad \inf_{x \in C} \langle -x^*, x \rangle = \inf_{x \in \widetilde{X}, \, \gamma_C(x) \le 1} \langle -x^*, x \rangle,$$

with its Lagrange dual problem (see (4.18))

$$(\widetilde{D}_L^{\gamma^0}) \quad \sup_{\lambda \ge 0}\{-\lambda - (\lambda\gamma_C)^*(x^*)\}.$$

For $\lambda \ge 0$ one has

$$(\lambda\gamma_C)^*(x^*) = \sup_{x \in \widetilde{X}}\{\langle x^*, x \rangle - \lambda\gamma_C(x)\} = \sup_{\substack{x_i \in Y_i, \\ i=1,...,n}} \left\{\sum_{i=1}^{n}\langle x_i^*, x_i \rangle - \lambda \max_{1 \le i \le n}\{\gamma_{C_i}(x_i)\}\right\}.$$

Now, let $X_0 := \mathbb{R}^n$, $K_0 = \mathbb{R}_+^n$, $X_1 = \widetilde{X}$, the function $f : \mathbb{R}^n \to \overline{\mathbb{R}}$ be defined by

$$f(y_1^0, ..., y_n^0) := \begin{cases} \max\{y_1^0, ..., y_n^0\}, & \text{if } y_i^0 \in \mathbb{R}_+, \ i = 1, ..., n, \\ +\infty, & \text{otherwise}, \end{cases}$$

and the function $F^1 : X_1 \to \mathbb{R}^n$ by

$$F^1(x_1, ..., x_n) := (\gamma_{C_1}(x_1), ..., \gamma_{C_n}(x_n))^T$$

Hence, the gauge γ_C can be written as

$$\gamma_C(x_1, ..., x_n) = (f \circ F^1)(x_1, ..., x_n).$$

$$= \min_{\substack{\lambda>0,\ x\in X, \\ \mathcal{T}^{C_i}_{\Omega_i,\gamma_{G_i}}(x)\leq\lambda,\ i=1,\dots,n}} \lambda = \min_{x\in\mathcal{H}}\max_{1\leq i\leq n}\left\{\mathcal{T}^{C_i}_{\Omega_i,\gamma_G}(x)\right\}.$$

This shows on the one hand that the optimization problem $(P_{\gamma_G},\ \tau)$ *has an optimal solution and on the other hand that the Lagrange multipliers* $\lambda>0$ *and* $x\in X$ *characterize the optimal objective value and the optimal solution to the problem* $(P_{\gamma_G},\ \tau)$, *respectively. Furthermore, this fact also implies that the relation between the primal problem* $(P_{\gamma_G},\ \tau)$, *its dual problem* $(\widetilde{D}_{\gamma_G},\ \tau)$ *and its bidual problem* $(D\widetilde{D}_{\gamma_G},\ \tau)$ *is completely symmetric under the considered hypotheses.*

Remark 4.20. *As noted in Remark 4.14, the problem* $(P_{\gamma_G},\ \tau)$ *can also be written as*

$$(P_{\gamma_G},\ \tau)\quad \inf_{\substack{x\in X, t\in\mathbb{R}, \\ \mathcal{T}^{C_i}_{\Omega_i,\gamma_{G_i}}(x)\leq t,\ i=1,\dots,n}} t = \inf_{\substack{x\in X,\ t\in\mathbb{R}, \\ (x,t)\in\text{epi}\,\mathcal{T}^{C_i}_{\Omega_i,\gamma_{G_i}},\ i=1,\dots,n}} t = \inf_{x\in X,\ t\in\mathbb{R}}\left\{t+\sum_{i=1}^{n}\delta_{\text{epi}\,\mathcal{T}^{C_i}_{\Omega_i,\gamma_{G_i}}}(x,t)\right\},$$

to which one can assign the corresponding Fenchel dual problem that can be reduced to

$$(D^F_{\gamma_G},\ \tau)\quad \sup_{\substack{x_i^*\in X^*,t_i^*\in\mathbb{R},\ i=1,\dots,n, \\ \sum_{i=1}^{n}x_i^*=0_{X^*},\ \sum_{i=1}^{n}t_i^*=-1}}\left\{-\sum_{i=1}^{n}\sigma_{\text{epi}\,\mathcal{T}^{C_i}_{\Omega_i,\gamma_{G_i}}}(x_i^*,t_i^*)\right\}.$$

Now, let us take a careful look at

$$\sigma_{\text{epi}\,\mathcal{T}^{C_i}_{\Omega_i,\gamma_{G_i}}}(x_i^*,t_i^*)=\sup_{(x,t)\in X\times\mathbb{R},\ \mathcal{T}^{C_i}_{\Omega_i,\gamma_{G_i}}(x)\leq t}\left\{\langle x_i^*,x\rangle+t_i^*t\right\}, \tag{4.57}$$

and as for fixed $(x_i^*,t_i^*)\in X^*\times\mathbb{R}$ *the Slater constraint qualification is obviously fulfilled in the right-hand side of* (4.57) *one has*

$$\sigma_{\text{epi}\,\mathcal{T}^{C_i}_{\Omega_i,\gamma_{G_i}}}(x_i^*,t_i^*)=\min_{\lambda_i\geq 0}\sup_{(x,t)\in X\times\mathbb{R}}\left\{\langle x_i^*,x\rangle+t_i^*t-\lambda_i\left(\mathcal{T}^{C_i}_{\Omega_i,\gamma_{G_i}}(x)-t\right)\right\},\ i=1,\dots,n.$$

If for some $i\in\{1,\dots,n\}$, $\lambda_i=0$, *then*

$$\sigma_{\text{epi}\,\mathcal{T}^{C_i}_{\Omega_i,\gamma_{G_i}}}(x_i^*,t_i^*)=\begin{cases}0, & \text{if } x_i^*=0_{X^*},\ t_i^*=0 \\ +\infty, & \text{otherwise,}\end{cases}$$

otherwise, it holds for some $i\in\{1,\dots,n\}$,

$$\sigma_{\text{epi}\,\mathcal{T}^{C_i}_{\Omega_i,\gamma_{G_i}}}(x_i^*,t_i^*)=\min_{\lambda_i>0}\left\{\lambda_i\sup_{x\in X}\left\{\left\langle\frac{1}{\lambda_i}x_i^*,x\right\rangle-\mathcal{T}^{C_i}_{\Omega_i,\gamma_{G_i}}(x)\right\}+\sup_{t\in\mathbb{R}}\{t(t_i^*+\lambda_i)\}\right\}$$

$$=\min_{\substack{\lambda_i>0,\ t_i^*=-\lambda_i, \\ \gamma_{C_i^0}(x_i^*)\leq\lambda_i,\ \gamma_{G_i^0}(x_i^*)\leq\lambda_i}}\left\{-\sigma_{\Omega_i}(x_i^*)\right\}.$$

Therefore, the Fenchel dual problem transforms to

$$(D^F_{\gamma G}, \tau) \qquad \sup_{\substack{\lambda_i \in \mathbb{R}, \ x_i^* \in X^*, \ \gamma_{C_i^0}(x_i^*) \leq \lambda_i, \ \gamma_{G_i^0}(x_i^*) \leq \lambda_i, \\ i=1,\ldots,n, \ \sum_{i=1}^n \lambda_i = 1, \ \sum_{i=1}^n x_i^* = 0_{X^*}}} \left\{ -\sum_{i=1}^n \sigma_{\Omega_i}(x_i^*) \right\}.$$

Setting $x_j^* = w_j^*$, $j \in I$, and $\lambda_i = z_i^*$, $i = 1, \ldots, n$, allows to write the Fenchel dual problem as

$$(D^F_{\gamma G}, \tau) \qquad \sup_{\substack{z_i^* \in \mathbb{R}, \ w_i^* \in X^*, \ \gamma_{C_i^0}(w_i^*) \leq z_i^*, \ \gamma_{G_i^0}(w_i^*) \leq z_i^*, \\ i=1,\ldots,n, \ \sum_{i=1}^n \lambda_i = 1, \ \sum_{i=1}^n w_i^* = 0_{X^*}}} \left\{ -\sum_{i=1}^n \sigma_{\Omega_i}(w_i^*) \right\}.$$

More than that, one has by weak duality $v(D^F_{\gamma G}, \tau) \leq v(D_{\gamma G}, \tau) = v(\widetilde{D}_{\gamma G}, \tau) = v(P_{\gamma G}, \tau)$. But this approach has two drawbacks. First, for strong duality between $(P_{\gamma G}, \tau)$ and $(D^F_{\gamma G}, \tau)$ one needs to verify the fulfillment of a regularity condition, see, for instance, [14, Theorem 3.2.8]). Moreover, this Fenchel dual cannot be easily reduced to an optimization problem of the form $(\widetilde{D}_{\gamma G}, \tau)$.

Remark 4.21. *A further dual problem of interests is the (direct) Lagrange dual to $(P_{\gamma G}, \tau)$*

$$(D^L_{\gamma G}, \tau) \qquad \sup_{\lambda_i \geq 0, \ i=1,\ldots,n} \inf_{(x,t) \in X \times \mathbb{R}} \left\{ t + \sum_{i=1}^n \lambda_i \left(\mathcal{T}^{C_i}_{\Omega_i, \gamma_{G_i}}(x) - t \right) \right\}$$

$$= \sup_{\lambda_i \geq 0, \ i=1,\ldots,n} \left\{ \inf_{x \in X} \left\{ \sum_{i=1}^n \lambda_i \mathcal{T}^{C_i}_{\Omega_i, \gamma_{G_i}}(x) \right\} - \sup_{t \in \mathbb{R}} \left\{ t \left(1 - \sum_{i=1}^n \lambda_i \right) \right\} \right\}$$

$$= \sup_{\lambda_i \geq 0, \ i=1,\ldots,n, \ \sum_{i=1}^n \lambda_i = 1} \inf_{x \in X} \left\{ \sum_{i=1}^n \lambda_i \mathcal{T}^{C_i}_{\Omega_i, \gamma_{G_i}}(x) \right\}.$$

The Slater condition is for the problem $(\widetilde{P}_{\gamma G}, \tau)$ fulfilled, i.e. it holds strong duality for the primal-dual pair $(\widetilde{P}_{\gamma G}, \tau) - (D^L_{\gamma G}, \tau)$, and from the optimality conditions of Theorem 4.6 follows that $\overline{\lambda} \in \mathbb{R}^n_+$ with $\overline{\lambda}_i = (1/\beta_i)\|\overline{y}_i^*\|_{\mathcal{H}}$, $i \in \overline{I}$, and $\overline{\lambda}_j = 0$, $j \notin \overline{I}$, is an optimal solution to the Lagrange dual $(D^L_{\gamma G}, \tau)$.

4.2.2 Special case II

In this part of the paper, we analyze the special case where $a_i = 0$, $h_i(x) := x + \delta_{\mathbb{R}_+}(x)$, $x \in \mathbb{R}$, $f_i(x) := \delta_{L_i}(x)$ and $L_i \subseteq X$ is a nonempty, closed and convex set for all $i = 1, \ldots, n$, such that the minmax location problem $(P^S_{h,\mathcal{T}})$ turns into

$$(P_{\mathcal{T}}) \qquad \inf_{x \in S} \max_{1 \leq i \leq n} \left\{ \mathcal{T}^{C_i}_{\Omega_i, \delta_{L_i}}(x) \right\},$$

Remark 4.22. By construction $v(P_T) > 0$. If $S = X = \mathbb{R}^m$ and the gauges are taken to be the corresponding Euclidean norm, then the problem (P_T) can be seen as finding a point $\bar{x} \in \mathbb{R}^m$ such that the maximal distance to its Euclidean projections onto the target sets $\Omega_i + L_i$, $i = 1, \ldots, n$, is minimal. If $n = 3$, $X = \mathbb{R}^2$, $\Omega_i = \{0_{\mathbb{R}^2}\}$ and $L_i = \{x \in \mathbb{R}^2 : \|x - p_i\| \le a_i\}$, where $a_i > 0$ and $p_i \in \mathbb{R}^2$, $i = 1, 2, 3$, then this problem is also known as the classical *Apollonius problem* (see [7, 78, 83]).

The corresponding dual problem (D_T) to (P_T) becomes, via Proposition 4.1,

$$(D_T) \qquad \sup_{\substack{z_i^{0*} \ge 0,\ i=1,\ldots,n, \\ I=\{i\in\{1,\ldots,n\}:z_i^{0*}>0\},\ z_i^{1*}\in X^*, i\in I, \\ \gamma_{C_i^0}(z_i^{1*})\le z_i^{0*}, i\in I,\ \sum_{i\in I} z_i^{0*}\le 1}} \left\{ -\sigma_S \left(-\sum_{i\in I} z_i^{1*} \right) - \sum_{i\in I} \left[\sigma_{L_i} \left(z_i^{1*} \right) + \sigma_{\Omega_i} \left(z_i^{1*} \right) \right] \right\}.$$

Additionally to this dual problem, we consider the following one, that is equivalent to it in the sense that they share the same optimal objective value

$$(\widetilde{D}_T) \qquad \sup_{\substack{y_i^*\in X^*,\ i=1,\ldots,n, \\ I=\left\{i\in\{1,\ldots,n\}:\gamma_{C_i^0}(y_i^*)>0\right\},\ \sum_{i\in I} \gamma_{C_i^0}(y_i^*)\le 1}} \left\{ -\sigma_S \left(-\sum_{i\in I} y_i^* \right) - \sum_{i\in I} \left[\sigma_{L_i} \left(y_i^* \right) + \sigma_{\Omega_i} \left(y_i^* \right) \right] \right\}.$$

Theorem 4.8. *It holds* $v(D_T) = v(\widetilde{D}_T)$.

Proof. Let y_i^*, $i = 1, \ldots, n$, be a feasible element to $(\widetilde{D}_T)$ and set $z_i^{1*} = y_i^*$, $z_i^{0*} = \gamma_{C_i^0}(y_i^*)$ for $i \in I$ and $z_i^{0*} = 0$, $z_i^{1*} = 0_{X^*}$ for $i \notin I$. Then, it is obvious that z_i^{0*} and z_i^{1*}, $i = 1, \ldots, n$, are feasible elements to (D_T) and it holds

$$- \sigma_S \left(-\sum_{i\in I} y_i^* \right) - \sum_{i\in I} \left[\sigma_{L_i} \left(y_i^* \right) + \sigma_{\Omega_i} \left(y_i^* \right) \right]$$

$$= - \sigma_S \left(-\sum_{i\in I} z_i^{1*} \right) - \sum_{i\in I} \left[\sigma_{L_i} \left(z_i^{1*} \right) + \sigma_{\Omega_i} \left(z_i^{1*} \right) \right] \le v(D_T) \qquad (4.58)$$

for all y_i^*, $i = 1, \ldots, n$, feasible to $(\widetilde{D}_T)$, which implies $v(\widetilde{D}_T) \le v(D_T)$.

Vice versa, let z_i^{0*} and z_i^{1*} be feasible elements to (D_T) for $i = 1, \ldots, n$, then we have $\gamma_{C_i^0}(z_i^{1*}) \le z_i^{0*}$ for $i \in I$, $\sum_{i\in I} z_i^{0*} \le 1$ and $z_i^{0*} = 0$, $z_i^{1*} = 0_{X^*}$ for $i \notin I$, from which follows by setting $y_i^* = z_i^{1*}$ for $i \in I$ and $y_i^* = 0_{X^*}$ for $i \notin I$ that

$$\sum_{i\in I} \gamma_{C_i^0}(y_i^*) \le 1,$$

in other words y_i^*, $i = 1, \ldots, n$, is a feasible solution to $(\widetilde{D}_T)$.

Furthermore, we have that

$$
- \sigma_S \left(- \sum_{i \in I} z_i^{1*} \right) - \sum_{i \in I} \left[\sigma_{L_i} \left(z_i^{1*} \right) + \sigma_{\Omega_i} \left(z_i^{1*} \right) \right]
$$

$$
= - \sigma_S \left(- \sum_{i \in I} y_i^* \right) - \sum_{i \in I} \left[\sigma_{L_i} \left(y_i^* \right) + \sigma_{\Omega_i} \left(y_i^* \right) \right] \leq v(\widetilde{D}_T) \qquad (4.59)
$$

for all z_i^{0*} and z_i^{1*}, $i = 1, ..., n$, feasible to (D_T), which implies that $v(D_T) \leq v(\widetilde{D}_T)$. Bringing the statements (4.58) and (4.59) together reveals that it must hold $v(D_T) = v(\widetilde{D}_T)$. $\square$

The following statement is then a direct consequence of Theorem 4.3.

Theorem 4.9. *(strong duality) Between (P_T) and $(\widetilde{D}_T)$ holds strong duality, i.e. $v(P_T) = v(\widetilde{D}_T)$ and the dual problem has an optimal solution $\overline{y}^* \in X^n$.*

The necessary and sufficient optimality conditions for the primal-dual pair of optimization problems $(P_T) - (\widetilde{D}_T)$ can be derived by Theorem 4.9 similarly to the ones in Theorem 4.4.

Theorem 4.10. *(optimality conditions) (a) Let $\overline{x} \in X$ be an optimal solution to the problem (P_T). Then there exists an optimal solution to $(\widetilde{D}_T)$ $\overline{y}^* \in (X^*)^n$ with the corresponding index set $\overline{I} \subseteq \{1, \dots, n\}$ such that*

(i) $\displaystyle \max_{1 \leq j \leq n} \left\{ \mathcal{T}_{\Omega_j, \delta_{L_j}}^{C_j} (\overline{x}) \right\} = \sum_{i \in \overline{I}} \gamma_{C_i^0}(\overline{y}_i^*) \mathcal{T}_{\Omega_i, \delta_{L_i}}^{C_i} (\overline{x}),$

(ii) $\displaystyle \sigma_S \left(- \sum_{i \in \overline{I}} \overline{y}_i^* \right) = - \left\langle \sum_{i \in \overline{I}} \overline{y}_i^*, \overline{x} \right\rangle,$

(iii) $\gamma_{C_i^0}(\overline{y}_i^*) \mathcal{T}_{\Omega_i, \delta_{L_i}}^{C_i} (\overline{x}) + \sigma_{L_i} (\overline{y}_i^*) + \sigma_{\Omega_i} (\overline{y}_i^*) = \langle \overline{y}_i^*, \overline{x} \rangle,\ i \in \overline{I},$

(iv) $\displaystyle \sum_{j \in \overline{I}} \gamma_{C_i^0}(\overline{y}_i^*) = 1,\ \overline{y}_i^* \in X^* \setminus \{0_{X^*}\},\ i \in \overline{I},$ and $\overline{y}_i^* = 0_{X^*},\ i \notin \overline{I},$

(v) $\mathcal{T}_{\Omega_i, \delta_{L_i}}^{C_i} (\overline{x}) = \displaystyle \max_{1 \leq j \leq n} \left\{ \mathcal{T}_{\Omega_j, \delta_{L_j}}^{C_j} (\overline{x}) \right\},\ i \in \overline{I}.$

(b) If there exists $\overline{x} \in X$ such that for some $(\overline{y}_1^*, \dots, \overline{y}_n^*) \in (X^*)^n$ with the corresponding index set $\overline{I}$ the conditions (i)-(v) are fulfilled, then $\overline{x}$ is an optimal solution to (P_T), $(\overline{y}_1^*, \dots, \overline{y}_n^*)$ is an optimal solution to $(\widetilde{D}_T)$ and $v(P_T) = v(\widetilde{D}_T)$.

Proof. Let $\overline{x} \in S$ be an optimal solution to $(P_{\mathcal{T}})$, then by Theorem 4.9 there exists $\overline{y}^* \in (X^*)^n$ with the corresponding index set $\overline{I} \subseteq \{1, ..., n\}$ such that $v(P_{\mathcal{T}}) = v(\widetilde{D}_{\mathcal{T}})$, i.e.

$$\max_{1 \leq j \leq n} \left\{ \mathcal{T}^{C_j}_{\Omega_j, \delta_{L_j}}(\overline{x}) \right\} = -\sigma_S \left(-\sum_{i \in I} \overline{y}^*_i \right) - \sum_{i \in I} [\sigma_{L_i}(\overline{y}^*_i) + \sigma_{\Omega_i}(\overline{y}^*_i)]$$

$$\Leftrightarrow \max_{1 \leq j \leq n} \left\{ \mathcal{T}^{C_j}_{\Omega_j, \delta_{L_j}}(\overline{x}) \right\} + \sigma_S \left(-\sum_{i \in I} \overline{y}^*_i \right) + \sum_{i \in I} [\sigma_{L_i}(\overline{y}^*_i) + \sigma_{\Omega_i}(\overline{y}^*_i)] = 0$$

$$\Leftrightarrow \max_{1 \leq j \leq n} \left\{ \mathcal{T}^{C_j}_{\Omega_j, \delta_{L_j}}(\overline{x}) \right\} + \sigma_S \left(-\sum_{i \in I} \overline{y}^*_i \right) + \sum_{i \in I} [\sigma_{L_i}(\overline{y}^*_i) + \sigma_{\Omega_i}(\overline{y}^*_i)]$$
$$+ \sum_{i \in I} \gamma_{C^0_i}(\overline{y}^*_i) \mathcal{T}^{C_i}_{\Omega_i, \delta_{L_i}}(\overline{x}) - \sum_{i \in I} \gamma_{C^0_i}(\overline{y}^*_i) \mathcal{T}^{C_i}_{\Omega_i, \delta_{L_i}}(\overline{x}) + \sum_{i \in I} \langle \overline{y}^*_i, \overline{x} \rangle - \sum_{i \in I} \langle \overline{y}^*_i, \overline{x} \rangle = 0$$

$$\Leftrightarrow \left[\max_{1 \leq j \leq n} \left\{ \mathcal{T}^{C_j}_{\Omega_j, \delta_{L_j}}(\overline{x}) \right\} - \sum_{i \in \overline{I}} \gamma_{C^0_i}(\overline{y}^*_i) \mathcal{T}^{C_i}_{\Omega_i, \delta_{L_i}}(\overline{x}) \right]$$
$$+ \left[\delta_S(\overline{x}) + \sigma_S \left(-\sum_{i \in \overline{I}} \overline{y}^*_i \right) + \left\langle \sum_{i \in \overline{I}} \overline{y}^*_i, \overline{x} \right\rangle \right]$$
$$+ \sum_{i \in \overline{I}} \left[\gamma_{C^0_i}(\overline{y}^*_i) \mathcal{T}^{C_i}_{\Omega_i, \delta_{L_i}}(\overline{x}) + \sigma_{L_i}(\overline{y}^*_i) + \sigma_{\Omega_i}(\overline{y}^*_i) - \langle \overline{y}^*_i, \overline{x} \rangle \right] = 0.$$

By Lemma 4.2 holds that the term within the first bracket is non-negative and by the Young-Fenchel inequality we derive that the term within the second bracket is non-negative. Further, from $\gamma_{C^0_i}(\overline{y}^*_i) > 0$ for $i \in \overline{I}$, it follows by the Young-Fenchel inequality that

$$\gamma_{C^0_i}(\overline{y}^*_i) \mathcal{T}^{C_i}_{\Omega_i, \delta_{L_i}}(\overline{x}) + \left(\gamma_{C^0_i}(\overline{y}^*_i) \mathcal{T}^{C_i}_{\Omega_i, \delta_{L_i}} \right)^*(x^*) \geq \langle x^*, \overline{x} \rangle \; \forall x^* \in X^*, \tag{4.60}$$

and since (recall that $(\mathcal{T}^{C_i}_{\Omega_i, \delta_{L_i}})^* = \delta_{C^0} + \sigma_{L_i} + \sigma_{\Omega_i}$, for all $i \in \overline{I}$, and see Remark 4.5)

$$\left(\gamma_{C^0_i}(\overline{y}^*_i) \mathcal{T}^{C_i}_{\Omega_i, \delta_{L_i}} \right)^*(x^*) = \gamma_{C^0_i}(\overline{y}^*_i) \left(\mathcal{T}^{C_i}_{\Omega_i, \delta_{L_i}} \right)^* \left(\frac{1}{\gamma_{C^0_i}(\overline{y}^*_i)} x^* \right)$$

$$= \gamma_{C^0_i}(\overline{y}^*_i) \left[\delta_{C^0} \left(\frac{1}{\gamma_{C^0_i}(\overline{y}^*_i)} x^* \right) + \sigma_{L_i} \left(\frac{1}{\gamma_{C^0_i}(\overline{y}^*_i)} x^* \right) + \sigma_{\Omega_i} \left(\frac{1}{\gamma_{C^0_i}(\overline{y}^*_i)} x^* \right) \right]$$

$$= \begin{cases} \sigma_{L_i}(x^*_i) + \sigma_{\Omega_i}(x^*_i), & \text{if } \gamma_{C^0_i}(x^*) \leq \gamma_{C^0_i}(\overline{y}^*_i), \\ +\infty, & \text{otherwise,} \end{cases} \tag{4.61}$$

for all $x^* \in X^*$, $i \in \overline{I}$, one has that $\gamma_{C^0_i}(\overline{y}^*_i) \mathcal{T}^{C_i}_{\Omega_i, \delta_{L_i}}(\overline{x}) + \sigma_{L_i}(\overline{y}^*_i) + \sigma_{\Omega_i}(\overline{y}^*_i) \geq \langle \overline{y}^*_i, \overline{x} \rangle$ for all $i \in \overline{I}$. This means that the terms within the other brackets are also non-negative and therefore, all the terms inside the brackets must be equal to zero. This implies the cases (i)-(iii). Further, we obtain by the first bracket

$$\max_{1\leq j\leq n}\left\{\mathcal{T}^{C_j}_{\Omega_j,\delta_{L_j}}(\overline{x})\right\} = \sum_{i\in\overline{I}}\gamma_{C_i^0}(\overline{y}_i^*)\mathcal{T}^{C_i}_{\Omega_i,\delta_{L_i}}(\overline{x})$$

$$\leq \sum_{i\in\overline{I}}\gamma_{C_i^0}(\overline{z}_i^*)\max_{1\leq j\leq n}\left\{\mathcal{T}^{C_j}_{\Omega_j,\delta_{L_j}}(\overline{x})\right\} \leq \max_{1\leq j\leq n}\left\{\mathcal{T}^{C_j}_{\Omega_j,\delta_{L_j}}(\overline{x})\right\}$$

and from here follows that $\sum_{i\in\overline{I}}\gamma_{C_i^0}(\overline{y}_i^*) = 1$, which yields condition (iv) (recall that $0 \leq \gamma_{C_i^0}(y_i^*) = \sigma_C(y_i^*) \leq 0 \Leftrightarrow y_i^* = 0_{X^*}$, $i = 1,...,n$), as well as

$$\sum_{i\in\overline{I}}\gamma_{C_i^0}(\overline{y}_i^*)\max_{1\leq j\leq n}\left\{\mathcal{T}^{C_j}_{\Omega_j,\delta_{L_j}}(\overline{x})\right\} = \sum_{i\in\overline{I}}\gamma_{C_i^0}(\overline{y}_i^*)\mathcal{T}^{C_i}_{\Omega_i,\delta_{L_i}}(\overline{x})$$

$$\Leftrightarrow \sum_{i\in\overline{I}}\gamma_{C_i^0}(\overline{y}_i^*)\left[\max_{1\leq j\leq n}\left\{\mathcal{T}^{C_j}_{\Omega_j,\delta_{L_j}}(\overline{x})\right\} - \mathcal{T}^{C_i}_{\Omega_i,\delta_{L_i}}(\overline{x})\right] = 0. \tag{4.62}$$

As the brackets in (4.62) are non-negative and $\gamma_{C_i^0}(\overline{y}_i^*) > 0$, $i \in \overline{I}$, we get that

$$\max_{1\leq j\leq n}\left\{\mathcal{T}^{C_j}_{\Omega_j,\delta_{L_j}}(\overline{x})\right\} = \mathcal{T}^{C_i}_{\Omega_i,\delta_{L_i}}(\overline{x}), \ i \in \overline{I}.$$

which yields the condition (v) and completes the proof. $\qquad\square$

For the rest of this section, we consider the situation when $S = X$, which yields

$$(\widetilde{D}_T) \qquad \sup_{\substack{y_i^*\in X^*,\ i=1,...,n,\\ I=\left\{i\in\{1,...,n\}:\gamma_{C_i^0}(y_i^*)>0\right\},\ \sum_{i\in I}\gamma_{C_i^0}(y_i^*)\leq 1}} \left\{-\sigma_X\left(-\sum_{i\in I}y_i^*\right) - \sum_{i\in I}[\sigma_{L_i}(y_i^*) + \sigma_{\Omega_i}(y_i^*)]\right\}$$

$$= \sup_{\substack{y_i^*\in X^*,\ i=1,...,n,I=\left\{i\in\{1,...,n\}:\gamma_{C_i^0}(y_i^*)>0\right\},\\ \sum_{i\in I}\gamma_{C_i^0}(y_i^*)\leq 1,\ \sum_{i\in I}y_i^*=0_{X^*}}} \left\{-\sum_{i\in I}[\sigma_{L_i}(y_i^*) + \sigma_{\Omega_i}(y_i^*)]\right\}.$$

Remark 4.23. *Considering $(\widetilde{D}_T)$ in a finitely dimensional setting as a minimization problem, the following economical interpretation arises, where the objective function can be seen as a cost function. The components of the dual variables $y_i^* \in \mathbb{R}^m$, $i = 1,...,n$, express the expected expenditures on public goods and services, where i can be identified as one of n locations. More precisely, every location i has its own vector of m expenditures. Examples of public goods and services can be, for instance, parks, police stations, fire departments or highways. If a component of any vector is zero, then this means that the market (or the citizens) of this location is (are) saturated regarding this good or service and if a component is negative, then the market is supersaturated.*

The constraint $\sum_{i=1}^{n} \gamma_{C_i^0}(y_i^*) \leq 1$ defines then the limitation of the budget for the public goods and services, while the constraint $\sum_{i=1}^{n} y_i^* = 0_{\mathbb{R}^m}$ describes the substitution character of the goods and services. The latter means that if the market i has expected expenditures on a special good or service, then it is taken from an another location which is supersaturated. Therefore, the dual problem $(\widetilde{D}_T)$ can be understood as a cost minimization problem of the government of n locations, which has to find the optimal allocation of public goods and services $(\overline{y}_1^*, \ldots, \overline{y}_n^*)$ for n locations (that can be districts, towns or federal states) such that all expected expenditures can be financed, the demands on public goods and services of the citizens are saturated and the costs minimal. Another scenario appears by considering the dual problem $(\widetilde{D}_T)$ as a cost minimization problem of the World Health Organization (WHO), where the components of a vector $z_i^* \in \mathbb{R}^m$ represents the expected expenditures on medical treatment and health care for m diseases for a region i in the world, $i = 1, \ldots, n,$. Here also the constraint $\sum_{i=1}^{n} \gamma_{C_i^0}(y_i^*) \leq 1$ characterizes the budget restrictions of the WHO. Moreover, if in a region i for a disease $j \in \{1, \ldots, m\}$ no medical treatment is required and medical staff and products are no longer needed (which means that the associated component y_{ij}^* is negative), for instance because a certain disease was eradicated there, then these expenditures can be reallocated to other regions which need medical treatment for the disease j. It is important not to waste any expenditures, therefore their sum must be zero, i.e. the constraint $\sum_{i=1}^{n} y_i^* = 0_{\mathbb{R}^m}$ must be fulfilled. For an economical interpretation of (P_T) we refer to [7].

The next statement makes use of Theorem 4.10 in order to provide an exact characterization of the optimal solutions of the optimization problem (P_T).

Theorem 4.11. Let $\overline{x} \in X$ be an optimal solution to the optimization problem (P_T). If $(\overline{y}_1^*, \ldots, \overline{y}_n^*) \in (X^*)^n$ is an optimal solution to $(\widetilde{D}_T)$ with the corresponding $\overline{I} \subseteq \{1, \ldots, n\}$, then

$$\overline{x} \in \bigcap_{i \in \overline{I}} \left[\partial \left(v(\widetilde{D}_T) \gamma_{C_i^0} \right)(\overline{y}_i^*) + \partial \sigma_{L_i}(\overline{y}_i^*) + \partial \sigma_{\Omega_i}(\overline{y}_i^*) \right].$$

The proof of Theorem 4.11 is analogous to the one of Theorem 4.7, so we skip it.
In the rest of the paper we assume that $X = \mathcal{H}$, where $\mathcal{H}$ is a real Hilbert space, $\beta_i > 0$ and $\gamma_{C_i}(\cdot) = \beta_i \| \cdot \|_{\mathcal{H}}$. Note that in this situation $C_i = \{x \in \mathcal{H} : \beta_i \|x\|_{\mathcal{H}} \leq 1\}$, $X^* = \mathcal{H}$ and $\gamma_{C_i^0}(\cdot) = (1/\beta_i) \| \cdot \|_{\mathcal{H}}$, $i = 1, \ldots, n$.

Corollary 4.1. Let $\overline{x} \in \mathcal{H}$ be an optimal solution to the optimization problem (P_T). If $(\overline{y}_1^*, \ldots, \overline{y}_n^*) \in \mathcal{H}^n$ is an optimal solution to $(\widetilde{D}_T)$ with the corresponding $\overline{I} \subseteq \{1, \ldots, n\}$, then there exist $\phi_i \in \Omega_i$ and $\psi_i \in L_i$ fulfilling $\sigma_{L_i}(\overline{y}_i^*) = \langle \overline{y}_i^*, \psi_i \rangle$ and $\sigma_{\Omega_i}(\overline{y}_i^*) = \langle \overline{y}_i^*, \phi_i \rangle$, i.e. $\psi_i \in \partial \sigma_{L_i}(\overline{y}_i^*)$ and $\phi_i \in \partial \sigma_{\Omega_i}(\overline{y}_i^*)$, $i \in \overline{I}$, such that

$$\overline{x} = \frac{1}{\sum_{i \in \overline{I}} \beta_i \|\overline{y}_i^*\|_{\mathcal{H}}} \sum_{i \in \overline{I}} \beta_i \|\overline{y}_i^*\|_{\mathcal{H}} (\phi_i + \psi_i).$$

Proof. By Theorem 4.11 it holds for each $i \in \overline{I}$

$$\overline{x} \in \partial \left(v(\widetilde{D}_T) \frac{1}{\beta_i} \| \cdot \|_{\mathcal{H}} \right) (\overline{y}_i^*) + \partial \sigma_{L_i}(\overline{y}_i^*) + \partial \sigma_{\Omega_i}(\overline{y}_i^*),$$

which means that there exist $\phi_i \in \partial \sigma_{L_i}(\overline{y}_i^*)$ and $\psi_i \in \partial \sigma_{\Omega_i}(\overline{y}_i^*)$ such that

$$\overline{x} - \phi_i - \psi_i \in v(\widetilde{D}_T) \frac{1}{\beta_i} \partial \left(\| \cdot \|_{\mathcal{H}} \right) (\overline{y}_i^*).$$

As $\overline{y}_i^* \neq 0_{\mathcal{H}}$, it follows, as $v(\widetilde{D}_T) > 0$ due to Remark 4.18 and Theorem 4.10, that

$$\frac{\beta_i}{v(\widetilde{D}_T)} (\overline{x} - \phi_i - \psi_i) = \frac{1}{\|\overline{y}_i^*\|_{\mathcal{H}}} \overline{y}_i^*, \ i \in \overline{I}. \tag{4.63}$$

Now, we take the sum over all $i \in \overline{I}$ in (4.63) and get that

$$\frac{1}{v(\widetilde{D}_T)} \sum_{i \in \overline{I}} \beta_i \|\overline{y}_i^*\|_{\mathcal{H}} \overline{x} = \frac{1}{v(\widetilde{D}_T)} \sum_{i \in \overline{I}} \beta_i \|\overline{y}_i^*\|_{\mathcal{H}} (\phi_i + \psi_i) + \sum_{i \in \overline{I}} \overline{y}_i^*. \tag{4.64}$$

From the optimality condition (*ii*) of Theorem 4.10 follows that the last term of (4.64) is equal to zero, which finally yields that

$$\overline{x} = \frac{1}{\sum_{i \in \overline{I}} \beta_i \|\overline{y}_i^*\|_{\mathcal{H}}} \sum_{i \in \overline{I}} \beta_i \|\overline{y}_i^*\|_{\mathcal{H}} (\phi_i + \psi_i). \tag{4.65}$$

$\square$

Remark 4.24. *If $\Omega_i = \mathcal{H}$ and $L_i = \{p_i\}$, $i = 1, \ldots, n$, where $p_1, \ldots, p_n$ are distinct points in $\mathcal{H}$, then*

$$\overline{x} = \frac{1}{\sum_{i \in \overline{I}} \beta_i \|\overline{y}_i^*\|_{\mathcal{H}}} \sum_{i \in \overline{I}} \beta_i \|\overline{y}_i^*\|_{\mathcal{H}} p_i.$$

Corollary 4.2. *Let $\beta_s := \max_{1 \leq i \leq n}\{\beta_i\}$ and $y_i^* \in \mathcal{H}$, $i = 1, ..., n$, and $I \subseteq \{1, ..., n\}$ be a feasible solution to $(\widetilde{D}_T)$, then it holds*

$$\|y_i^*\|_{\mathcal{H}} \leq \frac{\beta_s \beta_i}{\beta_s + \beta_i}, \ i \in I.$$

Proof. Assume that $y_i^* \in \mathcal{H}$, $i = 1, ..., n$ and $I \subseteq \{1, ..., n\}$ are feasible elements of the dual problem $(\widetilde{D}_T)$, then one has for $j \in I$,

$$\sum_{i \in I} y_i^* = 0_{\mathcal{H}} \Leftrightarrow y_j^* = - \sum_{\substack{i \in I \\ i \neq j}} y_i^*$$

and hence,

$$\|y_j^*\|_{\mathcal{H}} = \|\sum_{\substack{i \in I \\ i \neq j}} y_i^*\|_{\mathcal{H}} \leq \sum_{\substack{i \in I \\ i \neq j}} \|y_i^*\|_{\mathcal{H}}, \ j \in I. \tag{4.66}$$

Moreover, from the feasibility of y_i^*, $i \in I$, to $(\widetilde{D}_\mathcal{T})$ and by (4.66), we have

$$1 \geq \sum_{i \in I} \frac{1}{\beta_i} \|y_i^*\|_{\mathcal{H}} = \frac{1}{\beta_j} \|y_j^*\|_{\mathcal{H}} + \sum_{\substack{i \in I \\ i \neq j}} \frac{1}{\beta_i} \|y_i^*\|_{\mathcal{H}}$$

$$\geq \frac{1}{\beta_j} \|y_j^*\|_{\mathcal{H}} + \frac{1}{\beta_s} \sum_{\substack{i \in I \\ i \neq j}} \|y_i^*\|_{\mathcal{H}} \geq \frac{1}{\beta_j} \|y_j^*\|_{\mathcal{H}} + \frac{1}{\beta_s} \|y_j^*\|_{\mathcal{H}} = \frac{\beta_s + \beta_j}{\beta_s \beta_j} \|y_j^*\|_{\mathcal{H}}, \ j \in I,$$

and so,

$$\|y_j^*\|_{\mathcal{H}} \leq \frac{\beta_s \beta_j}{\beta_s + \beta_j}, \ j \in I.$$

$\square$

Remark 4.25. *Under the assumption that $\beta_1 = \ldots = \beta_n = 1$ and $\cap_{i=1}^n (\Omega_i + L_i) = \emptyset$, any two nonzero components of an optimal solution $\overline{y}^*$ to $(\widetilde{D}_\mathcal{T})$ are linearly independent. To see this, let us assume that there exist $i, j \in \overline{I}$, $i \neq j$, and $k_j > 0$ such that $\overline{y}_i^* = k_j \overline{y}_j^*$. Further, it holds by (4.63) that*

$$\overline{y}_i^* = \frac{\|\overline{y}_i^*\|_{\mathcal{H}}}{v(\widetilde{D}_\mathcal{T})}(\overline{x} - \phi_i - \psi_i) = k_j \overline{y}_j^* = k_j \frac{\|\overline{y}_j^*\|_{\mathcal{H}}}{v(\widetilde{D}_\mathcal{T})}(\overline{x} - \phi_j - \psi_j),$$

from which follows that

$$\overline{x} - \phi_i - \psi_i = k_j \frac{\|\overline{y}_j^*\|_{\mathcal{H}}}{\|\overline{y}_i^*\|_{\mathcal{H}}}(\overline{x} - \phi_j - \psi_j),$$

thus one gets that $\overline{x} - \phi_i - \psi_i = \overline{x} - \phi_j - \psi_j \Leftrightarrow \phi_i + \psi_i = \phi_j + \psi_j$, which contradicts the assumption that $\cap_{i=1}^n (\Omega_i + L_i) = \emptyset$. Therefore, $\overline{y}_i^* \neq k_j \overline{y}_j^*$, $k_j > 0$, $i \neq j$, for all $i, j \in \overline{I}$.

Remark 4.26. *(i) Clearly, if $\partial \sigma_{\Omega_i + L_i}(\overline{y}_i^*)$ is a singleton for some $i \in \overline{I}$ (which is the situation, when for instance the set $\Omega_i + L_i$ is strictly convex or its indicator function is Gâteaux-differentiable at $\overline{y}_i^*$), then the optimal solution $\overline{x}$ of $(P_\mathcal{T})$ can be determined immediately by the formula (4.63), i.e.*

$$\overline{x} = \frac{v(\widetilde{D}_\mathcal{T})}{\beta_i \|\overline{y}_i^*\|_{\mathcal{H}}} \overline{y}_i^* + \phi_i + \psi_i.$$

(ii) Recall moreover that $v(\widetilde{D}_{\mathcal{T}}) = v(P_{\mathcal{T}})$ due to Theorem 4.9. For instance, take for an index $i \in \overline{I}$ the sets $\Omega_i := \{0_{\mathcal{H}}\}$ and $L_i := \{x \in \mathcal{H} : \|x - p_i\|_{\mathcal{H}} \leq a_i\}$, where $p_i \in \mathcal{H}$ and $a_i > 0$, then

$$\psi_i \in \partial \sigma_{L_i}(\overline{y}_i^*) \Leftrightarrow \psi_i \in \partial \left(\frac{1}{a_i}\|\cdot\|_{\mathcal{H}}\right)(\overline{y}_i^*) + p_i \Leftrightarrow a_i(\psi_i - p_i) \in \partial (\|\cdot\|_{\mathcal{H}})(\overline{y}_i^*)$$

$$\Leftrightarrow a_i(\psi_i - p_i) = \frac{1}{\|\overline{y}_i^*\|_{\mathcal{H}}}\overline{y}_i^* \Leftrightarrow \psi_i = \frac{1}{a_i\|\overline{y}_i^*\|_{\mathcal{H}}}\overline{y}_i^* + p_i,$$

hence

$$\overline{x} = \left(\frac{v(\widetilde{D}_{\mathcal{T}})}{\beta_i} + \frac{1}{a_i}\right)\frac{1}{\|\overline{y}_i^*\|_{\mathcal{H}}}\overline{y}_i^* + p_i.$$

(iii) Let us consider another situation where $\mathcal{H} = \mathbb{R}^m$, $\|\cdot\|_{\infty}$ is the ∞-norm, $\|\cdot\|_1$ is the l_1-norm, $\Omega_i := \{0_{\mathbb{R}^m}\}$ and $L_i := \{x \in \mathbb{R}^m : \|x - p_i\|_{\infty} \leq a_i\}$. Further, let $\overline{x} = (\overline{x}_1, \ldots, \overline{x}_m)^T \in \mathbb{R}^m$, $\overline{y}_i^* = (\overline{y}_{i1}^*, \ldots, \overline{y}_{im}^*)^T \in \mathbb{R}^m$, $\psi_i = (\psi_{i1}, \ldots, \psi_{im})^T \in \mathbb{R}^m$ and $p_i = (p_{i1}, \ldots, p_{im})^T \in \mathbb{R}^m$. One has

$$\psi_i \in \partial \sigma_{L_i}(\overline{y}_i^*) = \partial (a_i\|\cdot\|_1)(\overline{y}_i^*) + p_i$$

$$\Leftrightarrow \frac{1}{a_i}(\psi_i - p_i) \in \partial(\|\cdot\|_1)(\overline{y}_i^*) \Leftrightarrow \frac{1}{a_i}(\psi_{ij} - p_{ij}) \in \begin{cases} \{1\}, & \text{if } \overline{y}_{ij}^* > 0, \\ \{-1\}, & \text{if } \overline{y}_{ij}^* < 0, \\ [-1,1], & \text{if } \overline{y}_{ij}^* = 0, \end{cases} \tag{4.67}$$

$i \in \overline{I}$, $j = 1, \ldots, m$. Now, we define the following index set $J = \{j \in \{1, \ldots, m\} : \overline{y}_{ij}^* = 0 \text{ for all } i \in \overline{I}\}$, then it follows for $j \in J$ from (4.67) that

$$\frac{1}{a_i}(\psi_{ij} - p_{ij}) \in [-1,1] \Leftrightarrow \psi_{ij} \in p_{ij} + [-a_i, a_i]. \tag{4.68}$$

Combining (4.63) and (4.68) implies that under the corresponding hypotheses it holds

$$\overline{x}_j \in \bigcap_{i \in \overline{I}}(p_{ij} + [-a_i, a_i]), \ j \in J.$$

Otherwise, if $j \notin J$, there exists $i \in \overline{I}$ such that $\overline{y}_{ij}^* > 0$ and by (4.67) it holds

$$\psi_{ij} = \begin{cases} a_i + p_{ij}, & \text{if } \overline{y}_{ij}^* > 0, \\ -a_i + p_{ij}, & \text{if } \overline{y}_{ij}^* < 0, \end{cases} \tag{4.69}$$

and hence, one gets by (4.63) and (4.69) that

$$\overline{x}_j = \begin{cases} \frac{v(\widetilde{D}_{\mathcal{T}})}{\beta_i\|\overline{y}_i^*\|}\overline{y}_{ij}^* + a_i + p_{ij}, & \text{if } \overline{y}_{ij}^* > 0, \\ \frac{v(\widetilde{D}_{\mathcal{T}})}{\beta_i\|\overline{y}_i^*\|}\overline{y}_{ij}^* - a_i + p_{ij}, & \text{if } \overline{y}_{ij}^* < 0. \end{cases}$$

(iv) Finally, when $m = 2$, $|\overline{I}| > 2$ and $\beta_1 = \ldots = \beta_n = 1$, it follows by Remark 4.25 that $J = \emptyset$ and hence, there exist $i, j \in \overline{I}$ with $\overline{y}_{i1}^* \neq 0$ and $\overline{y}_{j2}^* \neq 0$ such that

$$\overline{x}_1 = \begin{cases} \frac{v(\widetilde{D}_\mathcal{T})}{\|\overline{y}_i^*\|} \overline{y}_{i1}^* + a_i + p_{i1}, & \text{if } \overline{y}_{i1}^* > 0, \\ \frac{v(\widetilde{D}_\mathcal{T})}{\|\overline{y}_i^*\|} \overline{y}_{i1}^* - a_i + p_{i1}, & \text{if } \overline{y}_{i1}^* < 0, \end{cases} \quad \text{and} \quad \overline{x}_2 = \begin{cases} \frac{v(\widetilde{D}_\mathcal{T})}{\|\overline{y}_j^*\|} \overline{y}_{j2}^* + a_j + p_{j2}, & \text{if } \overline{y}_{j2}^* > 0, \\ \frac{v(\widetilde{D}_\mathcal{T})}{\|\overline{y}_j^*\|} \overline{y}_{j2}^* - a_j + p_{j2}, & \text{if } \overline{y}_{j2}^* < 0. \end{cases}$$

Remark 4.27. *The optimality conditions of Theorem 4.10 allow to give the following geometrical interpretation of the set of optimal solutions to the conjugate dual problem $(\widetilde{D}_\mathcal{T})$ in the situation when $\mathcal{H} = \mathbb{R}^m$. When $\gamma_{C_i}(\cdot) = \|\cdot\|$ one gets from condition (iii) of Theorem 4.10 via the Young-Fenchel inequality $\|\overline{y}_i^*\|\|\overline{x} - \phi_i - \psi_i\| = \langle \overline{y}_i^*, \overline{x} - \phi_i - \psi_i \rangle$, which means that the vectors $\overline{y}_i^*$ and $\overline{x} - \phi_i - \psi_i$, $i \in \overline{I}$, are parallel and directed to $\overline{x}$. Further, from the optimal condition (v) of Theorem 4.10 one gets that $i \in \overline{I}$, i.e. $\overline{y}_i^* \neq 0_\mathcal{H}$, if the points $\phi_i + \psi_i \in \Omega_i + L_i$ are lying on the boundary of the ball centered at $\overline{x}$ with radius $v(\widetilde{D}_\mathcal{T})$. If $i \notin \overline{I}$, i.e. $\overline{y}_i^* = 0_\mathcal{H}$, then the points $\phi_i + \psi_i \in \Omega_i + L_i$ are lying on the border of a ball centered at $\overline{x}$ with radius $t < v(\widetilde{D}_\mathcal{T})$. Hence, the vectors $\overline{y}_i^*$, $i \in \overline{I}$, can be interpreted as force vectors, which pull the sets $\Omega_i + L_i$, $i \in \overline{I}$, in direction of the center, the gravity point $\overline{x}$, to reduce the minimal time needed to reach the farthest set(s) (see Figure 6.7).*

4.3 Single minmax location problems with gauges

4.3.1 Location problems with set-up costs in Banach spaces

Let us now focus our discussion on the location problem $(P_{h,\mathcal{T}}^S)$ considered in Section 4.2 where p_i, $i = 1, ..., n$ $(n \geq 2)$, are distinct points, $f_i = \delta_{\{0_X\}}$ and $\Omega_i = \{p_i\}$ such that $\mathcal{T}_{\Omega_i, f_i}^{C_i}(x) = \gamma_{C_i}(x - p_i)$ for all $x \in X$, $i = 1, ..., n$. Hence, $(P_{h,\mathcal{T}}^S)$ transforms into the following geometrically constrained minmax location problem

$$(P_{h,a}^S) \quad \inf_{x \in S} \max_{1 \leq i \leq n} \{h_i(\gamma_{C_i}(x - p_i)) + a_i\}.$$

Recall that

- $a_i \in \mathbb{R}_+$ are given non-negative set-up costs,

- S is a non-empty, closed and convex subset of the Banach space X,

- C_i is a closed and convex subset of X such that $0_X \in \operatorname{int} C_i$ and

- $h_i : \mathbb{R} \to \overline{\mathbb{R}}$ with $h_i(x) \in \mathbb{R}_+$, if $x \in \mathbb{R}_+$, and $h_i(x) = +\infty$, otherwise, is a proper, convex, lower semicontinuous and increasing function on $\mathbb{R}_+$, $i = 1, ..., n$.

The case where the set-up costs are arbitrary, i.e. $a_i \in \mathbb{R}$, will be discussed in Remark 4.31. Notice moreover that the following investigations can be extended to a Fréchet space, too.

Since $\sigma_{\Omega_i}(\cdot) = \sigma_{\{p_i\}}(\cdot) = \langle \cdot, p_i \rangle$, $\sigma_S(\cdot) = -\inf_{x \in S} \langle \cdot, x \rangle$ and $f_i^*(\cdot) = 0$, $i = 1, ..., n$, the conjugate dual problem $(D_{h,a}^S)$ to $(P_{h,a}^S)$ becomes, by using the dual $(D_{h,\mathcal{T}}^S)$ given in (4.36), to

$$(D_{h,a}^S) \qquad \sup_{\substack{\lambda_i,\ z_i^{0*} \geq 0,\ z_i^{1*} \in X^*,\ i=1,...,n, \\ I=\{i \in \{1,...,n\}: z_i^{0*}>0\} \subseteq R=\{r \in \{1,...,n\}: \lambda_r>0\}, \\ \gamma_{C_i^0}(z_i^{1*}) \leq z_i^{0*}, i \in I,\ z_j^{1*}=0_{X^*},\ j \notin I,\ \sum_{r \in R} \lambda_r \leq 1}} \left\{ \inf_{x \in S} \left\{ \sum_{i \in I} \langle z_i^{1*}, x - p_i \rangle \right\} \right.$$

$$\left. - \sum_{r \in R} \lambda_r \left[h_r^* \left(\frac{z_r^{0*}}{\lambda_r} \right) - a_r \right] \right\}. \qquad (4.70)$$

Remark 4.28. *If $h_i(\cdot) = \cdot + \delta_{\mathbb{R}_+}(\cdot)$, then the conjugate function of h_i is $h_i^* = \delta_{(-\infty,1]}$, $i = 1, ..., n$, and the conjugate dual problem $(D_{h,a}^S)$ transforms to*

$$(D_{h,a}^S) \qquad \sup_{\substack{\lambda_i,\ z_i^{0*} \geq 0,\ z_i^{1*} \in X^*,\ i=1,...,n, \\ I=\{i \in \{1,...,n\}: z_i^{0*}>0\} \subseteq R=\{r \in \{1,...,n\}: \lambda_r>0\}, \\ z_r^{0*} \leq \lambda_r,\ r \in R,\ \gamma_{C_i^0}(z_i^{1*}) \leq z_i^{0*}, i \in I,\ z_j^{1*}=0_{X^*},\ j \notin I,\ \sum_{r \in R} \lambda_r \leq 1}} \left\{ \inf_{x \in S} \left\{ \sum_{i \in I} \langle z_i^{1*}, x - p_i \rangle \right\} + \sum_{r \in R} \lambda_r a_r \right\}.$$

Moreover, by using the same ideas as in the proof of Proposition 4.1, one can show that the dual problem $(D_{h,a}^S)$ can be reduced to the following equivalent one

$$(\widetilde{D}_{h,a}^S) \qquad \sup_{\substack{y_i^{0*} \geq 0, y_i^{1*} \in X^*, i=1,...,n, \widetilde{I}=\{i \in \{1,...,n\}: y_i^{0*}>0\}, \\ y_j^{1*}=0_{X^*}, j \notin \widetilde{I}, \gamma_{C_i^0}(y_i^{1*}) \leq y_i^{0*}, i \in \widetilde{I}, \sum_{i \in \widetilde{I}} y_i^{0*} \leq 1}} \left\{ \inf_{x \in S} \left\{ \sum_{i \in \widetilde{I}} \langle y_i^{1*}, x - p_i \rangle \right\} + \sum_{i \in \widetilde{I}} y_i^{0*} a_i \right\}. \qquad (4.71)$$

Remark 4.29. *The index sets I and R of the dual problem $(D_{h,a}^S)$ in (4.70) give a detailed characterization of the set of feasible solutions and are very useful in the further approach. But from the numerical aspect, these index sets make the dual in (4.70) very hard to solve, as they transform it into a discrete optimization problem.*

For this reason we prefer to use for theoretical approaches the dual $(D_{h,a}^S)$ in the form of (4.70) and for numerical studies its equivalent dual problem $(\widehat{D}_{h,a}^S)$ (see Proposition 4.2):

$$(\widehat{D}_{h,a}^S) \qquad \sup_{\substack{\lambda_i,\ z_i^{0*} \geq 0,\ z_i^{1*} \in X^*, \\ \gamma_{C_i^0}(z_i^{1*}) \leq z_i^{0*},\ i=1,...,n, \sum_{i=1}^{n} \lambda_i \leq 1}} \left\{ \inf_{x \in S} \left\{ \sum_{i=1}^{n} \langle z_i^{1*}, x - p_i \rangle \right\} - \sum_{i=1}^{n} \left[(\lambda_i h_i)^* \left(z_i^{0*} \right) - \lambda_i a_i \right] \right\}.$$

$$(4.72)$$

In this context, the dual of $(\widetilde{D}_{h,a}^S)$ in (4.71) looks like

$$(\widetilde{D}_{h,a}^S) \qquad \sup_{\substack{z_i^{0*} \geq 0, z_i^{1*} \in X^*,\ \gamma_{C_i^0}(y_i^{1*}) \leq y_i^{0*}, \\ i=1,...,n, \sum_{i=1}^{n} z_i^{0*} \leq 1}} \left\{ \inf_{x \in S} \left\{ \sum_{i=1}^{n} \langle z_i^{1*}, x - p_i \rangle \right\} + \sum_{i=1}^{n} z_i^{0*} a_i \right\}.$$

Finally, we can state the following result as a consequence of Theorem 4.3.

Theorem 4.12. *(strong duality) Between $(P_{h,a}^S)$ and $(D_{h,a}^S)$ strong duality holds, i.e. $v(P_{h,a}^S) = v(D_{h,a}^S)$ and the conjugate dual problem has an optimal solution.*

The following necessary and sufficient optimality conditions are a consequence of Theorem 4.4.

Theorem 4.13. *(optimality conditions) (a) Let $\overline{x} \in S$ be an optimal solution to the problem $(P_{h,a}^S)$. Then there exists an optimal solution to $(D_{h,a}^S)$ $(\overline{\lambda}_1, ..., \overline{\lambda}_n, \overline{z}_1^{0*}, ..., \overline{z}_n^{0*}, \overline{z}_1^{1*}, ..., \overline{z}_n^{1*}) \in \mathbb{R}_+^n \times \mathbb{R}_+^n \times (X^*)^n$ with the corresponding index sets $\overline{I} \subseteq \overline{R} \subseteq \{1, ..., n\}$ such that*

(i) $\displaystyle \max_{1 \le j \le n} \{h_j(\gamma_{C_j}(\overline{x} - p_j)) + a_j\} = \sum_{i \in \overline{I}} \overline{z}_i^{0*} \gamma_{C_i}(\overline{x} - p_i) - \sum_{r \in \overline{R}} \overline{\lambda}_r \left[h_r^* \left(\frac{\overline{z}_r^{0*}}{\overline{\lambda}_r} \right) - a_r \right]$

$\displaystyle \qquad = \sum_{r \in \overline{R}} \overline{\lambda}_r [h_r(\gamma_{C_r}(\overline{x} - p_r)) + a_r],$

(ii) $\overline{\lambda}_r h_r^* \left(\frac{\overline{z}_r^{0*}}{\overline{\lambda}_r} \right) + \overline{\lambda}_r h_r(\gamma_{C_r}(\overline{x} - p_r)) = \overline{z}_r^{0*} \gamma_{C_r}(\overline{x} - p_r) \ \forall r \in \overline{R},$

(iii) $\overline{z}_i^{0*} \gamma_{C_i}(\overline{x} - p_i) = \langle \overline{z}_i^{1*}, \overline{x} - p_i \rangle \ \forall i \in \overline{I},$

(iv) $\sum_{i \in \overline{I}} \langle \overline{z}_i^{1*}, \overline{x} \rangle = -\sigma_S \left(-\sum_{i \in \overline{I}} \overline{z}_i^{1*} \right),$

(v) $\displaystyle \max_{1 \le j \le n} \{h_j(\gamma_{C_j}(\overline{x} - p_j)) + a_j\} = h_r(\gamma_{C_r}(\overline{x} - p_r)) + a_r \ \forall r \in \overline{R},$

(vi) $\sum_{r \in \overline{R}} \overline{\lambda}_r = 1, \ \overline{\lambda}_k > 0, \ k \in \overline{R}, \ \overline{\lambda}_l = 0, \ l \notin \overline{R}, \ \overline{z}_i^{0*} > 0, \ i \in \overline{I}, \ and \ \overline{z}_j^{0*} = 0, \ j \notin \overline{I},$

(vii) $\gamma_{C_i^0}(\overline{z}_i^{1*}) = \overline{z}_i^{0*}, \ \overline{z}_i^{1*} \in X^* \setminus \{0_{X^*}\}, \ i \in \overline{I} \ and \ \overline{z}_j^{1*} = 0_{X^*}, \ j \notin \overline{I}.$

(b) If there exists $\overline{x} \in S$ such that for some $(\overline{\lambda}_1, ..., \overline{\lambda}_n, \overline{z}_1^{0*}, ..., \overline{z}_n^{0*}, \overline{z}_1^{1*}, ..., \overline{z}_n^{1*}) \in \mathbb{R}_+^n \times \mathbb{R}_+^n \times (X^*)^n$ with the corresponding index sets $\overline{I} \subseteq \overline{R} \subseteq \{1, ..., n\}$ the conditions (i)-(vii) are fulfilled, then $\overline{x}$ is an optimal solution to $(P_{h,a}^S)$, $(\overline{\lambda}_1, ..., \overline{\lambda}_n, \overline{z}_1^{0*}, ..., \overline{z}_n^{0*}, \overline{z}_1^{1*}, ..., \overline{z}_n^{1*})$ is an optimal solution to $(D_{h,a}^S)$ and $v(P_{h,a}^S) = v(D_{h,a}^S)$.

Remark 4.30. *The optimality conditions (i)-(iv) of the previous theorem can also be expressed by using subdifferentials. As*

$$f(y^0) = \begin{cases} \displaystyle \max_{1 \le i \le n} \{h_i(y_i^0) + a_i\}, & if \ y^0 = (y_1^0, ..., y_n^0)^T \in \mathbb{R}_+^n, \ i = 1, ..., n, \\ +\infty_{\mathbb{R}_+^n}, & otherwise, \end{cases}$$

and

$$f^*(z_1^{0*}, ..., z_n^{0*}) = \min_{\substack{\sum_{i=1}^n \lambda_i \le 1, \ \lambda_i \ge 0, \\ i = 1, ..., n}} \left\{ \sum_{i=1}^n [(\lambda_i h_i)^*(z_i^{0*}) - \lambda_i a_i] \right\},$$

we have by the optimal condition (i) of Theorem 4.13 that

$$f(\gamma_{C_1}(\overline{x} - p_1), ..., \gamma_{C_n}(\overline{x} - p_n)) + f^*(z_1^{0*}, ..., z_n^{0*}) = \sum_{i \in \overline{I}} \overline{z}_i^{0*} \gamma_{C_i}(\overline{x} - p_i).$$

By (2.4) the last equality is equivalent to

$$(z_1^{0*}, ..., z_n^{0*}) \in \partial f(\gamma_{C_1}(\overline{x} - p_1), ..., \gamma_{C_n}(\overline{x} - p_n)).$$

Therefore, the condition (i) of Theorem 4.13 can equivalently be written as

$$(i) \ (\overline{z}_1^{0*}, ..., \overline{z}_n^{0*}) \in \partial \left(\max_{1 \le j \le n} \{h_j(\cdot) + a_j\} \right) (\gamma_{C_1}(\overline{x} - p_1), ..., \gamma_{C_n}(\overline{x} - p_n)),$$

In the same way, we can rewrite the conditions (ii)-(iv)

(ii) $\overline{z}_r^{0*} \in \partial(\overline{\lambda}_r h_r)(\gamma_{C_r}(\overline{x} - p_r))$, $r \in \overline{R}$,

(iii) $\overline{z}_i^{1*} \in \partial(\overline{z}_i^{0*} \gamma_{C_i})(\overline{x} - p_i)$, $i \in \overline{I}$,

(iv) $-\sum_{i \in \overline{I}} \overline{z}_i^{1*} \in \partial \delta_S(\overline{x}) = N_S(\overline{x})$.

Bringing the optimality conditions (i) and (ii) together yields

$$(\overline{z}_1^{0*}, ..., \overline{z}_n^{0*}) \in \partial \left(\max_{1 \le j \le n} \{h_j(\cdot) + a_j\} \right) (\gamma_{C_1}(\overline{x} - p_1), ..., \gamma_{C_n}(\overline{x} - p_n))$$
$$\cap \left(\partial(\overline{\lambda}_1 h_1)(\gamma_{C_1}(\overline{x} - p_1)) \times ... \times \partial(\overline{\lambda}_n h_n)(\gamma_{C_n}(\overline{x} - p_n)) \right).$$

Moreover, summarizing the optimality conditions (iii) and (iv) reveals that

$$\sum_{i \in \overline{I}} \overline{z}_i^{1*} \in \sum_{i \in \overline{I}} \partial(\overline{z}_i^{0*} \gamma_{C_i})(\overline{x} - p_i) \cap (-N_S(\overline{x})).$$

Finally, take also note that the optimality conditions (iii) and (vii) of Theorem 4.13 give a detailed characterization of the subdifferential of $\overline{z}_i^{0*} \gamma_{C_i}$ at $\overline{x} - p_i$, $i = 1, ..., n$. More precisely,

$$\partial(\overline{z}_i^{0*} \gamma_{C_i})(\overline{x} - p_i) = \left\{ \overline{z}_i^{1*} \in X^* : \overline{z}_i^{0*} \gamma_{C_i}(\overline{x} - p_i) = \langle \overline{z}_i^{1*}, \overline{x} - p_i \rangle \text{ and } \gamma_{C_i^0}(\overline{z}_i^{1*}) = \overline{z}_i^{0*} \right\}, \ i \in \overline{I}.$$

Remark 4.31. *If we consider the situation when the set-up costs are arbitrary, i.e. a_i can also be negative, $i = 1, ..., n$, then the conjugate function of f looks like (see Remark 4.2)*

$$f^*(z_1^{0*}, ..., z_n^{0*}) = \min_{\substack{\sum_{i=1}^n \lambda_i = 1, \ \lambda_i \ge 0, \\ i=1,...,n}} \left\{ \sum_{i=1}^n [(\lambda_i h_i)^*(z_i^{0*}) - \lambda_i a_i] \right\}.$$

As a consequence, we derive the following corresponding dual problem

$$(D_{h,a}^S) \qquad \sup_{\substack{\lambda_i,\ z_i^{0*} \geq 0,\ z_i^{1*} \in X^*,\ i=1,\dots,n, \\ I=\{i \in \{1,\dots,n\}: z_i^{0*} > 0\} \subseteq R=\{r \in \{1,\dots,n\}: \lambda_r > 0\}, \\ \gamma_{C_i^0}(z_i^{1*}) \leq z_i^{0*},\, i \in I,\ z_j^{1*}=0_{X^*},\, j \notin I,\ \sum_{r \in R} \lambda_r = 1}} \left\{ \inf_{x \in S} \left\{ \sum_{i \in I} \langle z_i^{1*}, x - p_i \rangle \right\} \right. $$

$$\left. - \sum_{r \in R} \lambda_r \left[h_r^* \left(\frac{z_r^{0*}}{\lambda_r} \right) - a_r \right] \right\}.$$

Therefore, all the statements given in this subsection are also true in the case of arbitrary set-up costs with the difference that $\sum_{r \in R} \lambda_r = 1$ in the constraint set.

Minmax location problems with arbitrary set-up costs were considered for example in [40] and [77]. For readers who are also interested in minmax location problems with nonlinear set-up costs, we refer to [38] and [49].

4.3.2 Location problems with set-up costs in Hilbert spaces

This subsection is devoted to the case where $S = X = \mathcal{H}$, where $\mathcal{H}$ is a Hilbert space, $a_i \geq 0$ and $\gamma_{C_i} : \mathcal{H} \to \mathbb{R}$ is defined by $\gamma_{C_i}(x) := \|x\|_{\mathcal{H}}$, $i = 1, \dots, n$, such that the minmax location problem $(P_{h,a}^S)$ turns into

$$(P_{h,a}^{S,N}) \qquad \inf_{x \in \mathcal{H}} \max_{1 \leq i \leq n} \{ h_i(\|x - p_i\|_{\mathcal{H}}) + a_i \}.$$

Its corresponding dual problem $(D_{h,a}^{S,N})$ transforms by (4.70) to

$$\sup_{\substack{\lambda_i,\ z_i^{0*} \geq 0,\ z_i^{1*} \in \mathcal{H},\ i=1,\dots,n, \\ I=\{i \in \{1,\dots,n\}: z_i^{0*} > 0\} \subseteq R=\{r \in \{1,\dots,n\}: \lambda_r > 0\}, \\ \|z_i^{1*}\|_{\mathcal{H}} \leq z_i^{0*},\, i \in I,\ \sum_{r \in R} \lambda_r \leq 1}} \left\{ \inf_{x \in \mathcal{H}} \left\{ \sum_{i \in I} \langle z_i^{1*}, x - p_i \rangle_{\mathcal{H}} \right\} \right.$$

$$\left. - \sum_{r \in R} \lambda_r \left[h_r^* \left(\frac{z_r^{0*}}{\lambda_r} \right) - a_r \right] \right\}$$

$$= \sup_{\substack{\lambda_i,\ z_i^{0*} \geq 0,\ z_i^{1*} \in \mathcal{H},\ i=1,\dots,n, \\ I=\{i \in \{1,\dots,n\}: z_i^{0*} > 0\} \subseteq R=\{r \in \{1,\dots,n\}: \lambda_r > 0\}, \\ \|z_i^{1*}\|_{\mathcal{H}} \leq z_i^{0*},\, i \in I,\ \sum_{r \in R} \lambda_r \leq 1}} \left\{ - \sup_{x \in \mathcal{H}} \left\{ \left\langle -\sum_{i \in I} z_i^{1*}, x \right\rangle_{\mathcal{H}} \right\} \right.$$

$$\left. - \sum_{i \in I} \langle z_i^{1*}, p_i \rangle_{\mathcal{H}} - \sum_{r \in R} \lambda_r \left[h_r^* \left(\frac{z_r^{0*}}{\lambda_r} \right) - a_r \right] \right\}$$

$$= \sup_{\substack{\lambda_i,\ z_i^{0*}\geq 0,\ z_i^{1*}\in\mathcal{H},\ i=1,...,n, \\ I=\{i\in\{1,...,n\}:z_i^{0*}>0\}\subseteq R=\{r\in\{1,...,n\}:\lambda_r>0\}, \\ \|z_i^{1*}\|_{\mathcal{H}}\leq z_i^{0*},i\in I,\ \sum_{i\in I}z_i^{1*}=0_{\mathcal{H}},\ \sum_{r\in R}\lambda_r\leq 1}} \left\{ -\sum_{i\in I}\langle z_i^{1*},p_i\rangle_{\mathcal{H}} - \sum_{r\in R}\lambda_r\left[h_r^*\left(\frac{z_r^{0*}}{\lambda_r}\right)-a_r\right]\right\}.$$

The following duality statements are direct consequences of Theorem 4.12 and 4.13.

Theorem 4.14. *(strong duality) Between* $(P_{h,a}^{S,N})$ *and* $(D_{h,a}^{S,N})$ *holds strong duality, i.e.* $v(P_{h,a}^{S,N}) = v(D_{h,a}^{S,N})$ *and the dual problem has an optimal solution.*

Theorem 4.15. *(optimality conditions) (a) Let* $\overline{x}\in\mathcal{H}$ *be an optimal solution to the problem* $(P_{h,a}^{S,N})$. *Then there exists an optimal solution to* $(D_{h,a}^{S,N})$ $(\overline{\lambda}_1,...,\overline{\lambda}_n,\overline{z}^{0*},\overline{z}^{1*})\in\mathbb{R}_+^n\times\mathbb{R}_+^n\times\mathcal{H}^n$ *with the corresponding index sets* $\overline{I}\subseteq\overline{R}\subseteq\{1,...,n\}$ *such that*

(i) $\displaystyle\max_{1\leq j\leq n}\{h_j(\|\overline{x}-p_j\|_{\mathcal{H}})+a_j\} = \sum_{i\in\overline{I}}\overline{z}_i^{0*}\|\overline{x}-p_i\|_{\mathcal{H}} - \sum_{r\in\overline{R}}\overline{\lambda}_r\left[h_r^*\left(\frac{\overline{z}_r^{0*}}{\overline{\lambda}_r}\right)-a_r\right]$

$\displaystyle = \sum_{r\in\overline{R}}\overline{\lambda}_r[h_r(\|\overline{x}-p_r\|_{\mathcal{H}})+a_r],$

(ii) $\overline{\lambda}_r h_r^*\left(\frac{\overline{z}_r^{0*}}{\overline{\lambda}_r}\right) + \overline{\lambda}_r h_r(\|\overline{x}-p_r\|_{\mathcal{H}}) = \overline{z}_r^{0*}\|\overline{x}-p_r\|_{\mathcal{H}}\ \forall r\in\overline{R},$

(iii) $\overline{z}_i^{0*}\|\overline{x}-p_i\|_{\mathcal{H}} = \langle\overline{z}_i^{1*},\overline{x}-p_i\rangle_{\mathcal{H}}\ \forall i\in\overline{I},$

(iv) $\displaystyle\sum_{i\in\overline{I}}\overline{z}_i^{1*} = 0_{\mathcal{H}},$

(v) $\displaystyle\max_{1\leq j\leq n}\{h_j(\|\overline{x}-p_j\|_{\mathcal{H}})+a_j\} = h_r(\|\overline{x}-p_r\|_{\mathcal{H}})+a_r\ \forall r\in\overline{R},$

(vi) $\displaystyle\sum_{r\in\overline{R}}\overline{\lambda}_r = 1,\ \overline{\lambda}_k>0,\ k\in\overline{R},\ \lambda_l=0,\ l\notin\overline{R},\ \overline{z}_i^{0*}>0,\ i\in\overline{I},$ *and* $\overline{z}_j^{0*}=0,\ j\notin\overline{I},$

(vii) $\|\overline{z}_i^{1*}\|_{\mathcal{H}} = \overline{z}_i^{0*},\ \overline{z}_i^{1*}\in\mathcal{H}\setminus\{0_{\mathcal{H}}\},\ i\in\overline{I}$ *and* $\overline{z}_j^{1*}=0_{\mathcal{H}},\ j\notin\overline{I}.$

(b) If there exists $\overline{x}\in\mathcal{H}$ *such that for some* $(\overline{\lambda}_1,...,\overline{\lambda}_n,\overline{z}^{0*},\overline{z}^{1*})\in\mathbb{R}_+^n\times\mathbb{R}_+^n\times\mathcal{H}^n$ *with the corresponding index sets* $\overline{I}\subseteq\overline{R}$ *the conditions (i)-(vii) are fulfilled, then* $\overline{x}$ *is an optimal solution to* $(P_{h,a}^{S,N})$, $(\overline{\lambda}_1,...,\overline{\lambda}_n,\overline{z}^{0*},\overline{z}^{1*})$ *is an optimal solution to* $(D_{h,a}^{S,N})$ *and* $v(P_{h,a}^{S,N}) = v(D_{h,a}^{S,N})$.

Regarding the relation between the optimal solutions to the primal and the dual problem the following corollary can be given under the additional assumption that the function h_i is continuous and strictly increasing for all $i=1,...,n$.

Corollary 4.3. *Let the function*

$$h_i:\mathbb{R}\to\overline{\mathbb{R}},\ h_i(x):=\begin{cases}h_i(x)\in\mathbb{R}_+, & \text{if } x\in\mathbb{R}_+, \\ +\infty, & \text{otherwise,}\end{cases}$$

be convex, continuous and strictly increasing for all $i = 1, ..., n$, and $\overline{x} \in \mathcal{H}$ an optimal solution to the problem $(P_{h,a}^{S,N})$. If $(\overline{\lambda}_1, ..., \overline{\lambda}_n, \overline{z}^{0*}, \overline{z}^{1*}) \in \mathbb{R}_+^n \times \mathbb{R}_+^n \times \mathcal{H}^n$ with the corresponding index sets $\overline{I} \subseteq \overline{R} \subseteq \{1, ..., n\}$ is an optimal solutions to the dual problem $(D_{h,a}^{S,N})$, then it holds

$$\overline{x} = \frac{1}{\displaystyle\sum_{i \in \overline{I}} \frac{\|\overline{z}_i^{1*}\|_{\mathcal{H}}}{h_i^{-1}\left(v(D_{h,a}^{S,N}) - a_i\right)}} \sum_{i \in \overline{I}} \frac{\|\overline{z}_i^{1*}\|_{\mathcal{H}}}{h_i^{-1}\left(v(D_{h,a}^{S,N}) - a_i\right)} p_i.$$

Proof. The optimality conditions (iii) and (vii) of Theorem 4.15 imply that

$$\|\overline{z}_i^{1*}\|_{\mathcal{H}}\|\overline{x} - p_i\|_{\mathcal{H}} = \langle \overline{z}_i^{1*}, \overline{x} - p_i \rangle_{\mathcal{H}}, \ i \in \overline{I},$$

By [3, Fact 2.10] there exists $\alpha_i > 0$ such that

$$\overline{z}_i^{1*} = \alpha_i \left(\overline{x} - p_i\right), \ i \in \overline{I}, \tag{4.73}$$

and so, $\|\overline{z}_i^{1*}\|_{\mathcal{H}} = \alpha_i\|\overline{x} - p_i\|_{\mathcal{H}}, \ i \in \overline{I}$. Therefore, it follows from the optimality condition (v) of Theorem 4.15 that (note that $\overline{I} \subseteq \overline{R}$)

$$\max_{1 \le j \le n} \{h_j(\|\overline{x} - p_j\|_{\mathcal{H}}) + a_j\} = h_i\left(\frac{1}{\alpha_i}\|\overline{z}_i^{1*}\|_{\mathcal{H}}\right) + a_i$$

$$\Leftrightarrow h_i^{-1}\left(\max_{1 \le j \le n} \{h_j(\|\overline{x} - p_j\|_{\mathcal{H}}) + a_j\} - a_i\right) = \frac{1}{\alpha_i}\|\overline{z}_i^{1*}\|_{\mathcal{H}}$$

$$\Leftrightarrow \alpha_i = \frac{\|\overline{z}_i^{1*}\|_{\mathcal{H}}}{h_i^{-1}\left(\max_{1 \le j \le n} \{h_j(\|\overline{x} - p_j\|_{\mathcal{H}}) + a_j\} - a_i\right)} = \frac{\|\overline{z}_i^{1*}\|_{\mathcal{H}}}{h_i^{-1}\left(v(D_{h,a}^{S,N}) - a_i\right)}, \ i \in \overline{I}.$$
$$\tag{4.74}$$

Now, we take in (4.73) the sum over all $i \in \overline{I}$, which yields by condition (iv) of Theorem 4.15

$$0_{\mathcal{H}} = \sum_{i \in \overline{I}} \overline{z}_i^{1*} = \sum_{i \in \overline{I}} \alpha_i \left(\overline{x} - p_i\right) \Leftrightarrow \overline{x} = \frac{1}{\displaystyle\sum_{i \in \overline{I}} \alpha_i} \sum_{i \in \overline{I}} \alpha_i p_i. \tag{4.75}$$

Finally, bringing (4.74) and (4.75) together implies

$$\overline{x} = \frac{1}{\displaystyle\sum_{i \in \overline{I}} \frac{\|\overline{z}_i^{1*}\|_{\mathcal{H}}}{h_i^{-1}\left(v(D_{h,a}^{S,N}) - a_i\right)}} \sum_{i \in \overline{I}} \frac{\|\overline{z}_i^{1*}\|_{\mathcal{H}}}{h_i^{-1}\left(v(D_{h,a}^{S,N}) - a_i\right)} p_i.$$

$\square$

Example 4.1. *(a) Let α_{is}, $\beta_{is} \geq 0$, $s = 1, ..., v$, and $h_i : \mathbb{R} \to \overline{\mathbb{R}}$ be defined by*

$$h_i(x) := \begin{cases} \max\limits_{1 \leq s \leq v} \{\alpha_{is} x + \beta_{is}\}, & \text{if } x \in \mathbb{R}_+, \\ +\infty, & \text{otherwise,} \end{cases}$$

$i = 1, ..., n$, then the corresponding location problem looks like

$$(P_{h,a}^{S,N}) \quad \inf_{x \in \mathcal{H}} \max_{1 \leq i \leq n} \left\{ \max_{1 \leq s \leq v} \{\alpha_{is} \|x - p_i\|_{\mathcal{H}} + \beta_{is}\} + a_i \right\}$$

$$= \inf_{x \in \mathcal{H}} \max_{\substack{1 \leq i \leq n, \\ 1 \leq s \leq v}} \{\alpha_{is} \|x - p_i\|_{\mathcal{H}} + \beta_{is} + a_i\}.$$

Moreover, we define the function

$$f_s : \mathbb{R} \to \overline{\mathbb{R}}, \ f_s(x) := \begin{cases} \alpha_{is} x + \beta_{is}, & \text{if } x \in \mathbb{R}_+, \\ +\infty, & \text{otherwise,} \end{cases}$$

then we derive by [91, Theorem 3.2]

$$h_i^*(x^*) = \left(\max_{1 \leq s \leq v} \{f_s(\cdot)\} \right)^* (x^*) = \inf_{\substack{\sum\limits_{s=1}^{v} x_s^* = x^*, \ \sum\limits_{s=1}^{v} \tau_s = 1, \\ \tau_s \geq 0, \ s=1,...,v}} \left\{ \sum_{s=1}^{v} (\tau_s f_s)^*(x_s^*) \right\}.$$

As the conjugate of the function $\tau_s f_s$ is

$$(\tau_s f_s)^*(x_s^*) = \sup_{x \in \mathbb{R}} \{x_s^* x - \tau_s f_s(x)\} = \sup_{x \geq 0} \{x_s^* x - \tau_s \alpha_{is} x - \tau_s \beta_{is}\}$$

$$= -\tau_s \beta_{is} + \sup_{x \geq 0} \{(x_s^* - \tau_s \alpha_{is}) x\} = \begin{cases} -\tau_s \beta_{is}, & \text{if } x_s^* \leq \tau_s \alpha_{is}, \\ +\infty, & \text{otherwise,} \end{cases}$$

$s = 1, ..., v$, we have

$$h_i^*(x^*) = \inf_{\substack{\sum\limits_{s=1}^{v} x_s^* = x^*, \ \sum\limits_{s=1}^{v} \tau_s = 1, \\ \tau_s \geq 0, \ x_s \leq \tau_s \alpha_{is}, \ s=1,...,v}} \left\{ -\sum_{s=1}^{n} \tau_s \alpha_{is} \right\}, \ i = 1, ..., n,$$

and hence, the dual problem is given by

$$(D_{h,a}^{S,N}) \quad \sup_{\substack{\lambda_i, \ z_i^{0*} \geq 0, \ z_i^{1*} \in \mathcal{H}, \ i=1,...,n, \\ I = \{i \in \{1,...,n\}: z_i^{0*} > 0\} \subseteq R = \{r \in \{1,...,n\}: \lambda_r > 0\}, \\ \|z_i^{1*}\|_{\mathcal{H}} \leq z_i^{0*}, i \in I, \sum\limits_{i \in I} z_i^{1*} = 0_{\mathcal{H}}, \sum\limits_{r \in R} \lambda_r \leq 1 \\ \sum\limits_{s=1}^{v} x_s^* = \frac{z_r^{0*}}{\lambda_r}, \ \sum\limits_{s=1}^{v} \tau_s = 1, \ \tau_s \geq 0, \ x_s \leq \tau_s \alpha_{rs}, \ s=1,...,v}} \left\{ -\sum_{i \in I} \langle z_i^{1*}, p_i \rangle_{\mathcal{H}} + \sum_{r \in R} \lambda_r \left[\sum_{s=1}^{n} \tau_s \alpha_{rs} - a_r \right] \right\}.$$

Furthermore, $h_i^{-1}(y) = \min_{1 \le s \le v} \left\{ \frac{1}{\alpha_{is}} (y - \beta_{is}) \right\}$ for all $i = 1, ..., n$, and thus, we have by Corollary 4.3

$$\bar{x} = \frac{1}{\sum_{i \in \bar{I}} \frac{\|\bar{z}_i^{1*}\|_{\mathcal{H}}}{\min_{1 \le s \le v} \left\{ \frac{1}{\alpha_{is}} (v(D_{h,a}^{S,N}) - a_i - \beta_{is}) \right\}}} \sum_{i \in \bar{I}} \frac{\|\bar{z}_i^{1*}\|_{\mathcal{H}}}{\min_{1 \le s \le v} \left\{ \frac{1}{\alpha_{is}} (v(D_{h,a}^{S,N}) - a_i - \beta_{is}) \right\}} p_i.$$

(b) Let $h_i : \mathbb{R} \to \overline{\mathbb{R}}$ be defined by

$$h_i(x) := \begin{cases} w_i x^{\beta_i}, & \text{if } x \in \mathbb{R}_+, \\ +\infty, & \text{otherwise}, \end{cases}$$

with $w_i > 0$, $\beta_i > 1$, $i = 1, ..., n$, then

$$(P_{h,a}^{S,N}) \qquad \inf_{x \in \mathcal{H}} \max_{1 \le i \le n} \left\{ w_i \|x - p_i\|_{\mathcal{H}}^{\beta_i} + a_i \right\}$$

and since the conjugate function of h_i is given by (see [3, Example 13.2 (i)])

$$h_i^*(x^*) = w_i \frac{\beta_i - 1}{\beta_i} \left(\frac{1}{w_i} x^* \right)^{\frac{\beta_i}{\beta_i - 1}} = \frac{\beta_i - 1}{\beta_i w_i^{\frac{1}{\beta_i - 1}}} (x^*)^{\frac{\beta_i}{\beta_i - 1}}, \quad i = 1, ..., n,$$

the associated dual problem $(D_{h,a}^{S,N})$ is

$$(D_{h,a}^{S,N}) \qquad \sup_{\substack{\lambda_i, \ z_i^{0*} \ge 0, \ z_i^{1*} \in \mathcal{H}, \ i=1,...,n, \\ I = \{i \in \{1,...,n\} : z_i^{0*} > 0\} \subseteq R = \{r \in \{1,...,n\} : \lambda_r > 0\}, \\ \|z_i^{1*}\|_{\mathcal{H}} \le z_i^{0*}, i \in I, \sum_{i \in I} z_i^{1*} = 0_{\mathcal{H}}, \sum_{r \in R} \lambda_r \le 1}} \left\{ - \sum_{i \in I} \langle z_i^{1*}, p_i \rangle_{\mathcal{H}} \right.$$

$$\left. - \sum_{r \in R} \lambda_r \left[\frac{\beta_r - 1}{\beta_r (\lambda_r w_r)^{\frac{1}{\beta_r - 1}}} (z_r^{0*})^{\frac{\beta_r}{\beta_r - 1}} - a_r \right] \right\}.$$

In addition, as $h_i^{-1}(y) = (y/w_i)^{\frac{1}{\beta_i}}$ for all $i = 1, ..., n$, it holds

$$\bar{x} = \frac{1}{\sum_{i \in \bar{I}} \frac{w_i^{\frac{1}{\beta_i}} \|\bar{z}_i^{1*}\|_{\mathcal{H}}}{\left(v(D_{h,a}^{S,N}) - a_i \right)^{\frac{1}{\beta_i}}}} \sum_{i \in \bar{I}} \frac{w_i^{\frac{1}{\beta_i}} \|\bar{z}_i^{1*}\|_{\mathcal{H}}}{\left(v(D_{h,a}^{S,N}) - a_i \right)^{\frac{1}{\beta_i}}} p_i.$$

(c) Let $h_i : \mathbb{R} \to \overline{\mathbb{R}}$ be defined by

$$h_i(x) := \begin{cases} w_i x, & \text{if } x \in \mathbb{R}_+, \\ +\infty, & \text{otherwise}, \end{cases}$$

where $w_i > 0$, then $h_i^{-1}(y) = \frac{1}{w_i}y$ for all $i = 1, ..., n$, and hence,

$$\overline{x} = \frac{1}{\sum\limits_{i \in \overline{I}} \frac{w_i \|\overline{z}_i^{1*}\|_{\mathcal{H}}}{v(D_{h,a}^{S,N}) - a_i}} \sum\limits_{i \in \overline{I}} \frac{w_i \|\overline{z}_i^{1*}\|_{\mathcal{H}}}{v(D_{h,a}^{S,N}) - a_i} p_i. \tag{4.76}$$

If $a_i = 0$, $i = 1, ..., n$, then formula in (4.76) reduces to (see Remark 4.24)

$$\overline{x} = \frac{1}{\sum\limits_{i \in \overline{I}} w_i \|\overline{z}_i^{1*}\|_{\mathcal{H}}} \sum\limits_{i \in \overline{I}} w_i \|\overline{z}_i^{1*}\|_{\mathcal{H}} p_i. \tag{4.77}$$

Remark 4.32. *Let us note that all the results in this section hold also for negative set-up costs. Like already mentioned in Remark 4.31, we have in this case in the constraint set of the dual problem $\sum_{r \in R} \lambda_r = 1$.*

4.3.3 Constrained location problems without set-up costs in Banach spaces

In this section we discuss single minmax location problems without set-up costs (i.e. $a_i = 0$, $i = 1, ..., n$), where X is a Banach space, $S \subseteq X$ and $h_i : \mathbb{R} \to \overline{\mathbb{R}}$ is defined by

$$h_i(x) := \begin{cases} x, & \text{if } x \in \mathbb{R}_+, \\ +\infty, & \text{otherwise.} \end{cases}$$

Hence, the location problem $(P_{h,a}^S)$ turns into

$$(P^S) \qquad \inf_{x \in S} \max_{1 \le i \le n} \{\gamma_{C_i}(x - p_i)\}$$

and by (4.71) we can write the corresponding dual (D^S) as

$$(D^S) \qquad \sup_{\substack{y_i^{0*} \ge 0, y_i^{1*} \in X^*, i=1,...,n, \ I = \left\{i \in \{1,...,n\}: y_i^{0*} > 0\right\}, \\ y_j^{1*} = 0_{X^*}, j \notin I, \gamma_{C_i^0}(y_i^{1*}) \le y_i^{0*}, i \in I, \sum\limits_{i \in I} y_i^{0*} \le 1}} \left\{ \inf_{x \in S} \left\{ \sum\limits_{i \in I} \langle z_i^{1*}, x - p_i \rangle \right\} \right\}.$$

Let us now introduce the following optimization problem

$$(\widetilde{D}^S) \qquad \sup_{\substack{z_i^* \in X^*, \ i=1,...,n, \ I = \left\{i \in \{1,...,n\}: \gamma_{C_i^0}(z_i^*) > 0\right\}, \\ z_j^* = 0_{X^*}, \ j \notin I, \ \sum\limits_{i \in I} \gamma_{C_i^0}(z_i^*) \le 1}} \left\{ \inf_{x \in S} \left\{ \sum\limits_{i \in I} \langle z_i^*, x - p_i \rangle \right\} \right\}, \tag{4.78}$$

then the following theorem can be formulated and proven similarly to the proof of Theorem 4.8.

Theorem 4.16. *It holds $v(D^S) = v(\widetilde{D}^S)$.*

Motivated by Theorem 4.16 it follows immediately the following one.

Theorem 4.17. *(strong duality) Between (P^S) and $(\widetilde{D}^S)$ holds strong duality, i.e. $v(P^S) = v(\widetilde{D}^S)$ and the dual problem $v(\widetilde{D}^S)$ has an optimal solution.*

Now, it is possible to formulate the following optimality conditions for the primal-dual pair (P^S)-$(\widetilde{D}^S)$ by using Theorem 4.10 (recall that $L_i = \{0_X\}$ and $\Omega_i = \{p_i\}$ such that $\mathcal{T}^{C_i}_{\Omega_i,\,\delta_{L_i}}(x) = \gamma_{C_i}(x - p_i)$ for all $x \in X$, $i = 1, ..., n$).

Theorem 4.18. *(optimality conditions) (a) Let $\overline{x} \in S$ be an optimal solution to the problem (P^S). Then there exists an optimal solution to $(\widetilde{D}^S)$ $\overline{z}^* \in (X^*)^n$ with the corresponding index set $\overline{I} \subseteq \{1, ..., n\}$ such that*

(i) $\max\limits_{1 \leq j \leq n}\{\gamma_{C_j}(\overline{x} - p_j)\} = \sum\limits_{i \in \overline{I}} \gamma_{C_i^0}(\overline{z}_i^*)\gamma_{C_i}(\overline{x} - p_i),$

(ii) $\sum\limits_{i \in \overline{I}}\langle \overline{z}_i^*, \overline{x}\rangle = -\sigma_S\left(-\sum\limits_{i \in \overline{I}} \overline{z}_i^*\right),$

(iii) $\gamma_{C_i^0}(\overline{z}_i^*)\gamma_{C_i}(\overline{x} - p_i) = \langle \overline{z}_i^*, \overline{x} - p_i\rangle,\ i \in \overline{I},$

(iv) $\sum\limits_{j \in \overline{I}} \gamma_{C_j^0}(\overline{z}_j^*) = 1,\ \overline{z}_i^* \in X^* \setminus \{0_{X^*}\},\ i \in \overline{I},\ and\ \overline{z}_i^* = 0_{X^*},\ i \notin \overline{I},$

(v) $\gamma_{C_i}(\overline{x} - p_i) = \max\limits_{1 \leq j \leq n}\{\gamma_{C_j}(\overline{x} - p_j)\},\ i \in \overline{I}.$

(b) If there exists $\overline{x} \in S$ such that for some $\overline{z}^* \in (X^*)^n$ with the corresponding index set $\overline{I}$ the conditions (i)-(v) are fulfilled, then $\overline{x}$ is an optimal solution to (P^S), $\overline{z}^*$ is an optimal solution to $(\widetilde{D}^S)$ and $v(P^S) = v(\widetilde{D}^S)$.

4.3.4 Unconstrained location problems without set-up costs in the Euclidean space

Now, we turn our attention to the case where $S = X = \mathbb{R}^d$. Furthermore, we use as the gauge functions the weighted Euclidean norm, i.e. $\gamma_{C_i}(\cdot) = w_i\|\cdot\|$ where $w_i > 0$, $i = 1, ..., n$. By these settings, the minmax location problem (P^S) transforms into the following one

$$(P_N^S) \qquad \inf_{x \in \mathbb{R}^d} \max_{1 \leq i \leq n}\{w_i\|x - p_i\|\}.$$

By using (4.78) we obtain the following dual problem corresponding to (P_N^S),

$$
(\widetilde{D}_N^S) \quad \sup_{\substack{z_i^* \in \mathbb{R}^d,\ i=1,\dots,n,\ I=\left\{i \in \{1,\dots,n\}:\|z_i^*\|>0\right\}, \\ z_j^*=0_{\mathbb{R}^d},\ j \notin I,\ \sum_{i \in I}\frac{1}{w_i}\|z_i^*\|\leq 1}} \inf_{x \in \mathbb{R}^d} \left\{ \sum_{i \in I} \langle z_i^*, x - p_i \rangle \right\}
$$

$$
= \sup_{\substack{z_i^* \in \mathbb{R}^d,\ i=1,\dots,n,\ I=\left\{i \in \{1,\dots,n\}:\|z_i^*\|>0\right\}, \\ z_j^*=0_{\mathbb{R}^d},\ j \notin I,\ \sum_{i \in I}\frac{1}{w_i}\|z_i^*\|\leq 1}} \left\{ -\sigma_{\mathbb{R}^d}\left(-\sum_{i \in I} z_i^* \right) - \sum_{i \in I} \langle z_i^*, p_i \rangle \right\}
$$

$$
= \sup_{\substack{z_i^* \in \mathbb{R}^d,\ i=1,\dots,n,\ I=\left\{i \in \{1,\dots,n\}:\|z_i^*\|>0\right\}, \\ z_j^*=0_{\mathbb{R}^d},\ j \notin I,\ \sum_{i \in I}\frac{1}{w_i}\|z_i^*\|\leq 1,\ \sum_{i \in I} z_i^*=0_{\mathbb{R}^d}}} \left\{ -\sum_{i \in I} \langle z_i^*, p_i \rangle \right\}. \tag{4.79}
$$

Remark 4.33. *Note that for simplicity it is also possible to substitute $z_i^* = -z_i^*$ for all $i = 1, \dots, n$, whence it follows*

$$
(\widetilde{D}_N^S) \quad \sup_{\substack{z_i^* \in \mathbb{R}^d,\ i=1,\dots,n,\ I=\left\{i \in \{1,\dots,n\}:\|z_i^*\|>0\right\}, \\ z_j^*=0_{\mathbb{R}^d},\ j \notin I,\ \sum_{i \in I}\frac{1}{w_i}\|z_i^*\|\leq 1,\ \sum_{i \in I} z_i^*=0_{\mathbb{R}^d}}} \left\{ \sum_{i \in I} \langle z_i^*, p_i \rangle \right\}. \tag{4.80}
$$

Theorem 4.19. *(strong duality) Between (P_N^S) and $(\widetilde{D}_N^S)$ holds strong duality, i.e. $v(P_N^S) = v(\widetilde{D}_N^S)$ and the dual problem has an optimal solution.*

By Theorem 4.18 and 4.19 we derive the following necessary and sufficient optimality conditions.

Theorem 4.20. *(optimality conditions) (a) Let $\overline{x} \in \mathbb{R}^d$ be an optimal solution to the problem (P_N^S). Then there exists an optimal solution to $(\widetilde{D}_N^S)$ $\overline{z}_i^* \in \mathbb{R}^d$, $i = 1, \dots, n$, with the corresponding index set $\overline{I}$ such that*

(i) $\displaystyle \max_{1 \leq j \leq n} \{w_j \|\overline{x} - p_j\|\} = \sum_{i \in \overline{I}} \|\overline{z}_i^*\| \|\overline{x} - p_i\|$,

(ii) $\displaystyle \sum_{i \in \overline{I}} \overline{z}_i^* = 0_{\mathbb{R}^d}$,

(iii) $\|\overline{z}_i^*\| \|\overline{x} - p_i\| = \langle \overline{z}_i^*, \overline{x} - p_i \rangle$, $i \in \overline{I}$,

(iv) $\displaystyle \sum_{j \in \overline{I}} \frac{1}{w_j} \|\overline{z}_j^*\| = 1$, $\overline{z}_i^* \in \mathbb{R}^d \setminus \{0_{\mathbb{R}^d}\}$ for $i \in I$ and $\overline{z}_i^* = 0_{\mathbb{R}^d}$ for $i \notin \overline{I}$,

(v) $w_i \|\overline{x} - p_i\| = \displaystyle \max_{1 \leq j \leq n} \{w_j \|\overline{x} - p_j\|\}$, $i \in \overline{I}$.

(b) If there exists $\overline{x} \in \mathbb{R}^d$ such that for some $\overline{z}_i^* \in \mathbb{R}^d$, $i = 1, ..., n$, with the corresponding index set $\overline{I}$ the conditions (i)-(v) are fulfilled, then $\overline{x}$ is an optimal solution to (P_N^S), $\overline{z}^*$ is an optimal solution to $(\widetilde{D}_N^S)$ and $v(P_N^S) = v(\widetilde{D}_N^S)$.

For the length of the vectors z_i^*, $i \in I$, feasible to $(\widetilde{D}_N^S)$, the following estimation from above can be made by Corollary 4.2.

Corollary 4.4. *Let $w_s := \max_{1 \leq i \leq n}\{w_i\}$ and $z_i^* \in \mathbb{R}^d$, $i = 1, ..., n$, and $I \subseteq \{1, ..., n\}$ be a feasible solution to $(\widetilde{D}_N^S)$, then it holds*

$$\|z_i^*\| \leq \frac{w_s w_i}{w_s + w_i}, \ i \in I.$$

By the next remark we point out the relation between the minmax and minsum problems.

Remark 4.34. *The optimal solution $\overline{x}$ to the problem (P_N^S) is also a solution to the following generalized Fermat-Torricelli problem*

$$(P_N^{FT}) \quad \min_{x \in \mathbb{R}^d} \sum_{i \in \overline{I}} \widetilde{w}_i \|x - p_i\|,$$

where $\widetilde{w}_i = \|\overline{z}_i^*\|$, $i \in \overline{I}$.
This can be seen like follows: It is well known that $\overline{x}$ is an optimal solution to the problem (P_N^{FT}) with $\overline{x} \neq p_i$, $i \in I$, if and only if the resultant force R at $\overline{x}$, defined by

$$R(\overline{x}) := \sum_{i \in \overline{I}} \widetilde{w}_i \frac{\overline{x} - p_i}{\|\overline{x} - p_i\|},$$

is zero (see [71]). As $\overline{x}$ is an optimal solution to (P_N^S), we have by (4.73) that

$$\sum_{i \in \overline{I}} \widetilde{w}_i \frac{\overline{x} - p_i}{\|\overline{x} - p_i\|} = \sum_{i \in \overline{I}} \|\overline{z}_i^*\| \frac{\overline{x} - p_i}{\|\overline{x} - p_i\|} = \sum_{i \in \overline{I}} \alpha_i(\overline{x} - p_i) = \sum_{i \in \overline{I}} \overline{z}_i^* = 0_{\mathbb{R}^d},$$

which implies that $\overline{x}$ is also an optimal solution to the problem (P_N^{FT}). In this context, pay attention also to the fact that for the optimal solution $\overline{x}$ to the problem (P_N^S) it holds $\overline{x} \neq p_i$, $i \in \overline{I}$. Because if there exists $j \in \overline{I}$ such that $\overline{x} = p_j$, then $\overline{x} = p_i$ for all $i \in \overline{I}$, which contradicts the assumption that the given points are distinct.

Geometrical interpretation.
For simplicity let us suppose that $w_1 = ... = w_n = 1$, then it is well-known that the problem (P_N^N) can be interpreted as the finding of a ball with center $\overline{x}$ and minimal radius such that all given points p_i, $i = 1, ..., n$ are covered by this ball. This problem is also known as the minimum covering ball problem.

Our plan is now to give a geometrical interpretation of the set of optimal solutions of the dual problem $(\widetilde{D}_N^S)$ by using Theorem 4.20 (see also Remark 4.27). By condition (iii) we see that for $i \in \overline{I}$ the dual problem can geometrically be understood as the finding of vectors $\overline{z}_i^*$, which are parallel to the vectors $\overline{x} - p_i$ and directed to $\overline{x}$ fulfilling $\sum_{i \in \overline{I}} \overline{z}_i^* = 0_{\mathbb{R}^d}$ and $\sum_{i \in \overline{I}} \|\overline{z}_i^*\| = 1$. Especially, conditions (iv) and (v) are telling us that for $i \in \overline{I}$, i.e. $\overline{z}_i^* \neq 0_{\mathbb{R}^d}$, the corresponding point p_i is lying on the border of the minimal covering ball and for $i \notin \overline{I}$, i.e. $\overline{z}_i^{1*} = 0_{\mathbb{R}^d}$, the corresponding point p_i is lying inside the mentioned ball. Therefore, for $i \in \overline{I}$ the elements $\overline{z}_i^*$ can be interpreted as force vectors, which pull the points p_i lying on the border of the minimum covering ball inside of this ball in direction to the center, the gravity point $\overline{x}$, where the resultant force of the sum of these force vectors is zero. For illustration see Example 4.2 and Figure 4.1.

Another well-known geometrical characterization of the location problem (P_S^N) is to find the minimum radius of balls centered at the points p_i, $i = 1, ..., n$, such that their intersection is non-empty. In this situation, the set of optimal solutions of the dual problem can be described as force vectors fulfilling the optimality conditions of Theorem 4.20 and increasing these balls until their intersection is non-empty and the radius of the largest ball is minimal. From the conditions (iv) and (v) we obtain that a force vector $\overline{z}_i^*$ is equal to the zero vector if $\overline{x}$ is an element of the interior of the ball centered at point p_i with radius $v(P_N^S)$, which is exactly the case when $i \notin \overline{I}$. If $i \in \overline{I}$, which is exactly the case when $\overline{x}$ is lying on the border of the ball centered at point p_i with radius $v(P_N^S)$, then the corresponding force vector $\overline{z}_i^*$ is unequal to the zero vector and moreover, by the optimality condition (iii) follows that $\overline{z}_i^*$ is parallel to the vector $\overline{x} - p_i$ and has the same direction.

To demonstrate the statements we made above, let us discuss the following example.

Example 4.2. *Consider the unconstrained single minmax location problem in $\mathbb{R}^2$ defined by the given points:*

$$p_1 = (-5, -2.5)^T; \ p_2 = (-2, 1)^T; \ p_3 = (2.5, 3)^T; \ p_4 = (3.5, -2)^T \ and \ p_5 = (0, -3)^T.$$

The primal problem looks in this case like follows

$$(\overline{P}_N^S) \qquad \inf_{x \in \mathbb{R}^2} \max_{1 \leq i \leq 5} \{\|x - p_i\|\}$$

and by using the MATLAB *Optimization Toolbox we get the solution* $\overline{x} = (-0.866, -0.273)^T$ *with the objective function value* $\max_{1 \leq i \leq 5} \{\|\overline{x} - p_i\|\} = 4.695$.

For the dual problem we have the formulation (see Remark 4.29)

$$(\widetilde{D}_N^S) \qquad \sup_{\substack{z_i^* \in \mathbb{R}^2, \ i=1,...,5, \\ \sum_{i=1}^5 \|z_i^*\| \leq 1, \ \sum_{i=1}^5 z_i^* = 0_{\mathbb{R}^2}}} \left\{ -\sum_{i=1}^5 \langle z_i^*, p_i \rangle \right\}. \tag{4.81}$$

with the solution

$$\overline{z}_1^* = (0.412, 0.222)^T; \ \overline{z}_2^* = (0, 0)^T; \ \overline{z}_3^* = (-0.281, -0.273)^T;$$
$$\overline{z}_4^* = (-0.131, 0.052)^T; \ \overline{z}_5^* = (0, 0)^T.$$

The dual problem was also solved by using the MATLAB *Optimization Toolbox.*

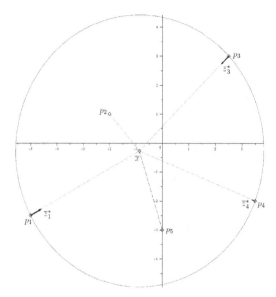

Figure 4.1: Geometrical illustration of the Example 4.2

In fact, it holds $\overline{I} = \{1, 3, 4\}$, $\langle \overline{z}_1^*, p_1 \rangle + \langle \overline{z}_3^*, p_3 \rangle + \langle \overline{z}_4^*, p_4 \rangle = 4.695$, $\overline{x} = \|\overline{z}_1^*\| p_1 + \|\overline{z}_3^*\| p_3 + \|\overline{z}_4^*\| p_4 = 0.468 \cdot (-5, -2.5)^T + 0.392 \cdot (2.5, 3)^T + 0.14 \cdot (3.5, -2)^T = (-0.866, -0.273)^T$ (see (4.77)) and the points p_1, p_3 and p_4 are lying on the border of the minimum covering circle as Figure 4.1 verifies.

Remark 4.35. *Let $w_i = 1$, $i=1,...,n$. Then, for the case $n = 2$ it follows immediately by condition (iv) of Theorem 4.20 and Corollary 4.4 the well-known fact that $\overline{x} = (1/2)(p_1 + p_2)$.*

Remark 4.36. *Let $w_i = 1$, $i=1,...,n$. If we consider the case $d = 1$, we can write the dual problem $(\widetilde{D}_N^S)$ as*

$$(\widetilde{D}_N^S) \quad \sup_{\substack{z_i^* \in \mathbb{R}, \ i=1,...,n, \ I = \{i \in \{1,...,n\}: |z_i^*| > 0\}, \\ z_j^* = 0, \ j \notin I, \ \sum_{i \in I} |z_i^*| \leq 1, \ \sum_{i \in I} z_i^* = 0}} \left\{ -\sum_{i \in I} z_i^* p_i \right\} = \sup_{\substack{z^* \in \mathbb{R}^n, \ \langle z^*, 1 \rangle = 0, \\ \|z^*\|_1 \leq 1}} \left\{ -\langle z^*, p \rangle \right\},$$

where $z^* = (z_1^*, ..., z_n^*)^T \in \mathbb{R}^n$, $p = (p_1, ..., p_n)^T \in \mathbb{R}^n$, $1 = (1, ..., 1)^T \in \mathbb{R}^n$ and $\| \cdot \|_1$ is the Manhattan norm. From the second formulation of the problem $(\widetilde{D}_N^S)$ it is clear that the set of the feasible elements is the intersection of a hyperplane orthogonal to the vector 1 and a cross-polytope (or hyperoctahedron), i.e. a convex polytope. Further, it is clear that the optimal solution to this problem can get immediately by the following consideration.

Let us assume that $p_1 < ... < p_n$, then it holds $p_1 < \overline{x} < p_n$ and by condition (v) of Theorem 4.20 one gets

$$\max_{1 \leq j \leq n} \{|\overline{x} - p_j|\} = |\overline{x} - p_1| = |\overline{x} - p_n|,$$

i.e. $\overline{I} = \{1, n\}$. By Remark 4.35 this means $\overline{x} = (1/2)(p_1 + p_n)$. Moreover, by Corollary 4.4 we have that $|\overline{z}_1^*| = |\overline{z}_n^*| = 0.5$ and by condition (iii) of Theorem 4.20 finally follows that $\overline{z}_1^* = 0.5$ and $\overline{z}_n^* = -0.5$. A more detailed analysis of location problems using rectilinear distances was given in [39].

By the next remark, we discover that the Lagrange multiplier associated with the linear equation constraint of the dual problem $(\widetilde{D}_N^S)$ is the optimal solution to the primal problem (P_N^S) and moreover, the Lagrange multiplier associated with the inequality constraint of the dual $(\widetilde{D}_N^S)$ is the optimal objective value. A similar result was shown in [69] for minsum location problems.

Remark 4.37. *First, let us notice that the dual problem $(\widetilde{D}_N^S)$ can be written as (see Remark 4.29)*

$$(\widetilde{D}_N^S) \qquad \sup_{\substack{z_i^* \in \mathbb{R}^d, \ i=1,...,n, \\ \sum_{i=1}^n \frac{1}{w_i}\|z_i^*\| \leq 1, \ \sum_{i=1}^n z_i^* = 0_{\mathbb{R}^d}}} \left\{ -\sum_{i=1}^n \langle z_i^*, p_i \rangle \right\},$$

then the Lagrange dual of the dual $(\widetilde{D}_N^S)$ looks like

$$(D\widetilde{D}_N^S) \qquad \inf_{\lambda \geq 0, \ x \in \mathbb{R}^d} \sup_{z_i^* \in \mathbb{R}^d, \ i=1,...,n} \left\{ -\sum_{i=1}^n \langle z_i^*, p_i \rangle + \left\langle x, \sum_{i=1}^n z_i^* \right\rangle - \lambda \left(\sum_{i=1}^n \frac{1}{w_i}\|z_i^*\| - 1 \right) \right\}$$

$$= \inf_{\lambda \geq 0, \ x \in \mathbb{R}^d} \left\{ \lambda + \sum_{i=1}^n \sup_{z_i^* \in \mathbb{R}^d} \left\{ \langle x - p_i, z_i^* \rangle - \frac{\lambda}{w_i}\|z_i^*\| \right\} \right\}. \qquad (4.82)$$

If $\lambda = 0$, then we get

$$\sup_{z_i^* \in \mathbb{R}^d} \langle x - p_i, z_i^* \rangle = \begin{cases} 0, & \text{if } x = p_i, \\ +\infty, & \text{otherwise}, \end{cases}$$

$i = 1, ..., n$, which contradicts the assumption from the beginning that the given points p_i, $i = 1, ..., n$ are distinct. Therefore, we can write for (4.82)

$$(D\widetilde{D}_N^S) \qquad \inf_{\lambda > 0, \ x \in \mathbb{R}^d} \left\{ \lambda + \lambda \sum_{i=1}^n \frac{1}{w_i} \sup_{z_i^* \in \mathbb{R}^d} \left\{ \left\langle \frac{w_i}{\lambda}(x - p_i), z_i^* \right\rangle - \|z_i^*\| \right\} \right\}$$

$$= \inf_{\substack{\lambda > 0, \ x \in \mathbb{R}^d, \\ w_i\|x-p_i\| \leq \lambda, \ i=1,...,n}} \lambda = \inf_{x \in \mathbb{R}^d} \max_{1 \leq i \leq n} \{w_i\|x - p_i\|\}.$$

We conclude, on the one hand, that the Lagrange dual of the dual problem $(\widetilde{D}_N^S)$ (i.e. the bidual of the primal location problem (P^S)) is the problem (P^S). On the other hand, we see that the Lagrange multipliers of the dual $(D\widetilde{D}_N^S)$ characterize the optimal solution and the optimal objective value of the primal problem (P^S). Therefore, we have a complete symmetry between the primal problem (P_N^S), the dual problem $(\widetilde{D}_N^S)$ and its Lagrange dual problem $(D\widetilde{D}_N^S)$.

The described symmetry between the primal, its dual and bidual problem can also be applied to general optimization problems where the underlying space is more general than the finite-dimensional one. In this context, we refer the interested reader to [42] where in addition criteria were given which secure this kind of symmetry.

4.4 Extended multifacility minmax location problems

4.4.1 Unconstraint location problems with set-up costs

The location problem, which we investigate in a more general setting as suggested by Drezner in [40] and studied by Michelot and Plastria in [35, 77], is

$$(EP_a^M) \qquad \inf_{(x_1,...,x_m)\in X^m} \max_{1\leq i\leq n} \left\{ \sum_{j=1}^{m} \gamma_{C_{ij}}(x_j - p_i) + a_i \right\},$$

where X is a Banach space (notice that most of the following investigations can be extended to a Fréchet space, too), $a_i \in \mathbb{R}_+$ are non-negative set-up costs, $p_i \in X$ are distinct points and $\gamma_{C_{ij}} : X \to \mathbb{R}$ are gauges defined by closed and convex subsets C_{ij} of X such that $0_X \in \operatorname{int} C_{ij}$, $i = 1,...,n$, $j = 1,...,m$.

Now, set $\widetilde{X} = X^m$, $x = (x_1,...,x_m) \in \widetilde{X}$, $\widetilde{p}_i = (p_i,...,p_i) \in \widetilde{X}$ and define the gauge $\gamma_{C_i} : \widetilde{X} \to \mathbb{R}$ by

$$\gamma_{C_i}(x) := \sum_{j=1}^{m} \gamma_{C_{ij}}(x_j), \ x = (x_1,...,x_m) \in \widetilde{X},$$

where $C_i = \{x \in \widetilde{X} : \gamma_{C_i}(x) \leq 1\}$, $i = 1,...,n$. Note that, as defined in the proof of Lemma 4.5, $\langle x^*, x \rangle = \sum_{j=1}^{m} \langle x_j^*, x_j \rangle$ for $x \in \widetilde{X}$ and $x^* \in \widetilde{X}^*$. Then, it is obvious that the location problem (EP_a^M) can also be written in a slightly different form, namely, as a single minmax location problem

$$(EP_a^M) \qquad \inf_{x\in\widetilde{X}} \max_{1\leq i\leq n} \{\gamma_{C_i}(x - \widetilde{p}_i) + a_i\}.$$

We use (4.17) of Lemma 4.5 and (4.71) and get for the dual problem corresponding to (EP_a^M)

$$(ED_a^M) \qquad \sup_{\substack{z_i^{0*}\geq 0,\ z_{ij}^{1*}\in X^*,\ i=1,...,n,\ j=1,...,m,\ I=\{i\in\{1,...,n\}:z_i^{0*}>0\}, \\ z_{kj}^{1*}=0_{X^*},\ k\notin I,\ \gamma_{C_{ij}^0}(z_{ij}^{1*})\leq z_i^{0*},\ i\in I,\ j=1,...,m,\ \sum_{i\in I} z_i^{0*}\leq 1}} \left\{ \inf_{x\in\widetilde{X}} \left\{ \sum_{i\in I} \langle z_i^{1*}, x - \widetilde{p}_i \rangle \right\} + \sum_{i\in I} z_i^{0*} a_i \right\}.$$

Because

$$\inf_{x \in \tilde{X}} \left\{ \sum_{i \in I} \langle z_i^{1*}, x - \tilde{p}_i \rangle \right\} = \inf_{\substack{x_l \in X, \\ l=1,\ldots,m}} \left\{ \sum_{i \in I} \sum_{j=1}^{m} \langle z_{ij}^{1*}, x_j - p_i \rangle \right\}$$

$$= \sum_{j=1}^{m} \inf_{x_j \in X} \left\{ \sum_{i \in I} \langle z_{ij}^{1*}, x_j \rangle \right\} - \sum_{i \in I} \sum_{j=1}^{m} \langle z_{ij}^{1*}, p_i \rangle,$$

we obtain finally for the conjugate dual problem of (EP_a^M)

$$(ED_a^M) \qquad \sup_{(z_1^{0*},\ldots,z_n^{0*},z_1^{1*},\ldots,z_n^{1*}) \in \mathcal{C}} \left\{ -\sum_{i \in I} \left[\left\langle \sum_{j=1}^{m} z_{ij}^{1*}, p_i \right\rangle - z_i^{0*} a_i \right] \right\},$$

where

$$\mathcal{C} = \left\{ (z_1^{0*},\ldots,z_n^{0*},z_1^{1*},\ldots,z_n^{1*}) \in \mathbb{R}_+^n \times (X^*)^m \times \ldots \times (X^*)^m : I = \left\{ i \in \{1,\ldots,n\} : z_i^{0*} > 0 \right\} \right.$$

$$z_{kj}^{1*} = 0_{X^*}, \ k \notin I, \ \gamma_{C_{ij}^0}(z_{ij}^{1*}) \leq z_i^{0*}, \ i \in I, \ \sum_{i \in I} z_{ij}^{1*} = 0_{X^*}, \ j = 1,\ldots,m, \ \sum_{i \in I} z_i^{0*} \leq 1 \right\}.$$

Remark 4.38. *A similar dual problem was formulated by Michelot and Cornejo in [35] in the situation where X is the Euclidean space, $m = 2$ and the gauges are a norm. The authors construct in their paper a Fenchel duality scheme to solve extended minmax location problems by a proximal algorithm.*

Remark 4.39. *In the sense of Remark 4.29 the dual problem (ED_a^M) is equivalent to*

$$(\widehat{ED}_a^M) \qquad \sup_{(z_1^{0*},\ldots,z_n^{0*},z_1^{1*},\ldots,z_n^{1*}) \in \hat{\mathcal{C}}} \left\{ -\sum_{i=1}^{n} \left[\left\langle \sum_{j=1}^{m} z_{ij}^{1*}, p_i \right\rangle - z_i^{0*} a_i \right] \right\},$$

where

$$\hat{\mathcal{C}} = \left\{ (z_1^{0*},\ldots,z_n^{0*},z_1^{1*},\ldots,z_n^{1*}) \in \mathbb{R}_+^n \times (X^*)^m \times \ldots \times (X^*)^m : \gamma_{C_{kj}^0}(z_{kj}^{1*}) \leq z_k^{0*}, \right.$$

$$\sum_{i=1}^{n} z_{ij}^{1*} = 0_{X^*}, \ k = 1,\ldots,n, \ j = 1,\ldots,m, \ \sum_{i=1}^{n} z_i^{0*} \leq 1 \right\}.$$

Let $v(EP_a^M)$ be the optimal objective value of the location problem (EP_a^M) and $v(ED_a^M)$ be the optimal objective value of the dual problem (ED_a^M), then we obtain the following duality statement as a direct consequence of Theorem 4.12.

Theorem 4.21. *(strong duality) Between (EP_a^M) and (ED_a^M) holds strong duality, i.e. $v(EP_a^M) = v(ED_a^M)$ and the conjugate dual problem has an optimal solution.*

The following necessary and sufficient optimality conditions are a consequence of the previous theorem.

Theorem 4.22. *(optimality conditions) (a) Let $(\overline{x}_1, ..., \overline{x}_m) \in X^m$ be an optimal solution to the problem (EP_a^M). Then there exists an an optimal solution to (ED_a^M)*

$$(\overline{z}_1^{0*}, ..., \overline{z}_n^{0*}, \overline{z}_1^{1*}, ..., \overline{z}_n^{1*}) \in \mathbb{R}_+^n \times (X^*)^m \times ... \times (X^*)^m \qquad (4.83)$$

with the corresponding index set $\overline{I} \subseteq \{1, ..., n\}$ as such that

(i) $\displaystyle \max_{1 \leq u \leq n} \left\{ \sum_{j=1}^m \gamma_{C_{uj}}(\overline{x}_j - p_u) + a_u \right\} = \sum_{i \in \overline{I}} \overline{z}_i^{0*} \left(\sum_{j=1}^m \gamma_{C_{ij}}(\overline{x}_j - p_i) + a_i \right)$,

(ii) $\displaystyle \sum_{i \in \overline{I}} \overline{z}_{ij}^{1*} = 0_{X^*}$, $j = 1, ..., m$,

(iii) $\overline{z}_i^{0*} \gamma_{C_{ij}}(\overline{x}_j - p_i) = \langle \overline{z}_{ij}^{1*}, \overline{x}_j - p_i \rangle$, $i \in \overline{I}$, $j = 1, ..., m$,

(iv) $\displaystyle \sum_{j \in \overline{I}} \overline{z}_j^{0*} = 1$, $\overline{z}_i^{0*} > 0$, $i \in \overline{I}$, and $\overline{z}_k^{0*} = 0$, $k \notin \overline{I}$,

(v) $\displaystyle \sum_{j=1}^m \gamma_{C_{ij}}(\overline{x}_j - p_i) + a_i = \max_{1 \leq u \leq n} \left\{ \sum_{j=1}^m \gamma_{C_{uj}}(\overline{x}_j - p_u) + a_u \right\}$, $i \in \overline{I}$,

(vi) $\displaystyle \max_{1 \leq l \leq m} \left\{ \gamma_{C_{il}^0}(\overline{z}_{il}^{1*}) \right\} = \overline{z}_i^{0*}$, $(\overline{z}_{i1}^{1*}, ..., \overline{z}_{im}^{1*}) \in X^* \times ... \times X^* \setminus \{(0_{X^*}, ..., 0_{X^*})\}$, $i \in \overline{I}$, and $\overline{z}_{kj}^{1*} = 0_{X^*}$, $k \notin \overline{I}$, $j = 1, ..., m$.

(b) If there exists $(\overline{x}_1, ..., \overline{x}_m) \in X^m$ such that for some

$$(\overline{z}_1^{0*}, ..., \overline{z}_n^{0*}, \overline{z}_1^{1*}, ..., \overline{z}_n^{1*}) \in \mathbb{R}_+^n \times (X^*)^m \times ... \times (X^*)^m \qquad (4.84)$$

with the corresponding index set $\overline{I} \subseteq \{1, ..., n\}$ the conditions (i)-(vi) are fulfilled, then $(\overline{x}_1, ..., \overline{x}_m)$ is an optimal solution to (EP_a^M), $(\overline{z}_1^{0*}, ..., \overline{z}_n^{0*}, \overline{z}_1^{1*}, ..., \overline{z}_n^{1*})$ is an optimal solution to (ED_a^M) and $v(EP_a^M) = v(ED_a^M)$.

Proof. From Theorem 4.21 we have $v(EP_a^M) = v(ED_a^M)$, i.e. for $(\overline{x}_1, ..., \overline{x}_m) \in X^m$ and $(\overline{z}_1^{0*}, ..., \overline{z}_n^{0*}, \overline{z}_1^{1*}, ..., \overline{z}_n^{1*}) \in \mathbb{R}_+^n \times (X^*)^m \times ... \times (X^*)^m$ with the corresponding index set $\overline{I} \subseteq \{1, ..., n\}$ it holds

$$\max_{1 \leq u \leq n} \left\{ \sum_{j=1}^m \gamma_{C_{uj}}(\overline{x}_j - p_u) + a_u \right\} = -\sum_{i \in \overline{I}} \left[\left\langle \sum_{j=1}^m \overline{z}_{ij}^{1*}, p_i \right\rangle - \overline{z}_i^{0*} a_i \right]$$

$$\Leftrightarrow \max_{1 \leq u \leq n} \left\{ \sum_{j=1}^m \gamma_{C_{uj}}(\overline{x}_j - p_u) + a_u \right\} + \sum_{i \in \overline{I}} \left[\left\langle \sum_{j=1}^m \overline{z}_{ij}^{1*}, p_i \right\rangle - \overline{z}_i^{0*} a_i \right] = 0$$

$$\Leftrightarrow \max_{1 \leq u \leq n} \left\{ \sum_{j=1}^{m} \gamma_{C_{uj}}(\overline{x}_j - p_u) + a_u \right\} + \sum_{i \in \overline{I}} \left[\left\langle \sum_{j=1}^{m} \overline{z}_{ij}^{1*}, p_i \right\rangle - \overline{z}_i^{0*} a_i \right]$$

$$+ \sum_{i \in \overline{I}} \sum_{j=1}^{m} \overline{z}_i^{0*} \gamma_{C_{ij}}(\overline{x}_j - p_i) - \sum_{i \in \overline{I}} \sum_{j=1}^{m} \overline{z}_i^{0*} \gamma_{C_{ij}}(\overline{x}_j - p_i)$$

$$+ \left\langle \sum_{i \in \overline{I}} \sum_{j=1}^{m} \overline{z}_{ij}^{1*}, \overline{x}_j \right\rangle - \left\langle \sum_{i \in \overline{I}} \sum_{j=1}^{m} \overline{z}_{ij}^{1*}, \overline{x}_j \right\rangle = 0$$

$$\Leftrightarrow \left[\max_{1 \leq u \leq n} \left\{ \sum_{j=1}^{m} \gamma_{C_{uj}}(\overline{x}_j - p_u) + a_u \right\} - \sum_{i \in \overline{I}} \sum_{j=1}^{m} \overline{z}_i^{0*} \gamma_{C_{ij}}(\overline{x}_j - p_i) - \sum_{i \in \overline{I}} \overline{z}_i^{0*} a_i \right]$$

$$+ \sum_{i \in \overline{I}} \sum_{j=1}^{m} [\overline{z}_i^{0*} \gamma_{C_{ij}}(\overline{x}_j - p_i) - \langle \overline{z}_{ij}^{1*}, \overline{x}_j - p_i \rangle] + \sum_{j=1}^{m} \left\langle \sum_{i \in \overline{I}} \overline{z}_{ij}^{1*}, \overline{x}_j \right\rangle = 0. \tag{4.85}$$

If we define the function $h_i : \mathbb{R} \to \overline{\mathbb{R}}$ by

$$h_i(y) := \begin{cases} y, & \text{if } y \in \mathbb{R}_+, \\ +\infty, & \text{otherwise,} \end{cases} \tag{4.86}$$

then it follows by Lemma 4.2 that

$$g\left(\sum_{j=1}^{m} \gamma_{C_{1j}}(\overline{x}_j - p_1), ..., \sum_{j=1}^{m} \gamma_{C_{nj}}(\overline{x}_j - p_n) \right) = \max_{1 \leq u \leq n} \left\{ \sum_{j=1}^{m} \gamma_{C_{uj}}(\overline{x}_j - p_u) + a_u \right\}$$

$$\geq \sum_{i \in \overline{I}} \overline{z}_i^{0*} \left[\sum_{j=1}^{m} \gamma_{C_{ij}}(\overline{x}_j - p_i) + a_i \right],$$

which means that the term in the first bracket of (4.85) is equal to zero. Moreover, by the Young-Fenchel inequality as well as by the fact that $\sum_{i \in \overline{I}} \overline{z}_{ij}^{1*} = 0_{X^*}$, $j = 1, ..., m$, we get that the terms in the other brackets are also equal to zero. Hence, we derive the optimality conditions (i)-(iii).

By the feasibility condition, $\sum_{i \in \overline{I}} \overline{z}_i^{0*} \leq 1$, and the equality in the first bracket of (4.85) it holds

$$\max_{1 \leq u \leq n} \left\{ \sum_{j=1}^{m} \gamma_{C_{uj}}(\overline{x}_j - p_u) + a_u \right\} = \sum_{i \in \overline{I}} \overline{z}_i^{0*} \left(\sum_{j=1}^{m} \gamma_{C_{ij}}(\overline{x}_j - p_i) + a_i \right)$$

$$\leq \sum_{i \in \overline{I}} \overline{z}_i^{0*} \max_{1 \leq u \leq n} \left\{ \sum_{j=1}^{m} \gamma_{C_{uj}}(\overline{x}_j - p_u) + a_u \right\}$$

$$\leq \max_{1 \leq u \leq n} \left\{ \sum_{j=1}^{m} \gamma_{C_{uj}}(\overline{x}_j - p_u) + a_u \right\}$$

and from here follows on the one hand that

$$\sum_{i \in \overline{I}} \overline{z}_i^{0*} = 1, \tag{4.87}$$

and on the other hand that

$$\sum_{j=1}^{m} \gamma_{C_{ij}}(\overline{x}_j - p_i) + a_i = \max_{1 \le u \le n} \left\{ \sum_{j=1}^{m} \gamma_{C_{uj}}(\overline{x}_j - p_u) + a_u \right\}, \ i \in \overline{I}. \tag{4.88}$$

Moreover, as $\overline{z}_i^{0*} \gamma_{C_{ij}}(\overline{x}_j - p_i) = \langle \overline{z}_{ij}^{1*}, \overline{x}_j - p_i \rangle$, $i \in \overline{I}$, $j = 1, ..., m$, one gets by the feasibility condition,

$$\gamma_{C_{ij}^0}(\overline{z}_{ij}^{1*}) \le \overline{z}_i^{0*} \ \forall j = 1, ..., m, \ i \in \overline{I}, \Leftrightarrow \max_{1 \le l \le m} \left\{ \gamma_{C_{ij}^0}(\overline{z}_{ij}^{1*}) \right\} \le \overline{z}_i^{0*}, \ i \in \overline{I}. \tag{4.89}$$

Recall that $\gamma_{C_i}(\overline{x} - \widetilde{p}_i) = \sum_{j=1}^{m} \gamma_{C_{ij}}(\overline{x}_j - p_i)$ and that by Lemma 4.5 we have $\gamma_{C_i^0}(\overline{z}_i^{1*}) = \max_{1 \le j \le m} \{ \gamma_{C_{ij}^0}(\overline{z}_{ij}^{1*}) \}$, where $\widetilde{p}_i = (p_i, ..., p_i) \in X^m$ and $\overline{z}_i^{1*} = (\overline{z}_{i1}^{1*}, ..., \overline{z}_{im}^{1*}) \in (X^*)^m$, $i \in \overline{I}$. Then one has by Lemma 4.3 (the generalized Cauchy-Schwarz inequality) that

$$\gamma_{C_i^0}(\overline{z}_i^{1*}) \gamma_{C_i}(\overline{x} - \widetilde{p}_i) \ge \langle \overline{z}_i^{1*}, \overline{x} - \widetilde{p}_i \rangle, \tag{4.90}$$

i.e.

$$\max_{1 \le l \le m} \left\{ \gamma_{C_{ij}^0}(\overline{z}_{ij}^{1*}) \right\} \sum_{j=1}^{m} \gamma_{C_{ij}}(\overline{x}_j - p_i) \ge \sum_{j=1}^{m} \langle \overline{z}_{ij}^{1*}, \overline{x}_j - p_i \rangle, \ i \in \overline{I}. \tag{4.91}$$

From here follows that

$$\overline{z}_i^{0*} \gamma_{C_i}(\overline{x} - \widetilde{p}_i) = \overline{z}_i^{0*} \sum_{j=1}^{m} \gamma_{C_{ij}}(\overline{x}_j - p_i) = \sum_{j=1}^{m} \langle \overline{z}_{ij}^{1*}, \overline{x}_j - p_i \rangle = \langle \overline{z}_i^{1*}, \overline{x} - \widetilde{p}_i \rangle$$

$$\le \gamma_{C_i^0}(\overline{z}_i^{1*}) \gamma_{C_i}(\overline{x} - \widetilde{p}_i) = \max_{1 \le l \le m} \left\{ \gamma_{C_{ij}^0}(\overline{z}_{ij}^{1*}) \right\} \sum_{j=1}^{m} \gamma_{C_{ij}}(\overline{x}_j - p_i)$$

$$\le \overline{z}_i^{0*} \sum_{j=1}^{m} \gamma_{C_{ij}}(\overline{x}_j - p_i), \ i \in \overline{I},$$

and thus, the inequality in (4.89) holds as equality. Taking now (4.87), (4.88) and (4.89) as equality together yields the optimality conditions (iv)-(vi) and completes the proof. $\square$

Remark 4.40. *Let* $h_i : \mathbb{R} \to \overline{\mathbb{R}}$ *be defined by*

$$h_i(x_i) := \begin{cases} x_i, & \text{if } x_i \in \mathbb{R}_+, \\ +\infty, & \text{otherwise,} \end{cases}$$

then the conjugate function of $\lambda_i h_i$, $\lambda_i \geq 0$, is

$$(\lambda_i h_i)^*(x_i^*) = \begin{cases} 0, & \text{if } x_i^* \leq \lambda_i, \\ +\infty, & \text{otherwise,} \end{cases} \quad , \ i = 1, ..., n.$$

In addition, we consider the function $f : \mathbb{R} \to \overline{\mathbb{R}}$,

$$f(y^0) = \begin{cases} \max_{1 \leq i \leq n} \{y_i^0 + a_i\}, & \text{if } y^0 = (y_1^0, ..., y_n^0)^T \in \mathbb{R}_+^n, \ i = 1, ..., n, \\ +\infty, & \text{otherwise,} \end{cases}$$

and get by Lemma 4.1 that

$$f^*(z_1^{0*}, ..., z_n^{0*}) = \min_{\substack{\sum_{i=1}^n \lambda_i \leq 1, \ \lambda_i \geq 0, \ z_i^{0*} \leq \lambda_i, \\ i=1,...,n}} \left\{ -\sum_{i=1}^n \lambda_i a_i \right\} \leq -\sum_{i=1}^n z_i^{0*} a_i$$

for all $z_i^{0*} \leq \lambda_i$ with $\lambda_i \geq 0$, $i = 1, ..., n$, $\sum_{i=1}^n \lambda_i \leq 1$. Hence, we have by the Young-Fenchel inequality and the optimal condition (i) of Theorem 4.22 that

$$\sum_{i \in \overline{I}} \overline{z}_i^{0*} \sum_{j=1}^m \gamma_{C_{ij}}(\overline{x}_j - p_i)$$

$$\leq f \left(\sum_{j=1}^m \gamma_{C_{1j}}(\overline{x}_j - p_1), ..., \sum_{j=1}^m \gamma_{C_{nj}}(\overline{x}_j - p_n)) \right) + f^*(\overline{z}_1^{0*}, ..., \overline{z}_n^{0*})$$

$$\leq f \left(\sum_{j=1}^m \gamma_{C_{1j}}(\overline{x}_j - p_1), ..., \sum_{j=1}^m \gamma_{C_{nj}}(\overline{x}_j - p_n)) \right) - \sum_{i=1}^n \overline{z}_i^{0*} a_i = \sum_{i \in \overline{I}} \overline{z}_i^{0*} \sum_{j=1}^m \gamma_{C_{ij}}(\overline{x}_j - p_i),$$

i.e.

$$f \left(\sum_{j=1}^m \gamma_{C_{1j}}(\overline{x}_j - p_1), ..., \sum_{j=1}^m \gamma_{C_{nj}}(\overline{x}_j - p_n)) \right) + f^*(\overline{z}_1^{0*}, ..., \overline{z}_n^{0*}) = \sum_{i \in \overline{I}} \overline{z}_i^{0*} \sum_{j=1}^m \gamma_{C_{ij}}(\overline{x}_j - p_i)$$

and by (2.4) this equality is equivalent to

$$(z_1^{0*}, ..., z_n^{0*}) \in \partial f \left(\sum_{j=1}^m \gamma_{C_{1j}}(\overline{x}_j - p_1), ..., \sum_{j=1}^m \gamma_{C_{nj}}(\overline{x}_j - p_n)) \right).$$

In other words, the condition (i) of Theorem 4.22 can be written by means of the subdifferential, i.e.,

$$(i) \ (\overline{z}_1^{0*}, ..., \overline{z}_n^{0*}) \in \partial \left(\max_{1 \leq j \leq n} \{\cdot + a_j\} \right) \left(\sum_{j=1}^m \gamma_{C_{1j}}(\overline{x}_j - p_1), ..., \sum_{j=1}^m \gamma_{C_{nj}}(\overline{x}_j - p_n)) \right).$$

Similarly, we can rewrite the condition (iii) of Theorem 4.22 as follows

(iii) $\overline{z}_{ij}^{1*} \in \partial(\overline{z}_i^{0*}\gamma_{C_{ij}})(\overline{x}_j - p_i)$, $i \in \overline{I}$, $j = 1, ..., m$.

Moreover, combining this condition with the optimality condition (ii) of Theorem 4.22 yields that

$$0_{X^*} \in \sum_{i \in \overline{I}} \partial(\overline{z}_i^{0*}\gamma_{C_{ij}})(\overline{x}_j - p_i), \ j = 1, ..., m.$$

Notice also that the optimality conditions (iii) and (vi) of Theorem 4.22 give a detailed characterization of the subdifferential of $\overline{z}_i^{0*}\gamma_{C_{ij}}$ at $\overline{x}_j - p_i$ such that

$$\partial(\overline{z}_i^{0*}\gamma_{C_{ij}})(\overline{x}_j - p_i)$$
$$= \left\{ \overline{z}_{ij}^{1*} \in X^* : \overline{z}_i^{0*}\gamma_{C_{ij}}(\overline{x}_j - p_i) = \langle \overline{z}_{ij}^{1*}, \overline{x}_j - p_i \rangle, \ \max_{1 \leq l \leq m}\left\{\gamma_{C_{il}^0}(\overline{z}_{il}^{1*})\right\} = \overline{z}_i^{0*} \right\}$$

for all $i \in \overline{I}$, $j = 1, ..., m$.

Let us now consider the extended location problem (EP_a^M) in the following framework. We set $X = \mathcal{H}$, where $\mathcal{H}$ is a real Hilbert space and $\gamma_{C_{ij}} : \mathcal{H} \to \mathbb{R}$, $\gamma_{C_{ij}}(x) := w_{ij}\|x\|_{\mathcal{H}}$, where $w_{ij} > 0$ for $j = 1, ..., m$, $i = 1, ..., n$. Hence, the location problem looks like

$$(EP_{N,a}^M) \qquad \inf_{(x_1, ..., x_m) \in \mathcal{H} \times ... \times \mathcal{H}} \max_{1 \leq i \leq n}\left\{\sum_{j=1}^m w_{ij}\|x_j - p_i\|_{\mathcal{H}} + a_i\right\}.$$

For this situation, where the gauges are all identical and the distances are measured by a round norm, Michelot and Plastria examined in [77] under which conditions an optimal solution of coincidence type exists. The authors showed that if the weights have a multiplicative structure, i.e. $w_{ij} = \lambda_i \mu_j$ with λ_i, $\mu_j > 0$, $i = 1, ..., n$, $j = 1, ..., m$, and $\sum_{j=1}^m \mu_j = 1$, then there exists an optimal solution to $(EP_{N,a}^M)$ such that all new facilities coincide. Moreover, they described when the optimal solution of coincidence type is unique and presented a full characterization of the set of optimal solutions for extended multifacility location problems where the weights have a multiplicative structure.

The next statement is based on the idea of weights with a multiplicative structure and illustrates in this situation the relation between the extended location problem $(EP_{N,a}^M)$ and its corresponding conjugate dual problem.

Theorem 4.23. *Let $X = \mathcal{H}$, $\gamma_{C_{ij}} : \mathcal{H} \to \mathbb{R}$ be defined by $\gamma_{C_{ij}}(x) := w_{ij}\|x\|_{\mathcal{H}}$, $i = 1, ..., n$, $j = 1, ..., m$, and $w_{ij} = \lambda_i \mu_j$ with λ_i, $\mu_j > 0$, $i = 1, ..., n$, $j = 1, ..., m$, and $\sum_{j=1}^m \mu_j = 1$. Assume that $\Delta_{\overline{x}} = (\overline{x}, ..., \overline{x}) \in \underbrace{\mathcal{H} \times ... \times \mathcal{H}}_{m-times}$ is an optimal solution of coincidence type to*

$$(EP_{N,a}^M) \qquad \inf_{(x_1, ..., x_m) \in \mathcal{H} \times ... \times \mathcal{H}} \max_{1 \leq i \leq n}\left\{\sum_{j=1}^m w_{ij}\|x_j - p_i\|_{\mathcal{H}} + a_i\right\}.$$

and $(\overline{z}_1^{0*}, ..., \overline{z}_n^{0*}, \overline{z}_1^{1*}, ..., \overline{z}_n^{1*})$ *with the corresponding index set* $\overline{I} \subseteq \{1, ..., n\}$ *is an optimal solution to the conjugate dual problem*

$$(ED_{N,a}^M) \qquad \sup_{(z_1^{0*}, ..., z_n^{0*}, z_1^{1*}, ..., z_n^{1*}) \in \mathcal{C}} \left\{ -\sum_{i \in I} \left[\left\langle \sum_{j=1}^m z_{ij}^{1*}, p_i \right\rangle_{\mathcal{H}} - z_i^{0*} a_i \right] \right\},$$

where

$$\mathcal{C} = \left\{ (z_1^{0*}, ..., z_n^{0*}, z_1^{1*}, ..., z_n^{1*}) \in \mathbb{R}_+^n \times \underbrace{\mathcal{H}^m \times ... \times \mathcal{H}^m}_{n-times} : I = \left\{ i \in \{1, ..., n\} : z_i^{0*} > 0 \right\} \right.$$

$$\left. z_{kj}^{1*} = 0_{\mathcal{H}}, \ k \notin I, \ \|z_{ij}^{1*}\|_{\mathcal{H}} \leq z_i^{0*} w_{ij}, \ i \in I, \ \sum_{i \in I} z_{ij}^{1*} = 0_{\mathcal{H}}, \ j = 1, ..., m, \ \sum_{i \in I} z_i^{0*} \leq 1 \right\}.$$

Then, it holds

$$\overline{x} = \frac{1}{\displaystyle\sum_{i \in \overline{I}} \frac{\lambda_i \|\overline{z}_{ij}^{1*}\|_{\mathcal{H}}}{v(ED_{N,a}^M) - a_i}} \sum_{i \in \overline{I}} \frac{\lambda_i \|\overline{z}_{ij}^{1*}\|_{\mathcal{H}}}{v(ED_{N,a}^M) - a_i} p_i \ \forall j \in \overline{J},$$

where

$$\overline{J} := \left\{ j \in \{1, ..., m\} : \frac{1}{w_{ij}} \|\overline{z}_{ij}^{1*}\|_{\mathcal{H}} = \max_{1 \leq l \leq m} \left\{ \frac{1}{w_{il}} \|\overline{z}_{il}^{1*}\|_{\mathcal{H}} \right\} \right\}, \ i \in \overline{I}.$$

Proof. First, let us remark that the dual norm of the weighted norm $\gamma_{C_{ij}} = w_{ij} \| \cdot \|_{\mathcal{H}}$ is given by $\gamma_{C_{ij}^0} = (1/w_{ij}) \| \cdot \|_{\mathcal{H}}$.

Now, let $\Delta_{\overline{x}} = (\overline{x}, ..., \overline{x})$ be an optimal solution of coincidence type, then the optimality conditions (ii), (iii), (v) and (vi) of Theorem 4.22 can be written as

(ii) $\displaystyle\sum_{i \in \overline{I}} \overline{z}_{ij}^{1*} = 0_{\mathcal{H}}, \ j = 1, ..., m,$

(iii) $\overline{z}_i^{0*} w_{ij} \|\overline{x} - p_i\|_{\mathcal{H}} = \langle \overline{z}_{ij}^{1*}, \overline{x} - p_i \rangle_{\mathcal{H}}, \ i \in \overline{I}, \ j = 1, ..., m,$

(v) $\displaystyle\sum_{j=1}^m w_{ij} \|\overline{x} - p_i\|_{\mathcal{H}} + a_i = \max_{1 \leq u \leq n} \left\{ \sum_{j=1}^m w_{uj} \|\overline{x} - p_u\|_{\mathcal{H}} + a_u \right\}, \ i \in \overline{I},$

(vi) $\displaystyle\max_{1 \leq l \leq m} \left\{ \frac{1}{w_{il}} \|\overline{z}_{il}^{1*}\|_{\mathcal{H}} \right\} = \overline{z}_i^{0*}, \ (\overline{z}_{i1}^{1*}, ..., \overline{z}_{im}^{1*}) \in \mathcal{H} \times ... \times \mathcal{H} \setminus \{(0_{\mathcal{H}}, ..., 0_{\mathcal{H}})\}, \ i \in \overline{I}$ and $\overline{z}_{kj}^{1*} = 0_{\mathcal{H}}, \ k \notin \overline{I}, \ j = 1, ..., m.$

By combining the conditions (iii) and (vi), we get

$$\|\overline{z}_{ij}^{1*}\|_{\mathcal{H}} \|\overline{x} - p_i\|_{\mathcal{H}} = \langle \overline{z}_{ij}^{1*}, \overline{x} - p_i \rangle_{\mathcal{H}}, \ i \in \overline{I}, \ j \in \overline{J}. \tag{4.92}$$

Moreover, by Fact 2.10 in [3] there exists $\alpha_{ij} > 0$ such that

$$\bar{z}_{ij}^{1*} = \alpha_{ij} \left(\bar{x} - p_i\right) \tag{4.93}$$

and from here one gets that

$$\|\bar{z}_{ij}^{1*}\|_{\mathcal{H}} = \alpha_{ij}\|\bar{x} - p_i\|_{\mathcal{H}}, \tag{4.94}$$

$i \in \bar{I}, j \in \bar{J}$. By condition (v) follows

$$\sum_{j=1}^{m} w_{ij}\|\bar{x} - p_i\|_{\mathcal{H}} + a_i = \max_{1 \leq u \leq n}\left\{\sum_{j=1}^{m} w_{uj}\|\bar{x} - p_u\|_{\mathcal{H}} + a_u\right\}$$

$$\Leftrightarrow \lambda_i \sum_{j=1}^{m} \mu_j\|\bar{x} - p_i\|_{\mathcal{H}} + a_i = \max_{1 \leq u \leq n}\left\{\lambda_u \sum_{j=1}^{m} \mu_j\|\bar{x} - p_u\|_{\mathcal{H}} + a_u\right\}$$

$$\Leftrightarrow \lambda_i\|\bar{x} - p_i\|_{\mathcal{H}} + a_i = \max_{1 \leq u \leq n}\left\{\lambda_u\|\bar{x} - p_u\|_{\mathcal{H}} + a_u\right\}, \ i \in \bar{I}. \tag{4.95}$$

Bringing (4.94) and (4.95) together yields

$$\frac{\lambda_i}{\alpha_{ij}}\|\bar{z}_{ij}^{1*}\|_{\mathcal{H}} + a_i = \max_{1 \leq u \leq n}\left\{\lambda_u\|\bar{x} - p_u\|_{\mathcal{H}} + a_u\right\}$$

$$\Leftrightarrow \alpha_{ij} = \frac{\lambda_i}{\max_{1 \leq u \leq n}\left\{\lambda_u\|\bar{x} - p_u\|_{\mathcal{H}} + a_u\right\} - a_i}\|\bar{z}_{ij}^{1*}\|_{\mathcal{H}}, \ i \in \bar{I}, \ j \in \bar{J}. \tag{4.96}$$

Taking the sum overall $i \in \bar{I}$ in (4.96) gives

$$\sum_{i \in \bar{I}} \alpha_{ij} = \sum_{i \in \bar{I}} \frac{\lambda_i\|\bar{z}_{ij}^{1*}\|_{\mathcal{H}}}{\max_{1 \leq u \leq n}\left\{\lambda_u\|\bar{x} - p_u\|_{\mathcal{H}} + a_u\right\} - a_i}, \ j \in \bar{J}. \tag{4.97}$$

Now, consider condition (ii), by (4.93) follows

$$0_{\mathcal{H}} = \sum_{i \in \bar{I}} \bar{z}_{ij}^{1*} = \sum_{i \in \bar{I}} \alpha_{ij}\left(\bar{x} - p_i\right) \Leftrightarrow \bar{x} = \frac{1}{\sum_{i \in \bar{I}} \alpha_{ij}}\sum_{i \in \bar{I}} \alpha_{ij}p_i, \ j \in \bar{J}. \tag{4.98}$$

Putting (4.96), (4.97) and (4.98) together reveals

$$\bar{x} = \frac{1}{\sum_{i \in \bar{I}} \frac{\lambda_i\|\bar{z}_{ij}^{1*}\|_{\mathcal{H}}}{\max_{1 \leq u \leq n}\{\lambda_u\|\bar{x} - p_u\|_{\mathcal{H}} + a_u\} - a_i}}\sum_{i \in \bar{I}} \frac{\lambda_i\|\bar{z}_{ij}^{1*}\|_{\mathcal{H}}}{\max_{1 \leq u \leq n}\{\lambda_u\|\bar{x} - p_u\|_{\mathcal{H}} + a_u\} - a_i}p_i$$

$$= \frac{1}{\sum_{i \in \bar{I}} \frac{\lambda_i\|\bar{z}_{ij}^{1*}\|_{\mathcal{H}}}{v(ED_{N,a}^M) - a_i}}\sum_{i \in \bar{I}} \frac{\lambda_i\|\bar{z}_{ij}^{1*}\|_{\mathcal{H}}}{v(ED_{N,a}^M) - a_i}p_i, \ j \in \bar{J},$$

and the proof is finished. $\qquad\square$

Remark 4.41. *In the context of Theorem 4.23, it holds that $\overline{x} - p_i$ and $\overline{z}_{ij}^{1*}$ are parallel and so the vectors $(1/w_{ij})\overline{z}_{ij}^{1*}$, $j \in \overline{J}$, are all parallel to each other. In other words, the vectors $(1/w_{ij})\overline{z}_{ij}^{1*}$, $j \in \overline{J}$, are identical. In this sense, one can understand the optimal solution to the conjugate dual problem also as a solution of coincidence type.*

The next statement holds for any weights, not necessary of multiplicative structure.

Lemma 4.7. *Let $w_{sj} := \max_{1 \leq u \leq n}\{w_{uj}\}$, $X = \mathcal{H}$, $\gamma_{C_{ij}} : \mathcal{H} \to \mathbb{R}$ be defined by $\gamma_{C_{ij}}(x) := w_{ij}\|x\|_{\mathcal{H}}$, $i = 1, ..., n$, $j = 1, ..., m$, and $(z_1^{0*}, ..., z_n^{0*}, z_1^{1*}, ..., z_n^{1*})$ a feasible solution to the conjugate dual problem $(ED_{N,a}^M)$, then it holds*

$$\|z_{ij}^{1*}\|_{\mathcal{H}} \leq \frac{w_{sj}w_{ij}}{w_{sj} + w_{ij}}, \ i \in I, \ j = 1, ..., m.$$

Proof. Let

$$(z_1^{0*}, ..., z_n^{0*}, z_1^{1*}, ..., z_n^{1*}) \in \mathbb{R}^n \times \underbrace{\mathcal{H} \times ... \times \mathcal{H}}_{m-\text{times}} \times ... \times \underbrace{\mathcal{H} \times ... \times \mathcal{H}}_{m-\text{times}}$$

be a feasible solution to the conjugate dual problem $(ED_{N,a}^M)$, then we have

1. $\sum_{i \in I} z_i^{0*} \leq 1$,

2. $\|z_{ij}^{1*}\|_{\mathcal{H}} \leq z_i^{0*}w_{ij}$, $j = 1, ..., m$, $i \in I$,

3. $\sum_{i \in I} z_{ij}^{1*} = 0_{\mathcal{H}}$.

The inequalities (i) and (ii) imply the inequality

$$\sum_{i \in I} \frac{1}{w_{ij}}\|z_{ij}^{1*}\|_{\mathcal{H}} \leq 1, \ j = 1, ..., m. \tag{4.99}$$

Furthermore, by (iii) we have

$$\sum_{i \in I} z_{ij}^{1*} = 0_{\mathcal{H}} \Leftrightarrow z_{kj}^{1*} = -\sum_{\substack{i \in I \\ i \neq k}} z_{ij}^{1*}, \ k \in I, \ j = 1, ..., m, \tag{4.100}$$

and hence,

$$\|z_{kj}^{1*}\|_{\mathcal{H}} = \|\sum_{\substack{i \in I \\ i \neq k}} z_{ij}^{1*}\|_{\mathcal{H}} \leq \sum_{\substack{i \in I \\ i \neq k}} \|z_{ij}^{1*}\|_{\mathcal{H}}, \ k \in I, \ j = 1, ..., m. \tag{4.101}$$

By (4.101) we get in (4.99)

$$1 \geq \frac{1}{w_{kj}}\|z_{kj}^{1*}\|_{\mathcal{H}} + \sum_{\substack{i \in I \\ i \neq k}} \frac{1}{w_{ij}}\|z_{ij}^{1*}\|_{\mathcal{H}} \geq \frac{1}{w_{kj}}\|z_{kj}^{1*}\|_{\mathcal{H}} + \frac{1}{w_{sj}} \sum_{\substack{i \in I \\ i \neq k}} \|z_{ij}^{1*}\|_{\mathcal{H}}$$

$$\geq \frac{1}{w_{kj}}\|z_{kj}^{1*}\|_{\mathcal{H}} + \frac{1}{w_{sj}}\|z_{kj}^{1*}\|_{\mathcal{H}} = \frac{w_{sj} + w_{kj}}{w_{kj}w_{sj}}\|z_{kj}^{1*}\|_{\mathcal{H}}, \ j = 1, ..., m,$$

and finally,

$$\|z_{kj}^{1*}\|_{\mathcal{H}} \leq \frac{w_{sj}w_{kj}}{w_{sj} + w_{kj}}, \ k \in I, \ j = 1, ..., m.$$

$\square$

Remark 4.42. *If we allow also negative set-up costs, then we have in the constraint set, as stated in Remark 4.31, $\sum_{i \in I} z_i^{0*} = 1$ instead $\sum_{i \in I} z_i^{0*} \leq 1$. One can easy verify that the results we presented above also holds in this case.*

4.4.2 Unconstrained location problems without set-up costs

In the next, we study the case where X is a Banach space and $a_i = 0$ for all $i = 1, ..., n$. With this assumption the extended multifacility location problem (EP_a^M) can be stated as

$$(EP^M) \qquad \inf_{(x_1,...,x_m) \in X^m} \max_{1 \leq i \leq n} \left\{ \sum_{j=1}^{m} \gamma_{C_{ij}}(x_j - p_i) \right\}.$$

In this situation its corresponding conjugated dual problem (ED_a^M) transforms into

$$(ED^M) \qquad \sup_{(z_1^{0*},...,z_n^{0*},z_1^{1*},...,z_n^{1*}) \in \mathcal{C}} \left\{ -\sum_{i \in I} \left\langle \sum_{j=1}^{m} z_{ij}^{1*}, p_i \right\rangle \right\}.$$

Additionally, let us consider the following dual problem

$$(E\widetilde{D}^M) \qquad \sup_{(z_1^*,...,z_n^*) \in \widetilde{\mathcal{C}}} \left\{ -\sum_{i \in I} \left\langle \sum_{j=1}^{m} z_{ij}^*, p_i \right\rangle \right\}$$

where

$$\widetilde{\mathcal{C}} = \left\{ (z_1^*, ..., z_n^*) \in (X^*)^m \times ... \times (X^*)^m : I = \left\{ i \in \{1, ..., n\} : \max_{1 \leq j \leq m} \left\{ \gamma_{C_{ij}^0}(z_{ij}^*) \right\} > 0 \right\} \right.$$

$$z_{kj}^* = 0_{X^*}, \ k \notin I, \ \sum_{i \in I} z_{ij}^* = 0_{X^*}, \ j = 1, ..., m, \ \sum_{i \in I} \max_{1 \leq l \leq m} \left\{ \gamma_{C_{il}^0}(z_{il}^*) \right\} \leq 1 \right\}.$$

Let us denote by $v(ED_a^M)$ and $v(E\widetilde{D}^M)$ the optimal objective values of the dual problems (ED_a^M) and $(E\widetilde{D}^M)$, respectively, then we can state.

Theorem 4.24. *It holds $v(ED^M) = v(E\widetilde{D}^M)$.*

Proof. The statement follows immediately by Theorem 4.16 and (4.17). $\square$

The next duality statements follow as direct consequences of Theorem 4.21 and Theorem 4.24.

Theorem 4.25. *(strong duality) Between* (EP^M) *and* $(E\widetilde{D}^M)$ *strong duality holds, i.e.* $v(EP^M) = v(E\widetilde{D}^M)$ *and the dual problem* $v(E\widetilde{D}^M)$ *has an optimal solution.*

We define

$$J_\gamma := \left\{ j \in \{1, ..., m\} : \gamma_{C^0_{ij}}(z^*_{ij}) > 0 \right\}, \ i \in I,$$

and obtain as a result of Theorem 4.22 (especially by using the optimality condition (vi)), Theorem 4.24 and 4.25 the following optimality conditions.

Theorem 4.26. *(optimality conditions) (a) Let* $(\overline{x}_1, ..., \overline{x}_m) \in X^m$ *be an optimal solution to the problem* (EP^M). *Then there exists an optimal solution to* $(E\widetilde{D}^M)$ $(\overline{z}^*_1, ..., \overline{z}^*_n) \in (X^*)^m \times ... \times (X^*)^m$ *with the corresponding index set* $\overline{I} \subseteq \{1, ..., n\}$ *such that*

(i) $\displaystyle \max_{1 \leq u \leq n} \left\{ \sum_{j=1}^m \gamma_{C_{uj}}(\overline{x}_j - p_u) \right\} = \sum_{i \in \overline{I}} \sum_{j=1}^m \gamma_{C^0_{ij}}(\overline{z}^*_{ij})\gamma_{C_{ij}}(\overline{x}_j - p_i),$

(ii) $\displaystyle \sum_{i \in \overline{I}} \overline{z}^*_{ij} = 0_{X^*}, \ j = 1, ..., m,$

(iii) $\gamma_{C^0_{ij}}(\overline{z}^*_{ij})\gamma_{C_{ij}}(\overline{x}_j - p_i) = \langle \overline{z}^*_{ij}, \overline{x}_j - p_i \rangle, \ i \in \overline{I}, \ j = 1, ..., m,$

(iv) $\displaystyle \sum_{i \in \overline{I}} \max_{1 \leq l \leq m} \left\{ \gamma_{C^0_{il}}(\overline{z}^*_{il}) \right\} = 1,$

(v) $\displaystyle \max_{1 \leq u \leq n} \left\{ \sum_{j=1}^m \gamma_{C_{uj}}(\overline{x}_j - p_u) \right\} = \sum_{j=1}^m \gamma_{C_{ij}}(\overline{x}_j - p_i), \ i \in \overline{I},$

(vi) $\displaystyle \max_{1 \leq l \leq m} \left\{ \gamma_{C^0_{il}}(\overline{z}^*_{il}) \right\} = \gamma_{C^0_{ij}}(\overline{z}^*_{ij}) > 0, \ j \in \overline{J}_\gamma,$

$(\overline{z}^{1*}_{i1}, ..., \overline{z}^{1*}_{im}) \in X^* \times ... \times X^* \setminus \{(0_{X^*}, ..., 0_{X^*})\}, \ i \in \overline{I},$

and $\overline{z}^*_{ks} = 0_{X^*}, \ k \notin \overline{I}, \ s = 1, ..., m.$

(b) If there exists $(\overline{x}_1, ..., \overline{x}_m) \in X^m$ *such that for some* $(\overline{z}^*_1, ..., \overline{z}^*_n) \in (X^*)^m \times ... \times (X^*)^m$ *with the corresponding index set* $\overline{I}$ *the conditions* (i)-(vi) *are fulfilled, then* $(\overline{x}_1, ..., \overline{x}_m)$ *is an optimal solution to* (EP^M), $(\overline{z}^*_1, ..., \overline{z}^*_n)$ *is an optimal solution to* $(E\widetilde{D}^M)$ *and* $v(EP^M) = v(E\widetilde{D}^M).$

Now, our aim is to investigate the location problem (EP^M) from the geometrical point of view. For this purpose let $X = \mathbb{R}^d$ and the distances are measured by the Euclidean norm. Then, the problem (EP^M) turns into

$$(EP^M_N) \qquad \inf_{(x_1, ..., x_m) \in \mathbb{R}^d \times ... \times \mathbb{R}^d} \max_{1 \leq i \leq n} \left\{ \sum_{j=1}^m w_{ij}\|x_j - p_i\| \right\},$$

while its conjugate dual problem transforms into

$$(E\widetilde{D}_N^M) \qquad \sup_{(z_1^*,...,z_n^*)\in\widetilde{C}} \left\{ -\sum_{i\in I} \left\langle \sum_{j=1}^m z_{ij}^*, p_i \right\rangle \right\}$$

with

$$\widetilde{C} = \left\{ (z_1^*,...,z_n^*) \in (\mathbb{R}^d)^m \times ... \times (\mathbb{R}^d)^m : I = \left\{ i \in \{1,...,n\} : \max_{1\leq j\leq m} \left\{ \frac{1}{w_{ij}} \|z_{ij}^*\| \right\} > 0 \right\} \right.$$

$$\left. z_{kj}^* = 0_{\mathbb{R}^d}, \ k\notin I, \ \sum_{i\in I} z_{ij}^* = 0_{\mathbb{R}^d}, \ j = 1,...,m, \ \sum_{i\in I} \max_{1\leq l\leq m} \left\{ \frac{1}{w_{il}} \|z_{il}^*\| \right\} \leq 1 \right\}.$$

Via Theorem 4.25 and 4.26 the following statements follows immediately.

Theorem 4.27. *(strong duality) Between (EP_N^M) and $(E\widetilde{D}_N^M)$ holds strong duality, i.e.* $v(EP_N^M) = v(E\widetilde{D}_N^M)$ *and the dual problem has an optimal solution.*

Theorem 4.28. *(optimality conditions) (a) Let $(\overline{x}_1,...,\overline{x}_m) \in \mathbb{R}^d \times ... \times \mathbb{R}^d$ be an optimal solution to the problem (EP_N^M). Then there exists an optimal solution to $(E\widetilde{D}_N^M)$*

$$(\overline{z}_1^*,...,\overline{z}_n^*) \in \underbrace{\mathbb{R}^d \times ... \times \mathbb{R}^d}_{m-times} \times ... \times \underbrace{\mathbb{R}^d \times ... \times \mathbb{R}^d}_{m-times}$$

with the corresponding index set $\overline{I}$ such that

(i) $\displaystyle\max_{1\leq u\leq n} \left\{ \sum_{j=1}^m w_{uj}\|\overline{x}_j - p_u\| \right\} = \sum_{i\in\overline{I}} \sum_{j=1}^m \|\overline{z}_{ij}^*\|\|\overline{x}_j - p_i\|,$

(ii) $\displaystyle\sum_{i\in I} \overline{z}_{ij}^* = 0_{\mathbb{R}^d}, \ j = 1,...,m,$

(iii) $\|\overline{z}_{ij}^*\|\|\overline{x}_j - p_i\| = \langle \overline{z}_{ij}^*, \overline{x}_j - p_i \rangle, \ i\in\overline{I}, \ j = 1,...,m,$

(iv) $\displaystyle\sum_{i\in\overline{I}} \max_{1\leq l\leq m} \left\{ \frac{1}{w_{il}} \|\overline{z}_{il}^*\| \right\} = 1,$

(v) $\displaystyle\max_{1\leq u\leq n} \left\{ \sum_{j=1}^m w_{uj}\|\overline{x}_j - p_u\| \right\} = \sum_{j=1}^m w_{ij}\|\overline{x}_j - p_i\|, \ i\in\overline{I},$

(vi) $\displaystyle\max_{1\leq l\leq m} \left\{ \frac{1}{w_{il}}\|\overline{z}_{il}^*\| \right\} = \frac{1}{w_{ij}}\|\overline{z}_{ij}^*\|, \ j\in\overline{J}_\gamma = \left\{ j\in\{1,...,m\} : \|\overline{z}_{ij}^*\| > 0 \right\}, \ (\overline{z}_{i1}^{1*},...,\overline{z}_{im}^{1*}) \in \mathbb{R}^d \times ... \times \mathbb{R}^d \setminus \{(0_{\mathbb{R}^d},...,0_{\mathbb{R}^d})\}, \ i\in\overline{I}, \ and \ \overline{z}_{kj}^* = 0_{\mathbb{R}^d}, \ k\notin\overline{I}, \ j = 1,...,m.$

(b) If there exists $(\overline{x}_1, ..., \overline{x}_m) \in \mathbb{R}^d \times ... \times \mathbb{R}^d$ *such that for some*

$$(\overline{z}_1^*, ..., \overline{z}_n^*) \in \underbrace{\mathbb{R}^d \times ... \times \mathbb{R}^d}_{m-times} \times ... \times \underbrace{\mathbb{R}^d \times ... \times \mathbb{R}^d}_{m-times}$$

with the corresponding index set $\overline{I}$ the conditions (i)-(vi) are fulfilled, then $(\overline{x}_1, ..., \overline{x}_m)$ is an optimal solution to (EP_N^M), $(\overline{z}_1^*, ..., \overline{z}_n^*)$ is an optimal solution to $(E\widetilde{D}_N^M)$ and $v(EP_N^M) = v(E\widetilde{D}_N^M)$.

Geometrical interpretation.

We want now, in the concluding part of this section, to illustrate the results we presented above and describe the set of optimal solutions of the conjugate dual problem. For that end, let us first take a closer look at the optimality conditions stated in Theorem 4.28.

By the condition (iii) follows that the vectors $\overline{z}_{ij}^*$ and $\overline{x}_j - p_i$ are parallel and moreover, these vectors have the same direction, $i \in \overline{I}$, $j = 1, ..., m$. From the optimality condition (vi) we additionally deduce that the vectors $\overline{z}_{ij}^*$, $j = 1, ..., m$, are all unequal to the zero vector if $i \in \overline{I}$, which is the situation when the sum of the weighted distances in condition (v) is equal to the optimal objective value. In the reverse case, when $i \notin \overline{I}$, i.e. the sum of the weighted distances in condition (v) is less than the optimal objective value, the vectors $\overline{z}_{ij}^*$, $j = 1, ..., m$, are all equal to the zero vector.

Therefore, it is appropriate to interpret for $i \in \overline{I}$ the vectors $\overline{z}_{ij}^*$ fulfilling $\sum_{i \in \overline{I}} \overline{z}_{ij}^* = 0_{X^*}$ and $\sum_{i \in \overline{I}} \max_{1 \leq l \leq m} \left\{ \frac{1}{w_{il}} \|\overline{z}_{il}^*\| \right\} = 1$ as force vectors pulling the given point p_i in direction to the associated gravity points $\overline{x}_j$, $j = 1, ..., m$. As an illustration of the nature of the optimal solutions to the conjugate dual problem, let us consider the following example in the plane and especially, Figure 4.2.

Example 4.3. *Let us consider the points $p_1 = (0,0)^T$, $p_2 = (8,0)^T$ and $p_3 = (5,6)^T$ in the plane $(d = 2)$. For the given weights $w_{11} = 2$, $w_{12} = 3$, $w_{21} = 3$, $w_{22} = 3$, $w_{31} = 2$ and $w_{32} = 2$ we want to determine $m = 2$ new points minimizing the objective function of the location problem*

$$(EP_N^M) \quad \inf_{(x_1, x_2) \in \mathbb{R}^2 \times \mathbb{R}^2} \max\{2\|x_1 - p_1\| + 3\|x_2 - p1\|, 3\|x_1 - p_2\| + 3\|x_2 - p_2\|,$$

$$2\|x_1 - p_3\| + 2\|x_2 - p_3\|\}.$$

To solve this problem, we used the MATLAB Optimization Toolbox and obtained as optimal solution $\overline{x}_1 = (6.062, 0.858)^T$, $\overline{x}_2 = (2.997, 0.837)^T$ and as optimal objective value $(EP_N^M) = 21.578$.

The corresponding conjugate dual problem becomes to (see also Remark 4.39)

$$(E\widetilde{D}_N^M) \quad \sup_{(z_1^*, z_2^*, z_3^*) \in \widetilde{C}} \left\{ -\langle z_{11}^* + z_{12}^*, p_1 \rangle - \langle z_{21}^* + z_{22}^*, p_2 \rangle - \langle z_{31}^* + z_{32}^*, p_3 \rangle \right\},$$

where

$$\widetilde{\mathcal{C}} = \left\{ (z_1^*, z_2^*, z_3^*) \in (\mathbb{R}^2 \times \mathbb{R}^2) \times (\mathbb{R}^2 \times \mathbb{R}^2) \times (\mathbb{R}^2 \times \mathbb{R}^2) : \right.$$

$$z_{11}^* + z_{21}^* + z_{31}^* = 0_{\mathbb{R}^2}, \; z_{12}^* + z_{22}^* + z_{32}^* = 0_{\mathbb{R}^2},$$

$$\left. \max\left\{ \frac{1}{2}\|z_{11}^*\|, \frac{1}{3}\|z_{12}^*\| \right\} + \max\left\{ \frac{1}{3}\|z_{21}^*\|, \frac{1}{3}\|z_{22}^*\| \right\} + \max\left\{ \frac{1}{2}\|z_{31}^*\|, \frac{1}{2}\|z_{32}^*\| \right\} \le 1 \right\}.$$

The dual problem $(E\widetilde{D}_N^M)$ *was also solved with the* MATLAB *Optimization Toolbox. The optimal solution was*

$$\overline{z}_{11}^* = (0.803, 0.114)^T, \; \overline{z}_{12}^* = (1.171, 0.327)^T,$$

$$\overline{z}_{21}^* = (-0.909, 0.402)^T, \; \overline{z}_{22}^* = (-0.98, 0.164)^T,$$

$$\overline{z}_{31}^* = (0.106, -0.516)^T, \; \overline{z}_{32}^* = (-0.191, -0.491)^T$$

and the optimal objective function value $v(E\widetilde{D}_N^M) = 21.578 = v(EP_N^M)$. *See Figure 4.2 for an illustration of the relation between the optimal solutions of the primal and the conjugate dual problem.*

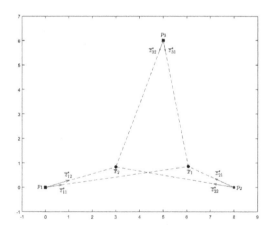

Figure 4.2: Illustration of the Example 4.3.

An alternative geometrical interpretation of the set of optimal solutions of the conjugate dual problem is based on the fact that the extended multifacility location problem (EP^M) can be reduced to a single minmax location problem as seen in the beginning of Section 4.4.1. This means precisely that the sum of distances in the objective function of the location problem (EP_N^M) can be understood as the finding the minimum value for n norms

d_i defined by the weighted sum of Euclidean norms, i.e. $d_i(y_1, ..., y_m) := \sum_{j=1}^{m} w_{ij} \|y_j\|$ with $y_j \in \mathbb{R}^d$, $w_{ij} > 0$, $j = 1, ..., m$, such that the associated norm balls centered at the points $\widetilde{p}_i = (p_i, ..., p_i)$ with $p_i \in \mathbb{R}^d$, $i = 1, ..., n$, have a non-empty intersection. In this case, it is possible to interpret the optimal solution to the corresponding conjugate dual problem as force vectors fulfilling the conditions in point (a) of Theorem 4.28 and increasing the norm balls until their intersection is non-empty. Notice that the optimality conditions (v) and (vi) imply that the vectors $\overline{z}_{ij}^*$, $j = 1, ..., m$, are equal to the zero vector if $i \notin \overline{I}$, which is exactly the case when $\overline{x}$ is an element of the interior of the ball associated to the norm d_i. But this also means that the vectors $\overline{z}_{ij}^*$, $j = 1, ..., m$, are all unequal to the zero vector if $i \in \overline{I}$, which exactly holds if $\overline{x}$ is lying on the border of the ball associated to the norm d_i.

For a better geometrical illustration of this interpretation, let us consider an example, where $d = 1$. In this case the Euclidean norm reduces to the absolute value.

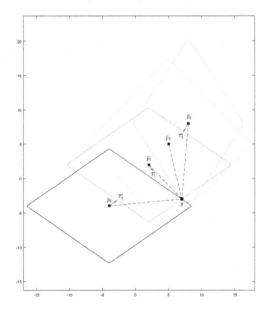

Figure 4.3: Illustration of the Example 4.4.

Example 4.4. *For the given points* $\widetilde{p}_1 = (p_1, p_1) = (2, 2)^T$, $\widetilde{p}_2 = (p_2, p_2) = (-4, -4)^T$, $\widetilde{p}_3 = (p_3, p_3) = (5, 5)^T$, $\widetilde{p}_4 = (p_4, p_4) = (8, 8)^T$ *and the weights* $w_{11} = 2$, $w_{12} = 3$, $w_{21} = 2$,

$w_{22} = 3$, $w_{31} = 2$, $w_{32} = 2$, $w_{41} = 3$, $w_{42} = 2$ we want to locate an optimal solution $x = (x_1, x_2)^T \in \mathbb{R}^2$ of the problem

$$(EP^M) \quad \inf_{(x_1,x_2)^T \in \mathbb{R}^2} \max\{2|x_1 - 2| + 3|x_2 - 2|, 2|x_1 + 4| + 3|x_2 + 4|,$$

$$2|x_1 - 5| + 2|x_2 - 5|, 3|x_1 - 8| + 2|x_2 - 8|\}.$$

We solved the problem (EP^M) with the MATLAB Optimization Toolbox and obtain as optimal solution $\overline{x} = (\overline{x}_1, \overline{x}_2)^T = (7, -3)^T$ and as optimal objective value $v(EP^M) = 25$.

For the corresponding conjugate dual problem (see also Remark 4.39)

$$(\widetilde{ED}^M) \quad \sup_{(z_1^*, z_2^*, z_3^*, z_4^*) \in \widetilde{C}} \left\{ -2(z_{11}^* + z_{12}^*) + 4(z_{21}^* + z_{22}^*) - 5(z_{31}^* + z_{32}^*) - 8(z_{41}^* + z_{42}^*) \right\},$$

where

$$\widetilde{C} = \Big\{ (z_1^*, z_2^*, z_3^*, z_4^*) \in \mathbb{R}^2 \times \mathbb{R}^2 \times \mathbb{R}^2 \times \mathbb{R}^2 :$$

$$z_{11}^* + z_{21}^* + z_{31}^* + z_{41}^* = 0, \ z_{12}^* + z_{22}^* + z_{32}^* + z_{42}^* = 0,$$

$$\max\left\{\frac{1}{2}|z_{11}^*|, \frac{1}{3}|z_{12}^*|\right\} + \max\left\{\frac{1}{2}|z_{21}^*|, \frac{1}{2}|z_{22}^*|\right\} + \max\left\{\frac{1}{2}|z_{31}^*|, \frac{1}{2}|z_{32}^*|\right\}$$

$$+ \max\left\{\frac{1}{3}|z_{41}^*|, \frac{1}{2}|z_{42}^*|\right\} \leq 1 \Big\},$$

we obtain by using again the MATLAB Optimization Toolbox the associated optimal solution

$$\overline{z}_1^* = (\overline{z}_{11}^*, \overline{z}_{12}^*)^T = (0.333, -0.5)^T, \ \overline{z}_2^* = (\overline{z}_{21}^*, \overline{z}_{22}^*)^T = (0.867, 1.3)^T,$$

$$\overline{z}_3^* = (\overline{z}_{31}^*, \overline{z}_{32}^*)^T = (0, 0)^T, \ \overline{z}_4^* = (\overline{z}_{41}^*, \overline{z}_{42}^*)^T = (-1.2, -0.8)^T$$

and the optimal objective value $v(\widetilde{ED}^M) = 25 = v(EP^M)$. The numerical results are illustrated in Figure 4.3. Take note that $\overline{x}$ is lying inside the norm ball centered at the point $\widetilde{p}_3$ and that for this reason $\overline{z}_3^*$ is equal to the zero vector.

4.5 Classical multifacility minmax location problems

4.5.1 Constrained location problems in Banach spaces

In this section we use the results of our previous approach to develop a conjugate dual problem of the multifacility minmax location problem with mixed gauges and geometric constraints. Furthermore, we show the validity of strong duality and derive optimality conditions for the corresponding primal-dual pair.

Let X be a Banach space (note that most of the following investigations can be extended to a Fréchet space, too), $J := \{jk : 1 \leq j \leq m, \ 1 \leq k \leq m, \ j \neq k\}$ and $\widetilde{J} := \{ji : 1 \leq j \leq m, \ 1 \leq i \leq t\}$. Further, let $V \subseteq J$ and $\widetilde{V} \subseteq \widetilde{J}$ be given index sets, the sets $C_{jk} \subseteq X$ with

$0_X \in \operatorname{int} C_{jk}$ for $jk \in V$ and $\widetilde{C}_{ji} \subseteq X$ with $0_X \in \operatorname{int} \widetilde{C}_{ji}$ for $ji \in \widetilde{V}$ be closed and convex as well as $S \subseteq X^m$ non-empty, closed and convex. Obviously, the gauges given by the sets C_{jk}, $jk \in V$, and $\widetilde{C}_{ji}$, $ji \in \widetilde{V}$, are convex, lower semicontinuous and well-defined.

For given distinct points $p_i \in X$, $1 \le i \le t$, the multifacility minmax location problem minimizes the maximum of gauges between pairs of m new facilities $x_1, ..., x_m$ and between pairs of m new and t existing facilities, concretely this means that

$$(P^M) \qquad \inf_{(x_1,...,x_m)\in S} \max \left\{ \left(\gamma_{C_{jk}}(x_j - x_k)\right)_{jk\in V}, \ \left(\gamma_{\widetilde{C}_{ji}}(x_j - p_i)\right)_{ji\in\widetilde{V}} \right\}.$$

The motivation of introducing the sets V and $\widetilde{V}$ is founded in the circumstance that some distances between pairs of new and pairs of new and existing facilities may not be relevant for the location problem (P^M) and thus, do not need to be considered in the objective function. In addition, take note that $|V| \le |J| = m(m-1)$ and $|\widetilde{V}| \le |\widetilde{J}| = mt$.

Now, we set $X_0 = \mathbb{R}^{|V|} \times \mathbb{R}^{|\widetilde{V}|}$ ordered by $K_0 = \mathbb{R}_+^{|V|} \times \mathbb{R}_+^{|\widetilde{V}|}$, $X_1 = X^{|V|} \times X^{|\widetilde{V}|}$ ordered by the trivial cone $K_1 = \{0_{X_1}\}$ and $X_2 = X^m$, where the corresponding dual spaces and dual variables are $(z^{0*}, \widetilde{z}^{0*}) = \left((z_{jk}^{0*})_{jk\in V}, (\widetilde{z}_{ji}^{0*})_{ji\in\widetilde{V}}\right) \in \mathbb{R}^{|V|} \times \mathbb{R}^{|\widetilde{V}|}$ and $(z^{1*}, \widetilde{z}^{1*}) = \left((z_{jk}^{1*})_{jk\in V}, (\widetilde{z}_{ji}^{1*})_{ji\in\widetilde{V}}\right) \in (X^*)^{|V|} \times (X^*)^{|\widetilde{V}|}$.

We continue with the decomposition of the objective function of the problem (P^M) into the following functions:

- $f : \mathbb{R}^{|V|} \times \mathbb{R}^{|\widetilde{V}|} \to \overline{\mathbb{R}}$ defined by

$$f(y^0, \widetilde{y}^0) = \max \left\{ \left(y_{jk}^0\right)_{jk\in V}, \ \left(\widetilde{y}_{ji}^0\right)_{ji\in\widetilde{V}} \right\}$$

if $y^0 = (y_{jk}^0)_{jk\in V} \in \mathbb{R}_+^{|V|}$ and $\widetilde{y}^0 = (\widetilde{y}_{ji}^0)_{ji\in\widetilde{V}} \in \mathbb{R}_+^{|\widetilde{V}|}$, otherwise $f(y^0, \widetilde{y}^0) = +\infty$,

- $F^1 : X^{|V|} \times X^{|\widetilde{V}|} \to \mathbb{R}^{|V|} \times \mathbb{R}^{|\widetilde{V}|}$ defined by

$$F^1(y^1, \widetilde{y}^1) = \left((\gamma_{C_{jk}}(y_{jk}^1))_{jk\in V}, (\gamma_{\widetilde{C}_{ji}}(\widetilde{y}_{ji}^1))_{ji\in\widetilde{V}}\right),$$

where $y^1 = (y_{jk}^1)_{jk\in V} \in X^{|V|}$ and $\widetilde{y}^1 = (\widetilde{y}_{ji}^1)_{ji\in\widetilde{V}} \in X^{|\widetilde{V}|}$,

- $F^2 : X^m \to X^{|V|} \times X^{|\widetilde{V}|}$ defined by

$$F^2(x) = \left((A_{jk}x)_{jk\in V}, (B_{ji}x - p_i)_{ji\in\widetilde{V}}\right),$$

where

$$A_{jk} = (\overset{1}{0}, ..., 0, \overset{j}{I}, 0, ..., \overset{k}{-I}, 0, ..., \overset{m}{0}), \ jk \in V, \ B_{ji} = (\overset{1}{0}, ..., 0, \overset{j}{I}, 0, ..., \overset{m}{0}), \ ji \in \widetilde{V},$$

0 is the zero mapping and I is the identity mapping, i.e. $0x_i = 0_X$ and $Ix_i = x_i \ \forall x_i \in X$, $i = 1, ..., m$. In particular, $A_{jk} : X^m \to X$ is defined as the mapping

$$x = (x_1, ..., x_m) \mapsto$$
$$0x_1 + ... + 0x_{j-1} + Ix_j + 0x_{j+1} + ... + 0x_{k-1} - Ix_k + 0x_{k+1} + ... + 0x_m,$$

i.e. $(x_1, ..., x_m) \mapsto x_j - x_k$, $jk \in V$, and $B_{ji} : X^m \to X$ is defined as the mapping

$$(x_1, ..., x_m) \mapsto 0x_1 + ... + 0x_{j-1} + Ix_j + 0x_{j+1} + ... + 0x_m = x_j, \ ji \in \widetilde{V}.$$

Thus, the problem (P^M) can be represented in the form

$$(P^M) \quad \inf_{x \in S} \max \left\{ \left(\gamma_{C_{jk}}(x_j - x_k) \right)_{jk \in V}, \ \left(\gamma_{\widetilde{C}_{ji}}(x_j - p_i) \right)_{ji \in \widetilde{V}} \right\}$$
$$= \inf_{x \in S} f \left(\left(\gamma_{C_{jk}}(A_{jk}x) \right)_{jk \in V}, \ \left(\gamma_{\widetilde{C}_{ji}}(B_{ji}x - p_i) \right)_{ji \in \widetilde{V}} \right)$$
$$= \inf_{x \in S} (f \circ F^1 \circ F^2)(x),$$

where the expression $F^1 \circ F^2$ in the last equality is defined as the component-wise composition of the vector-valued functions F^1 and F^2, i.e.

$$(F^1 \circ F^2)(x) = \left(\left(\left(\gamma_{C_{jk}} \circ A_{jk} \right)(x) \right)_{jk \in V}, \ \left(\left(\gamma_{\widetilde{C}_{ji}} \circ (B_{jk} \cdot -p_i) \right)(x) \right)_{ji \in \widetilde{V}} \right).$$

As mentioned in Remark 3.5, we do not need the monotonicity assumption for the function F^1, because F^2 is an affine function. Furthermore, it is clear that (P^M) is a convex optimization problem. Besides, it can easily be verified that f is proper, convex, $\mathbb{R}_+^{|V|} \times \mathbb{R}_+^{|\widetilde{V}|}$-increasing on $F^1(\mathrm{dom}\, F^1) + K_0 = \mathrm{dom}\, f = \mathbb{R}_+^{|V|} \times \mathbb{R}_+^{|\widetilde{V}|}$ and lower semicontinuous, and that F^1 is proper and $\mathbb{R}_+^{|V|} \times \mathbb{R}_+^{|\widetilde{V}|}$-convex as well as $\mathbb{R}_+^{|V|} \times \mathbb{R}_+^{|\widetilde{V}|}$-epi closed.

To use the formula from the previous section for the dual problem of (P^M), we set $Z = X^m$ ordered by the trivial cone $Q = X^m$ and define the function $g : X^m \to X^m$ by $g(x_1, ..., x_m) := (x_1, ..., x_m)$. As $Q^* = \{0_{(X^*)^m}\}$, which means that $z^{2*} = 0_{(X^*)^m}$, we derive for the dual problem

$$(D^M) \quad \sup_{\substack{(z^{0*}, \tilde{z}^{0*}) \in \mathbb{R}_+^{|V|} \times \mathbb{R}_+^{|\widetilde{V}|}, \\ (z^{1*}, \tilde{z}^{1*}) \in (X^*)^{|V|} \times (X^*)^{|\widetilde{V}|}}} \left\{ \inf_{x \in S} \left\{ \sum_{jk \in V} \langle z_{jk}^{1*}, A_{jk}x \rangle + \sum_{ji \in \widetilde{V}} \langle \tilde{z}_{ji}^{1*}, B_{ji}x - p_i \rangle \right\} \right.$$

$$\left. -f^*(z^{0*}, \tilde{z}^{0*}) - ((z^{0*}, \tilde{z}^{0*})F^1)^*(z^{1*}, \tilde{z}^{1*}) \right\},$$

and hence, we need to calculate the conjugate functions f^* and $((z^{0*}, \widetilde{z}^{0*})F^1)^*$. By Lemma 4.1 and Remark 4.40 we get for f^*,

$$f^*(z^{0*}, \widetilde{z}^{0*}) = \begin{cases} 0, & \text{if } z^{0*}_{jk} \leq \lambda_{jk}, \ \widetilde{z}^{0*}_{ji} \leq \widetilde{\lambda}_{ji}, \ \sum\limits_{jk \in V} \lambda_{jk} + \sum\limits_{ji \in \widetilde{V}} \widetilde{\lambda}_{ji} \leq 1 \\ & (\lambda_{jk})_{jk \in V} \in \mathbb{R}^{|V|}_+ \text{ and } (\lambda_{ji})_{ji \in \widetilde{V}} \in \mathbb{R}^{|\widetilde{V}|}_+, \\ +\infty, & \text{otherwise,} \end{cases}$$

$$= \begin{cases} 0, & \text{if } \sum\limits_{jk \in V} z^{0*}_{jk} + \sum\limits_{ji \in \widetilde{V}} \widetilde{z}^{0*}_{ji} \leq 1, \ z^{0*} \in \mathbb{R}^{|V|}_+, \ \widetilde{z}^{0*} \in \mathbb{R}^{|\widetilde{V}|}_+, \\ +\infty, & \text{otherwise,} \end{cases}$$

while for $((z^{0*}, \widetilde{z}^{0*})F^1)^*$ we obtain by using the definition of the conjugate function

$$((z^{0*}, \widetilde{z}^{0*})F^1)^*(z^{1*}, \widetilde{z}^{1*})$$

$$= \sup_{y^1 \in X^{|V|}, \ \widetilde{y}^1 \in X^{|\widetilde{V}|}} \left\{ \sum\limits_{jk \in V} \langle z^{1*}_{jk}, y^1_{jk} \rangle + \sum\limits_{ji \in \widetilde{V}} \langle \widetilde{z}^{1*}_{ji}, \widetilde{y}^1_{ji} \rangle - \sum\limits_{jk \in V} z^{0*}_{jk} \gamma_{C_{jk}}(y^1_{jk}) - \sum\limits_{ji \in \widetilde{V}} \widetilde{z}^{0*}_{ji} \gamma_{\widetilde{C}_{ji}}(\widetilde{y}^1_{ji}) \right\}$$

$$= \sum\limits_{jk \in V} \sup_{y^1_{jk} \in X} \left\{ \langle z^{1*}_{jk}, y^1_{jk} \rangle - z^{0*}_{jk} \gamma_{C_{jk}}(y^1_{jk}) \right\} + \sum\limits_{ji \in \widetilde{V}} \sup_{\widetilde{y}^1_{ji} \in X} \left\{ \langle \widetilde{z}^{1*}_{ji}, \widetilde{y}^1_{ji} \rangle - \widetilde{z}^{0*}_{ji} \gamma_{\widetilde{C}_{ji}}(\widetilde{y}^1_{ji}) \right\}$$

$$= \sum\limits_{jk \in V} (z^{0*}_{jk} \gamma_{C_{jk}})^*(z^{1*}_{jk}) + \sum\limits_{ji \in \widetilde{V}} (\widetilde{z}^{0*}_{ji} \gamma_{\widetilde{C}_{ji}})^*(\widetilde{z}^{1*}_{ji})$$

for all $(z^{0*}, \widetilde{z}^{0*}) \in \mathbb{R}^{|V|}_+ \times \mathbb{R}^{|\widetilde{V}|}_+$ and $z^{1*} = (z^{1*}_{jk})_{jk \in V} \in X^{|V|}$ and $\widetilde{z}^{1*} = (z^{1*}_{ji})_{ji \in \widetilde{V}} \in X^{|\widetilde{V}|}$. Hence, the dual problem may be written as

$$(D^M) \qquad \sup_{\substack{(z^{0*}, \widetilde{z}^{0*}, z^{1*}, \widetilde{z}^{1*}) \in \mathbb{R}^{|V|}_+ \times \mathbb{R}^{|\widetilde{V}|}_+ \times X^{|V|} \times X^{|\widetilde{V}|} \\ \sum\limits_{jk \in V} z^{0*}_{jk} + \sum\limits_{ji \in \widetilde{V}} \widetilde{z}^{0*}_{ji} \leq 1}} \inf_{x \in S} \Phi(z^{0*}, \widetilde{z}^{0*}, z^{1*}, \widetilde{z}^{1*}),$$

where

$$\Phi(z^{0*}, \widetilde{z}^{0*}, z^{1*}, \widetilde{z}^{1*}) = \inf_{x \in S} \left\{ \sum\limits_{jk \in V} \langle z^{1*}_{jk}, A_{jk}x \rangle + \sum\limits_{ji \in \widetilde{V}} \langle \widetilde{z}^{1*}_{ji}, B_{ji}x - p_i \rangle \right\}$$

$$- \sum\limits_{jk \in V} (z^{0*}_{jk} \gamma_{C_{jk}})^*(z^{1*}_{jk}) - \sum\limits_{ji \in \widetilde{V}} (\widetilde{z}^{0*}_{ji} \gamma_{\widetilde{C}_{ji}})^*(\widetilde{z}^{1*}_{ji}).$$

Let $I := \{jk \in V : z_{jk}^{0*} > 0\}$ and $\widetilde{I} := \{ji \in \widetilde{V}_{ji} : \widetilde{z}_{ji}^{0*} > 0\}$, then we separate in the objective function Φ the sum into the terms with $z_{jk}^{0*}, \widetilde{z}_{ji}^{0*} > 0$ and the terms with $z_{jk}^{0*}, \widetilde{z}_{ji}^{0*} = 0$:

$$
\Phi(z^{0*}, \widetilde{z}^{0*}, z^{1*}, \widetilde{z}^{1*}) = \inf_{x \in S} \left\{ \sum_{jk \in V} \langle z_{jk}^{1*}, A_{jk}x \rangle + \sum_{ji \in \widetilde{V}} \langle \widetilde{z}_{ji}^{1*}, B_{ji}x - p_i \rangle \right\}
$$
$$
- \sum_{jk \in I} (z_{jk}^{0*} \gamma_{C_{jk}})^*(z_{jk}^{1*}) - \sum_{ji \in \widetilde{I}} (\widetilde{z}_{ji}^{0*} \gamma_{\widetilde{C}_{ji}})^*(\widetilde{z}_{ji}^{1*})
$$
$$
- \sum_{jk \in V \setminus I} (0 \cdot \gamma_{C_{jk}})^*(z_{jk}^{1*}) - \sum_{ji \in \widetilde{V} \setminus \widetilde{I}} (0 \cdot \gamma_{\widetilde{C}_{ji}})^*(\widetilde{z}_{ji}^{1*}).
$$

Now, for $jk \in I$ we have (see Lemma 4.4 and Remark 4.5)

$$
(z_{jk}^{0*} \gamma_{C_{jk}})^*(z_{jk}^{1*}) = z_{jk}^{0*} \gamma_{C_{jk}}^* \left(\frac{z_{jk}^{1*}}{z_{jk}^{0*}} \right) = \begin{cases} 0, & \text{if } \sigma_{C_{jk}} \left(\frac{z_{jk}^{1*}}{z_{jk}^{0*}} \right) \leq 1, \\ +\infty, & \text{otherwise,} \end{cases}
$$
$$
= \begin{cases} 0, & \text{if } \sigma_{C_{jk}}(z_{jk}^{1*}) \leq z_{jk}^{0*}, \\ +\infty, & \text{otherwise,} \end{cases} = \begin{cases} 0, & \text{if } \gamma_{C_{jk}^0}(z_{jk}^{1*}) \leq z_{jk}^{0*}, \\ +\infty, & \text{otherwise,} \end{cases} \tag{4.102}
$$

and analogously, it follows for $ji \in \widetilde{I}$ that

$$
(\widetilde{z}_{ji}^{0*} \gamma_{\widetilde{C}_{ji}})^*(\widetilde{z}_{ji}^{1*}) = \begin{cases} 0, & \text{if } \gamma_{\widetilde{C}_{ji}^0}(\widetilde{z}_{ji}^{1*}) \leq \widetilde{z}_{ji}^{0*}, \\ +\infty, & \text{otherwise.} \end{cases} \tag{4.103}
$$

For $jk \in V \setminus I$ we have

$$
(0 \cdot \gamma_{C_{jk}})^*(z_{jk}^{1*}) = \sup_{y_{jk}^1 \in X} \left\{ \langle z_{jk}^{1*}, y_{jk}^1 \rangle \right\} = \begin{cases} 0, & \text{if } z_{jk}^{1*} = 0_{X^*}, \\ +\infty, & \text{otherwise,} \end{cases}
$$

and analogously, we get for $ji \in \widetilde{V} \setminus \widetilde{I}$,

$$
(0 \cdot \gamma_{\widetilde{C}_{ji}})^*(\widetilde{z}_{ji}^{1*}) = \begin{cases} 0, & \text{if } \widetilde{z}_{ji}^{1*} = 0_{X^*}, \\ +\infty, & \text{otherwise,} \end{cases}
$$

which implies that if $jk \in V \setminus I$, then $z_{jk}^{1*} = 0_{X^*}$ and if $ji \in \widetilde{V} \setminus \widetilde{I}$, then $\widetilde{z}_{ji}^{1*} = 0_{X^*}$. Therefore, we obtain for the dual problem of the location problem (P^M):

$$
(D^M) \qquad \sup_{(z^{0*}, \widetilde{z}^{0*}, z^{1*}, \widetilde{z}^{1*}) \in \mathcal{B}} \inf_{x \in S} \left\{ \sum_{jk \in I} \langle z_{jk}^{1*}, A_{jk}x \rangle + \sum_{ji \in \widetilde{I}} \langle \widetilde{z}_{ji}^{1*}, B_{ji}x - p_i \rangle \right\},
$$

where

$$\mathcal{B} = \Bigg\{ (z^{0*}, \widetilde{z}^{0*}, z^{1*}, \widetilde{z}^{1*}) \in \mathbb{R}_+^{|V|} \times \mathbb{R}_+^{|\widetilde{V}|} \times (X^*)^{|V|} \times (X^*)^{|\widetilde{V}|} : I \subseteq V, \ \widetilde{I} \subseteq \widetilde{V}, \ z_{jk}^{0*} > 0,$$

$$z_{jk}^{1*} \in X^*, \ \gamma_{C_{jk}^0}(z_{jk}^{1*}) \le z_{jk}^{0*}, \ jk \in I, \ \widetilde{z}_{ji}^{0*} > 0, \ \widetilde{z}_{ji}^{1*} \in X^*, \ \gamma_{\widetilde{C}_{ji}^0}(\widetilde{z}_{ji}^{1*}) \le \widetilde{z}_{ji}^{0*}, \ ji \in \widetilde{I}, \ z_{ef}^{0*} = 0,$$

$$z_{ef}^{1*} = 0_{X^*}, \ ef \in V \setminus I, \ \widetilde{z}_{ed}^{0*} = 0, \ \widetilde{z}_{ed}^{1*} \in 0_{X^*}, \ ed \in \widetilde{V} \setminus \widetilde{I}, \ \sum_{jk \in I} z_{jk}^{0*} + \sum_{ji \in \widetilde{I}} \widetilde{z}_{ji}^{0*} \le 1 \Bigg\}.$$

Since, the objective function of the conjugate dual problem (D^M) can also be written as

$$\inf_{x \in S} \Bigg\{ \sum_{jk \in I} \langle z_{jk}^{1*}, A_{jk}x \rangle + \sum_{ji \in \widetilde{I}} \langle \widetilde{z}_{ji}^{1*}, B_{ji}x - p_i \rangle \Bigg\}$$

$$= \inf_{x \in S} \Bigg\{ \Bigg\langle \sum_{jk \in I} A_{jk}^* z_{jk}^{1*} + \sum_{ji \in \widetilde{I}} B_{ji}^* \widetilde{z}_{ji}^{1*}, x \Bigg\rangle \Bigg\} - \sum_{ji \in I} \langle \widetilde{z}_{ji}^{1*}, p_i \rangle,$$

where

$$\langle A_{jk}^* z_{jk}^{1*}, x \rangle = \langle (\overset{1}{0_{X^*}}, ..., 0_{X^*}, \overset{j}{z_{jk}^{1*}}, 0_{X^*}, ..., 0_{X^*}, \overset{k}{-z_{jk}^{1*}}, 0_{X^*}, ..., \overset{m}{0_{X^*}}), (x_1, ..., x_m) \rangle$$

$$= \langle z_{jk}^{1*}, x_j - x_k \rangle$$

and

$$\langle B_{ji}^* \widetilde{z}_{ji}^{1*}, x \rangle = \langle (\overset{1}{0_{X^*}}, ..., 0_{X^*}, \overset{j}{\widetilde{z}_{ji}^{1*}}, 0_{X^*}, ..., \overset{m}{0_{X^*}}), (x_1, ..., x_m) \rangle = \langle \widetilde{z}_{ji}^{1*}, x_j \rangle,$$

we can express (D^M) as

$$(D^M) \qquad \sup_{(z^{0*}, \widetilde{z}^{0*}, z^{1*}, \widetilde{z}^{1*}) \in \mathcal{B}} \Bigg\{ -\sigma_S \Bigg(-\sum_{jk \in I} A_{jk}^* z_{jk}^{1*} - \sum_{ji \in \widetilde{I}} B_{ji}^* \widetilde{z}_{ji}^{1*} \Bigg) - \sum_{ji \in \widetilde{I}} \langle \widetilde{z}_{ji}^{1*}, p_i \rangle \Bigg\}.$$

Remark 4.43. *Take note that the problem (D^M) is equivalent to the following one*

$$(\widehat{D}^M) \qquad \sup_{(z^{0*}, \widetilde{z}^{0*}, z^{1*}, \widetilde{z}^{1*}) \in \mathcal{B}} \Bigg\{ -\sigma_S \Bigg(-\sum_{jk \in V} A_{jk}^* z_{jk}^{1*} - \sum_{ji \in \widetilde{V}} B_{ji}^* \widetilde{z}_{ji}^{1*} \Bigg) - \sum_{ji \in \widetilde{V}} \langle \widetilde{z}_{ji}^{1*}, p_i \rangle \Bigg\},$$

where

$$\widehat{\mathcal{B}} = \Bigg\{ (z^{0*}, \widetilde{z}^{0*}, z^{1*}, \widetilde{z}^{1*}) \in \mathbb{R}_+^{|V|} \times \mathbb{R}_+^{|\widetilde{V}|} \times (X^*)^{|V|} \times (X^*)^{|\widetilde{V}|} : \gamma_{C_{jk}^0}(z_{jk}^{1*}) \le z_{jk}^{0*}, \ jk \in V,$$

$$\gamma_{\widetilde{C}_{ji}^0}(\widetilde{z}_{ji}^{1*}) \le \widetilde{z}_{ji}^{0*}, \ ji \in \widetilde{V}, \ \sum_{jk \in V} z_{jk}^{0*} + \sum_{ji \in \widetilde{V}} \widetilde{z}_{ji}^{0*} \le 1 \Bigg\},$$

which can be proved as follows.

Let $(z^{0*}, \widetilde{z}^{0*}, z^{1*}, \widetilde{z}^{1*}) \in \widehat{\mathcal{B}}$ be a feasible solution to $(\widehat{D}^M)$, then it holds for $jk \in V \setminus I$ and $ji \in \widetilde{V} \setminus \widetilde{I}$ (see also [61, Proposition 2.2.3]),

$$0 \leq \gamma_{C_{jk}^0}(z_{jk}^{1*}) = \sup_{x \in C_{jk}} \langle z_{jk}^{1*}, x \rangle \leq 0 \Leftrightarrow \langle z_{jk}^{1*}, x \rangle = 0 \ \forall x \in C_{jk} \Leftrightarrow z_{jk}^{1*} = 0_{X^*}.$$

as well as

$$0 \leq \gamma_{\widetilde{C}_{ji}^0}(\widetilde{z}_{ji}^{1*}) = \sup_{x \in \widetilde{C}_{ji}} \langle \widetilde{z}_{ji}^{1*}, x \rangle \leq 0 \Leftrightarrow \langle \widetilde{z}_{ji}^{1*}, x \rangle = 0 \ \forall x \in \widetilde{C}_{ji} \Leftrightarrow \widetilde{z}_{ji}^{1*} = 0_{X^*}.$$

The latter implies that from $jk \in V \setminus I$, i.e. $z_{jk}^{0*} = 0$, follows $z_{jk}^{1*} = 0_{X^*}$ and from $ji \in \widetilde{V} \setminus \widetilde{I}$, i.e. $\widetilde{z}_{ji}^{0*} = 0$, $\widetilde{z}_{ji}^{1*} = 0_{X^*}$. This relation means that $\widehat{\mathcal{B}} = \mathcal{B}$, i.e. that $(z^{0*}, \widetilde{z}^{0*}, z^{1*}, \widetilde{z}^{1*})$ is also a feasible solution to (D^M) and as

$$\sigma_S \left(-\sum_{jk \in V} A_{jk}^* z_{jk}^{1*} - \sum_{ji \in \widetilde{V}} B_{ji}^* \widetilde{z}_{ji}^{1*} \right) + \sum_{ji \in \widetilde{V}} \langle \widetilde{z}_{ji}^{1*}, p_i \rangle$$

$$= \sigma_S \left(-\sum_{jk \in I} A_{jk}^* z_{jk}^{1*} - \sum_{ji \in \widetilde{I}} B_{ji}^* \widetilde{z}_{ji}^{1*} \right) + \sum_{ji \in \widetilde{I}} \langle \widetilde{z}_{ji}^{1*}, p_i \rangle,$$

one has immediately that $v(D^M) = v(\widehat{D}^M)$.

Vice versa, if we take a feasible solution $(z^{0*}, \widetilde{z}^{0*}, z^{1*}, \widetilde{z}^{1*})$ of the problem (D^M), then it is obvious that we have then also a feasible solution to $(\widehat{D}^M)$, which again implies that $v(D^M) = v(\widehat{D}^M)$.

From the theoretical aspect a dual problem of the form (D^M) is very useful, as one has a more detailed characterization of the set of feasible solutions. But from the numerical viewpoint it is complicate to solve, as the index sets I and $\widetilde{I}$ bring an undesirable discretization in the dual problem. For this reason it is preferable to use the dual problem $(\widehat{D}^M)$ for numerical and (D^M) for theoretical studies. It is also important to emphasize that the index sets I and $\widetilde{I}$ are determinable for a fixed and feasible solution $(z^{0*}, \widetilde{z}^{0*}, z^{1*}, \widetilde{z}^{1*})$ to the dual problem (D^M).

We know that the weak duality between the problem (P^M) and its corresponding dual problem (D^M) always holds. Now, we are interested to know whether we also can guarantee strong duality. For this purpose we use the results from Section 3.2. As $Z = X^m$ ordered by the trivial cone $Q = X^m$ and $g : X^m \to X^m$ is defined by $g(x_1, ..., x_m) = (x_1, ..., x_m)$, it is obvious that g is Q-epi closed and $0_{X^m} \in \text{sqri}(g(x) + Q) = \text{sqri}(X^m + Q) = X^m$.

More than that, recall that f is lower semicontinous, $K_0 = \mathbb{R}_+^{|V|} \times \mathbb{R}_+^{|\widetilde{V}|}$ is closed, S is closed and F^1 is $\mathbb{R}_+^{|V|} \times \mathbb{R}_+^{|\widetilde{V}|}$-epi closed. As

$$0_{\mathbb{R}_+^{|V|} \times \mathbb{R}_+^{|\widetilde{V}|}} \in \operatorname{sqri}(F^1(\operatorname{dom} F^1) - \operatorname{dom} f + K_0)$$
$$= \operatorname{sqri}(F^1(\operatorname{dom} F^1) - \mathbb{R}_+^{|V|} \times \mathbb{R}_+^{|\widetilde{V}|} + \mathbb{R}_+^{|V|} \times \mathbb{R}_+^{|\widetilde{V}|})$$
$$= \mathbb{R}^{|V|} \times \mathbb{R}^{|\widetilde{V}|},$$
$$0_{X^{|V|} \times X^{|\widetilde{V}|}} \in \operatorname{sqri}(F^2(\operatorname{dom} F^2) - \operatorname{dom} F^1 + K_1)$$
$$= \operatorname{sqri}(X^{|V|} \times X^{|\widetilde{V}|} - \operatorname{dom} F^1 + K_1) = X^{|V|} \times X^{|\widetilde{V}|}$$

and F^2 is $\{0_{X^{|V|} \times X^{|\widetilde{V}|}}\}$-epi closed, the generalized interior point regularity condition (RC) is fulfilled, it follows by Theorem 3.3 the following statement (note that we denote by $v(P^M)$ and $v(D^M)$ the optimal objective values of the problems (P^M) and (D^M), respectively).

Theorem 4.29. *(strong duality) Between (P^M) and (D^M) holds strong duality, i.e. $v(P^M) = v(D^M)$ and the conjugate dual problem has an optimal solution.*

The previous theorem implies the following necessary and sufficient optimality conditions for the primal-dual pair (P^M)-(D^M).

Theorem 4.30. *(optimality conditions) (a) Let $\overline{x} \in S$ be an optimal solution to the problem (P^M). Then there exists an optimal solution to (D^M) $(\overline{z}^{0*}, \overline{\widetilde{z}}^{0*}, \overline{z}^{1*}, \overline{\widetilde{z}}^{1*}) \in \mathbb{R}_+^{|V|} \times \mathbb{R}_+^{|\widetilde{V}|} \times (X^*)^{|V|} \times (X^*)^{|\widetilde{V}|}$ with corresponding index sets $\overline{I} \subseteq V$ and $\overline{\widetilde{I}} \subseteq \widetilde{V}$ such that*

(i) $\max\left\{ \left(\gamma_{C_{ef}}(\overline{x}_e - \overline{x}_f)\right)_{ef \in V}, \left(\gamma_{\widetilde{C}_{ed}}(\overline{x}_e - p_d)\right)_{ed \in \widetilde{V}} \right\}$
$$= \sum_{jk \in \overline{I}} \overline{z}_{jk}^{0*} \gamma_{C_{jk}}(\overline{x}_j - \overline{x}_k) + \sum_{ji \in \overline{\widetilde{I}}} \overline{\widetilde{z}}_{ji}^{0*} \gamma_{\widetilde{C}_{ji}}(\overline{x}_j - p_i),$$

(ii) $\left\langle \sum_{jk \in \overline{I}} A_{jk}^* \overline{z}_{jk}^{1*} + \sum_{ji \in \overline{\widetilde{I}}} B_{ji}^* \overline{\widetilde{z}}_{ji}^{1*}, \overline{x} \right\rangle = \inf_{x \in S} \left\{ \left\langle \sum_{jk \in \overline{I}} A_{jk}^* \overline{z}_{jk}^{1*} + \sum_{ji \in \overline{\widetilde{I}}} B_{ji}^* \overline{\widetilde{z}}_{ji}^{1*}, x \right\rangle \right\},$

(iii) $\sum_{jk \in \overline{I}} \overline{z}_{jk}^{0*} + \sum_{ji \in \overline{\widetilde{I}}} \overline{\widetilde{z}}_{ji}^{0*} = 1$, $\overline{z}_{jk}^{0*} > 0$, $jk \in \overline{I}$, $\overline{\widetilde{z}}_{ji}^{0*} > 0$, $ji \in \overline{\widetilde{I}}$ and $\overline{z}_{ef}^{0*} = 0$, $ef \in V \setminus \overline{I}$, $\overline{\widetilde{z}}_{ed}^{0*} = 0$, $ed \in \widetilde{V} \setminus \overline{\widetilde{I}}$,

(iv) $\overline{z}_{jk}^{0*} \gamma_{C_{jk}}(\overline{x}_j - \overline{x}_k) = \langle \overline{z}_{jk}^{1*}, \overline{x}_j - \overline{x}_k \rangle$, $jk \in \overline{I}$,

(v) $\overline{\widetilde{z}}_{ji}^{0*} \gamma_{\widetilde{C}_{ji}}(\overline{x}_j - p_i) = \langle \overline{\widetilde{z}}_{ji}^{1*}, \overline{x}_j - p_i \rangle$, $ji \in \overline{\widetilde{I}}$,

(vi) $\max\left\{ \left(\gamma_{C_{ef}}(\overline{x}_e - \overline{x}_f)\right)_{ef \in V}, \left(\gamma_{\widetilde{C}_{ed}}(\overline{x}_e - p_d)\right)_{ed \in \widetilde{V}} \right\} = \gamma_{C_{jk}}(\overline{x}_j - \overline{x}_k), jk \in \overline{I},$

(vii) $\max\left\{\left(\gamma_{C_{ef}}(\overline{x}_e - \overline{x}_f)\right)_{ef \in V}, \left(\gamma_{\widetilde{C}_{ed}}(\overline{x}_e - p_d)\right)_{ed \in \widetilde{V}}\right\} = \gamma_{\widetilde{C}_{ji}}(\overline{x}_j - p_i), \ ji \in \widetilde{\overline{I}},$

(viii) $\gamma_{C_{jk}^0}(\overline{z}_{jk}^{1*}) = \overline{z}_{jk}^{0*}, \ \overline{z}_{jk}^{1*} \in X^*, \ jk \in \overline{I} \ and \ \overline{z}_{ef}^{1*} = 0_{X^*}, \ ef \in V \setminus \overline{I},$

(ix) $\gamma_{\widetilde{C}_{ji}^0}(\overline{\overline{z}}_{ji}^{1*}) = \overline{\overline{z}}_{ji}^{0*}, \ \overline{\overline{z}}_{ji}^{1*} \in X^*, \ ji \in \widetilde{\overline{I}} \ and \ \overline{\overline{z}}_{ed}^{1*} = 0_{X^*}, \ ed \in \widetilde{\overline{V}} \setminus \widetilde{\overline{I}}.$

(b) If there exists $\overline{x} \in S$ such that for some $(\overline{z}^{0*}, \overline{\overline{z}}^{0*}, \overline{z}^{1*}, \overline{\overline{z}}^{1*})$ and the corresponding index sets $\overline{I}$ and $\widetilde{\overline{I}}$ the conditions (i)-(ix) are fulfilled, then $\overline{x}$ is an optimal solution to (P^C), $(\overline{z}^{0*}, \overline{\overline{z}}^{0*}, \overline{z}^{1*}, \overline{\overline{z}}^{1*})$ is an optimal solution to (D^M) and $v(P^M) = v(D^M)$.

Proof. (a) From Theorem 3.4 one gets

(i) $\max\left\{\left(\gamma_{C_{ef}}(\overline{x}_e - \overline{x}_f)\right)_{ef \in V}, \left(\gamma_{\widetilde{C}_{ed}}(\overline{x}_e - p_d)\right)_{ed \in \widetilde{V}}\right\}$
$= \sum_{jk \in \overline{I}} \overline{z}_{jk}^{0*} \gamma_{C_{jk}}(\overline{x}_j - \overline{x}_k) + \sum_{ji \in \widetilde{\overline{I}}} \overline{\overline{z}}_{ji}^{0*} \gamma_{\widetilde{C}_{ji}}(\overline{x}_j - p_i),$

(ii) $\sum_{jk \in \overline{I}} \overline{z}_{jk}^{0*} \gamma_{C_{jk}}(\overline{x}_j - \overline{x}_k) + \sum_{ji \in \widetilde{\overline{I}}} \overline{\overline{z}}_{ji}^{0*} \gamma_{\widetilde{C}_{ji}}(\overline{x}_j - p_i) = \sum_{jk \in \overline{I}} \langle \overline{z}_{jk}^{1*}, \overline{x}_j - \overline{x}_k \rangle + \sum_{ji \in \widetilde{\overline{I}}} \langle \overline{\overline{z}}_{ji}^{1*}, \overline{x}_j - p_i \rangle,$

(iii) $\left\langle \sum_{jk \in \overline{I}} A_{jk}^* \overline{z}_{jk}^{1*} + \sum_{ji \in \widetilde{\overline{I}}} B_{ji}^* \overline{\overline{z}}_{ji}^{1*}, \overline{x} \right\rangle = -\sigma_S \left(-\sum_{jk \in \overline{I}} A_{jk}^* \overline{z}_{jk}^{1*} - \sum_{ji \in \widetilde{\overline{I}}} B_{ji}^* \overline{\overline{z}}_{ji}^{1*} \right),$

(iv) $\sum_{jk \in \overline{I}} \overline{z}_{jk}^{0*} + \sum_{ji \in \widetilde{\overline{I}}} \overline{\overline{z}}_{ji}^{0*} \leq 1, \ \overline{z}_{jk}^{0*} > 0, \ jk \in \overline{I}, \ \overline{\overline{z}}_{ji}^{0*} > 0, \ ji \in \widetilde{\overline{I}} \ and \ \overline{z}_{ef}^{0*} = 0, \ ef \in V \setminus \overline{I},$
$\overline{\overline{z}}_{ed}^{0*} = 0, \ ed \in \widetilde{\overline{V}} \setminus \widetilde{\overline{I}},$

(v) $\gamma_{C_{jk}^0}(\overline{z}_{jk}^{1*}) \leq \overline{z}_{jk}^{0*}, \ \overline{z}_{jk}^{1*} \in X^*, \ jk \in \overline{I} \ and \ \overline{z}_{ef}^{1*} = 0_{X^*}, \ ef \in V \setminus \overline{I},$

(vi) $\gamma_{\widetilde{C}_{ji}^0}(\overline{\overline{z}}_{ji}^{1*}) \leq \overline{\overline{z}}_{ji}^{0*}, \ \overline{\overline{z}}_{ji}^{1*} \in X^*, \ ji \in \widetilde{\overline{I}} \ and \ \overline{\overline{z}}_{ed}^{1*} = 0_{X^*}, \ ed \in \widetilde{\overline{V}} \setminus \widetilde{\overline{I}}.$

Condition (ii) yields

$$\sum_{jk \in \overline{I}} [\overline{z}_{jk}^{0*} \gamma_{C_{jk}}(\overline{x}_j - \overline{x}_k) - \langle \overline{z}_{jk}^{1*}, \overline{x}_j - \overline{x}_k \rangle] + \sum_{ji \in \widetilde{\overline{I}}} [\overline{\overline{z}}_{ji}^{0*} \gamma_{\widetilde{C}_{ji}}(\overline{x}_j - p_i) - \langle \overline{\overline{z}}_{ji}^{1*}, \overline{x}_j - p_i \rangle] = 0$$
$$(4.104)$$

and by (4.102), (4.103) and the Young-Fenchel inequality it follows that the brackets in (4.104) are non-negative and must be equal to zero, i.e.

$$\overline{z}_{jk}^{0*} \gamma_{C_{jk}}(\overline{x}_j - \overline{x}_k) = \langle \overline{z}_{jk}^{1*}, \overline{x}_j - \overline{x}_k \rangle, \ jk \in \overline{I} \ and \ \overline{\overline{z}}_{ji}^{0*} \gamma_{\widetilde{C}_{ji}}(\overline{x}_j - p_i) = \langle \overline{\overline{z}}_{ji}^{1*}, \overline{x}_j - p_i \rangle, \ ji \in \widetilde{\overline{I}}.$$
$$(4.105)$$

Combining the condition (v) with (4.105) reveals by using the generalized Cauchy-Schwarz inequality (see Lemma 4.3) that

$$\overline{z}_{jk}^{0*}\gamma_{C_{jk}}(\overline{x}_j - \overline{x}_k) = \langle \overline{z}_{jk}^{1*}, \overline{x}_j - \overline{x}_k \rangle \leq \gamma_{C_{jk}^0}(\overline{z}_{jk}^{1*})\gamma_{C_{jk}}(\overline{x}_j - \overline{x}_k) \leq \overline{z}_{jk}^{0*}\gamma_{C_{jk}}(\overline{x}_j - \overline{x}_k), \ jk \in \overline{I},$$

which means that

$$\gamma_{C_{jk}^0}(\overline{z}_{jk}^{1*}) = \overline{z}_{jk}^{0*}, \ jk \in \overline{I}. \tag{4.106}$$

In the same way we get

$$\gamma_{\widetilde{C}_{ji}^0}(\overline{\overline{z}}_{ji}^{1*}) = \overline{\overline{z}}_{ji}^{0*}, \ ji \in \overline{\overline{I}}. \tag{4.107}$$

Moreover, by conditions (i) and (iv) we have

$$\max\left\{ \left(\gamma_{C_{ef}}(\overline{x}_e - \overline{x}_f)\right)_{ef \in V}, \ \left(\gamma_{\widetilde{C}_{ed}}(\overline{x}_e - p_d)\right)_{ed \in \widetilde{V}} \right\} \tag{4.108}$$

$$= \sum_{jk \in \overline{I}} \overline{z}_{jk}^{0*}\gamma_{C_{jk}}(\overline{x}_j - \overline{x}_k) + \sum_{ji \in \overline{\overline{I}}} \overline{\overline{z}}_{ji}^{0*}\gamma_{\widetilde{C}_{ji}}(\overline{x}_j - p_i)$$

$$\leq \sum_{jk \in \overline{I}} \overline{z}_{jk}^{0*} \max\left\{ \left(\gamma_{C_{ef}}(\overline{x}_e - \overline{x}_f)\right)_{ef \in V}, \ \left(\gamma_{\widetilde{C}_{ed}}(\overline{x}_e - p_d)\right)_{ed \in \widetilde{V}} \right\}$$

$$+ \sum_{ji \in \overline{\overline{I}}} \overline{\overline{z}}_{ji}^{0*} \max\left\{ \left(\gamma_{C_{ef}}(\overline{x}_e - \overline{x}_f)\right)_{ef \in V}, \ \left(\gamma_{\widetilde{C}_{ed}}(\overline{x}_e - p_d)\right)_{ed \in \widetilde{V}} \right\}$$

$$\leq \max\left\{ \left(\gamma_{C_{ef}}(\overline{x}_e - \overline{x}_f)\right)_{ef \in V}, \ \left(\gamma_{\widetilde{C}_{ed}}(\overline{x}_e - p_d)\right)_{ed \in \widetilde{V}} \right\}, \tag{4.109}$$

which implies that

$$\sum_{jk \in \overline{I}} \overline{z}_{jk}^{0*}\left[\max\left\{ \left(\gamma_{C_{ef}}(\overline{x}_e - \overline{x}_f)\right)_{ef \in V}, \ \left(\gamma_{\widetilde{C}_{ed}}(\overline{x}_e - p_d)\right)_{ed \in \widetilde{V}} \right\} - \gamma_{C_{jk}}(\overline{x}_j - \overline{x}_k) \right]$$

$$+ \sum_{ji \in \overline{\overline{I}}} \overline{\overline{z}}_{ji}^{0*}\left[\max\left\{ \left(\gamma_{C_{ef}}(\overline{x}_e - \overline{x}_f)\right)_{ef \in V}, \ \left(\gamma_{\widetilde{C}_{ed}}(\overline{x}_e - p_d)\right)_{ed \in \widetilde{V}} \right\} - \gamma_{\widetilde{C}_{ji}}(\overline{x}_j - p_i) \right] = 0$$

and as $\overline{z}_{jk}^{0*} > 0$, $jk \in \overline{I}$, and $\overline{\overline{z}}_{ji}^{0*} > 0$, $ji \in \overline{\overline{I}}$, it follows that

$$\max\left\{ \left(\gamma_{C_{ef}}(\overline{x}_e - \overline{x}_f)\right)_{ef \in V}, \ \left(\gamma_{\widetilde{C}_{ed}}(\overline{x}_e - p_d)\right)_{ed \in \widetilde{V}} \right\} = \gamma_{C_{jk}}(\overline{x}_j - \overline{x}_k), \ ik \in \overline{I}, \tag{4.110}$$

and

$$\max\left\{ \left(\gamma_{C_{ef}}(\overline{x}_e - \overline{x}_f)\right)_{ef \in V}, \ \left(\gamma_{\widetilde{C}_{ed}}(\overline{x}_e - p_d)\right)_{ed \in \widetilde{V}} \right\} = \gamma_{\widetilde{C}_{ji}}(\overline{x}_j - p_i), \ ji \in \overline{\overline{I}}. \tag{4.111}$$

Furthermore, we get by (4.109) that

$$\sum_{jk\in\overline{I}} \overline{z}_{jk}^{0*} \max\left\{ \left(\gamma_{C_{ef}}(\overline{x}_e - \overline{x}_f)\right)_{ef\in V}, \ \left(\gamma_{\widetilde{C}_{ed}}(\overline{x}_e - p_d)\right)_{ed\in\widetilde{V}} \right\}$$
$$+ \sum_{ji\in\overline{\widetilde{I}}} \overline{\overline{z}}_{ji}^{0*} \max\left\{ \left(\gamma_{C_{ef}}(\overline{x}_e - \overline{x}_f)\right)_{ef\in V}, \ \left(\gamma_{\widetilde{C}_{ed}}(\overline{x}_e - p_d)\right)_{ed\in\widetilde{V}} \right\}$$
$$= \max\left\{ \left(\gamma_{C_{ef}}(\overline{x}_e - \overline{x}_f)\right)_{ef\in V}, \ \left(\gamma_{\widetilde{C}_{ed}}(\overline{x}_e - p_d)\right)_{ed\in\widetilde{V}} \right\},$$

from which follows that

$$\sum_{jk\in\overline{I}} \overline{z}_{jk}^{0*} + \sum_{ji\in\overline{\widetilde{I}}} \overline{\overline{z}}_{ji}^{0*} = 1. \tag{4.112}$$

Combining now the conditions (i)-(vi) with (4.105), (4.106), (4.107), (4.110), (4.111) and (4.112) provides us the desired conclusion.

(b) The calculations made in (a) can also be done in the reverse direction, which completes the proof. $\qquad\square$

Remark 4.44. *We want to point out that the optimality condition (i) of the previous theorem can be expressed by means of the subdifferential. We have*

$$f(y^0, \widetilde{y}^0) = \begin{cases} \max\left\{ \left(y_{jk}^0\right)_{jk\in V}, \ \left(\widetilde{y}_{ji}^0\right)_{ji\in\widetilde{V}} \right\}, & \text{if } (y^0, \widetilde{y}^0) \in \mathbb{R}_+^{|V|} \times \mathbb{R}_+^{|\widetilde{V}|}, \\ +\infty, & \text{otherwise}, \end{cases}$$

and

$$f^*(z^{0*}, \widetilde{z}^{0*}) = \begin{cases} 0, & \text{if } \sum_{jk\in V} z_{jk}^{0*} + \sum_{ji\in\widetilde{V}} \widetilde{z}_{ji}^{0*} \le 1, \ z^{0*} \in \mathbb{R}_+^{|V|}, \ \widetilde{z}^{0*} \in \mathbb{R}_+^{|\widetilde{V}|}, \\ +\infty, & \text{otherwise}, \end{cases}$$

and by the optimality condition (i) of the previous theorem, it holds

$$f\left(\left(\gamma_{C_{ef}}(\overline{x}_e - \overline{x}_f)\right)_{ef\in V}, \left(\gamma_{\widetilde{C}_{ed}}(\overline{x}_e - p_d)\right)_{ed\in\widetilde{V}} \right) + f^*(\overline{z}^{0*}, \overline{\overline{z}}^{0*})$$
$$= \sum_{jk\in\overline{I}} \overline{z}_{jk}^{0*} \gamma_{C_{jk}}(\overline{x}_j - \overline{x}_k) + \sum_{ji\in\overline{\widetilde{I}}} \overline{\overline{z}}_{ji}^{0*} \gamma_{\widetilde{C}_{ji}}(\overline{x}_j - p_i),$$

in other words, the optimality condition (i) can be rewritten as

(i) $(\overline{z}^{0*}, \overline{\overline{z}}^{0*}) \in \partial f\left(\left(\gamma_{C_{ef}}(\overline{x}_e - \overline{x}_f)\right)_{ef\in V}, \left(\gamma_{\widetilde{C}_{ed}}(\overline{x}_e - p_d)\right)_{ed\in\widetilde{V}} \right).$

Moreover, for the optimality conditions (ii), (iv) and (v) we get by analogous considerations

(ii) $-\sum_{jk\in\overline{I}} A_{jk}^* \overline{z}_{jk}^{1*} - \sum_{ji\in\overline{\widetilde{I}}} B_{ji}^* \overline{\overline{z}}_{ji}^{1*} \in \partial\delta_S(\overline{x}) = N_S(\overline{x}),$

(iv) $\overline{z}_{jk}^{1*} \in \partial(\overline{z}_{jk}^{0*}\gamma_{C_{jk}})(\overline{x}_j - \overline{x}_k) = \partial(\overline{z}_{jk}^{0*}\gamma_{C_{jk}})(A_{jk}\overline{x})$
$\quad \Leftrightarrow A_{jk}^*\overline{z}_{jk}^{1*} \in A_{jk}^*\partial((\overline{z}_{jk}^{0*}\gamma_{C_{jk}}) \circ A_{jk})(\overline{x}), \; jk \in \overline{I},$

(v) $\overline{\overline{z}}_{ji}^{1*} \in \partial(\overline{\overline{z}}_{ji}^{0*}\gamma_{\widetilde{C}_{ji}})(\overline{x}_j - p_i) = \partial(\overline{\overline{z}}_{ji}^{0*}\gamma_{\widetilde{C}_{ji}})(B_{ji}\overline{x} - p_i)$
$\quad \Leftrightarrow B_{ji}^*\overline{\overline{z}}_{ji}^{1*} \in B_{ji}^*\partial\left(((\overline{\overline{z}}_{ji}^{0*}\gamma_{\widetilde{C}_{ji}}) \circ B_{ji})(\cdot - p_i)\right)(\overline{x}), \; ji \in \overline{\overline{I}}.$

Taking (ii), (iv) *and* (v) *together implies that*

$$\sum_{jk\in\overline{I}} A_{jk}^*\overline{z}_{jk}^{1*} + \sum_{ji\in\overline{\overline{I}}} B_{ji}^*\overline{\overline{z}}_{ji}^{1*} \in$$

$$\left(\sum_{jk\in\overline{I}} A_{jk}^*\partial((\overline{z}_{jk}^{0*}\gamma_{C_{jk}}) \circ A_{jk})(\overline{x}) + \sum_{ji\in\overline{\overline{I}}} B_{ji}^*\partial\left(((\overline{\overline{z}}_{ji}^{0*}\gamma_{\widetilde{C}_{ji}}) \circ B_{ji})(\cdot - p_i)\right)(\overline{x})\right) \bigcap(-N_S(\overline{x})).$$

Finally, notice that the optimality conditions (iv), (v), (viii) *and* (ix) *of the previous theorem give a detailed characterization of the subdifferentials of the associated gauges.*

Now, we show that the dual problem (D^M) is equivalent to the problem

$$(\widetilde{D}^M) \quad \sup_{(z^*,\widetilde{z}^*)\in\widetilde{\mathcal{B}}} \left\{ -\sigma_S\left(-\sum_{jk\in I} A_{jk}^*z_{jk}^* - \sum_{ji\in\widetilde{I}} B_{ji}^*\widetilde{z}_{ji}^*\right) - \sum_{ji\in\widetilde{I}}\langle\widetilde{z}_{ji}^*, p_i\rangle \right\}, \qquad (4.113)$$

where $(z^*, \widetilde{z}^*) = \left((z_{jk}^*)_{jk\in V}, (\widetilde{z}_{ji}^*)_{ji\in\widetilde{V}}\right)$ and

$$\widetilde{\mathcal{B}} = \left\{ \left((z_{jk}^*)_{jk\in V}, (\widetilde{z}_{ji}^*)_{ji\in\widetilde{V}}\right) \in (X^*)^{|V|} \times (X^*)^{|\widetilde{V}|} : I \subseteq V, \; \widetilde{I} \subseteq \widetilde{V}, \right.$$

$$\sum_{jk\in I}\gamma_{C_{jk}^0}(z_{jk}^*) + \sum_{ji\in\widetilde{I}}\gamma_{\widetilde{C}_{ji}^0}(\widetilde{z}_{ji}^*) \leq 1,$$

$$\left. z_{jk}^* \in X^*, \; jk \in I, \; \widetilde{z}_{ji}^* \in X^*, \; ji \in \widetilde{I} \text{ and } z_{ef}^* = 0_{X^*}, \; ef \in V\setminus I, \; \widetilde{z}_{ed}^* = 0_{X^*}, \; ed \in \widetilde{V}\setminus\widetilde{I} \right\},$$

in the sense of the next theorem, where $v(\widetilde{D}^M)$ denotes the optimal objective value of the problem $(\widetilde{D}^M)$.

Theorem 4.31. *It holds* $v(D^M) = v(\widetilde{D}^M)$.

Proof. Let $(z^*, \widetilde{z}^*)$ be a feasible element to $(\widetilde{D}^M)$ and set

$$z_{jk}^{1*} = z_{jk}^*, \; z_{jk}^{0*} = \gamma_{C_{jk}^0}(z_{jk}^*) \text{ for } jk \in I, \; z_{ef}^{1*} = 0_{X^*}, \; z_{ef}^{0*} = 0 \text{ for } ef \in V\setminus I,$$

and

$$\widetilde{z}_{ji}^{1*} = \widetilde{z}_{ji}^*, \ \widetilde{z}_{ji}^{0*} = \gamma_{\widetilde{C}_{ji}^0}(\widetilde{z}_{ji}^*) \text{ for } ji \in \widetilde{I}, \ \widetilde{z}_{ed}^{1*} = 0_{X^*}, \ \widetilde{z}_{ed}^{0*} = 0 \text{ for } ed \in \widetilde{V} \setminus \widetilde{I}.$$

Then, it is clear that $(z^{0*}, \widetilde{z}^{0*}, z^{1*}, \widetilde{z}^{1*})$ is a feasible element to (D^M). Furthermore, it holds

$$- \sigma_S \left(- \sum_{jk \in I} A_{jk}^* z_{jk}^* - \sum_{ji \in \widetilde{I}} B_{ji}^* \widetilde{z}_{ji}^* \right) - \sum_{ji \in \widetilde{I}} \langle \widetilde{z}_{ji}^*, p_i \rangle$$

$$= - \sigma_S \left(- \sum_{jk \in I} A_{jk}^* z_{jk}^{1*} - \sum_{ji \in \widetilde{I}} B_{ji}^* \widetilde{z}_{ji}^{1*} \right) - \sum_{ji \in \widetilde{I}} \langle \widetilde{z}_{ji}^{1*}, p_i \rangle \leq v(D^M),$$

for all $(z^*, \widetilde{z}^*)$ feasible to $(\widetilde{D}^M)$, from which follows that $v(\widetilde{D}^M) \leq v(D^M)$.

Now, let $(z^{0*}, \widetilde{z}^{0*}, z^{1*}, \widetilde{z}^{1*})$ be a feasible element to (D^M). By a careful look at the constraint set $\mathcal{B}$ we get by setting $z_{jk}^* = z_{jk}^{1*}$ for $jk \in I$, $\widetilde{z}_{ji}^* = \widetilde{z}_{ji}^{1*}$ for $ji \in \widetilde{I}$ and $z_{ef}^* = 0_{X^*}$ for $ef \in V \setminus I$, $\widetilde{z}_{ed}^* = 0_{X^*}$ for $ed \in \widetilde{V} \setminus \widetilde{I}$ that

$$\sum_{jk \in I} \gamma_{C_{jk}^0}(z_{jk}^*) + \sum_{ji \in \widetilde{I}} \gamma_{\widetilde{C}_{ji}^0}(\widetilde{z}_{ji}^*) \leq 1.$$

Therefore, $(z^*, \widetilde{z}^*)$ is feasible to $(\widetilde{D}^M)$ and we have

$$- \sigma_S \left(- \sum_{jk \in I} A_{jk}^* z_{jk}^{1*} - \sum_{ji \in \widetilde{I}} B_{ji}^* \widetilde{z}_{ji}^{1*} \right) - \sum_{ji \in \widetilde{I}} \langle \widetilde{z}_{ji}^{1*}, p_i \rangle$$

$$= - \sigma_S \left(- \sum_{jk \in I} A_{jk}^* z_{jk}^* - \sum_{ji \in \widetilde{I}} B_{ji}^* \widetilde{z}_{ji}^* \right) - \sum_{ji \in \widetilde{I}} \langle \widetilde{z}_{ji}^*, p_i \rangle \leq v(\widetilde{D}^M),$$

for all $(z^{0*}, \widetilde{z}^{0*}, z^{1*}, \widetilde{z}^{1*})$ feasible to (D^M), i.e. $v(D^M) \leq v(\widetilde{D}^M)$, which completes the proof. $\qquad \square$

The next two theorems are direct consequences of Theorem 4.31.

Theorem 4.32. *(strong duality) Between (P^M) and $(\widetilde{D}^M)$ holds strong duality, i.e. $v(P^M) = v(\widetilde{D}^M)$ and the dual problem has an optimal solution.*

Theorem 4.33. *(optimality conditions) (a) Let $\overline{x} \in S$ be an optimal solution to the problem (P^M). Then there exists an optimal solution to $(\widetilde{D}^M)$ $(\overline{z}^*, \overline{\widetilde{z}}^*) \in (X^*)^{|V|} \times (X^*)^{|\widetilde{V}|}$ with the corresponding index sets $\overline{I} \subseteq V$ and $\overline{\widetilde{I}} \subseteq \widetilde{V}$ such that*

$$(i) \ \max \left\{ \left(\gamma_{C_{ef}}(\overline{x}_e - \overline{x}_f) \right)_{ef \in V}, \ \left(\gamma_{\widetilde{C}_{ed}}(\overline{x}_e - p_d) \right)_{ed \in \widetilde{V}} \right\}$$

$$= \sum_{jk \in \overline{I}} \gamma_{C_{jk}^0}(\overline{z}_{jk}^*) \gamma_{C_{jk}}(\overline{x}_j - \overline{x}_k) + \sum_{ji \in \overline{\widetilde{I}}} \gamma_{\widetilde{C}_{ji}^0}(\overline{\widetilde{z}}_{ji}^*) \gamma_{\widetilde{C}_{ji}}(\overline{x}_j - p_i),$$

(ii) $\left\langle \sum\limits_{jk\in\overline{I}} A_{jk}^* \overline{z}_{jk}^* + \sum\limits_{ji\in\widetilde{\overline{I}}} B_{ji}^* \overline{\overline{z}}_{ji}^*, \overline{x} \right\rangle = -\sigma_S \left(\sum\limits_{jk\in\overline{I}} A_{jk}^* \overline{z}_{jk}^* + \sum\limits_{ji\in\widetilde{\overline{I}}} B_{ji}^* \overline{\overline{z}}_{ji}^* \right),$

(iii) $\gamma_{C_{jk}^0}(\overline{z}_{jk}^*)\gamma_{C_{jk}}(\overline{x}_j - \overline{x}_k) = \langle \overline{z}_{jk}^*, \overline{x}_j - \overline{x}_k \rangle, \; jk \in \overline{I},$

(iv) $\gamma_{\widetilde{C}_{ji}^0}(\overline{\overline{z}}_{ji}^*)\gamma_{\widetilde{C}_{ji}}(\overline{x}_j - p_i) = \langle \overline{\overline{z}}_{ji}^*, \overline{x}_j - p_i \rangle, \; ji \in \widetilde{\overline{I}},$

(v) $\max\left\{ \left(\gamma_{C_{ef}}(\overline{x}_e - \overline{x}_f)\right)_{ef\in V}, \; \left(\gamma_{\widetilde{C}_{ed}}(\overline{x}_e - p_d)\right)_{ed\in\widetilde{V}} \right\} = \gamma_{C_{jk}}(\overline{x}_j - \overline{x}_k), \; jk \in \overline{I},$

(vi) $\max\left\{ \left(\gamma_{C_{ef}}(\overline{x}_e - \overline{x}_f)\right)_{ef\in V}, \; \left(\gamma_{\widetilde{C}_{ed}}(\overline{x}_e - p_d)\right)_{ed\in\widetilde{V}} \right\} = \gamma_{\widetilde{C}_{ji}}(\overline{x}_j - p_i), \; ji \in \widetilde{\overline{I}},$

(vii) $\sum\limits_{jk\in\overline{I}} \gamma_{C_{jk}^0}(\overline{z}_{jk}^*) + \sum\limits_{ji\in\widetilde{\overline{I}}} \gamma_{\widetilde{C}_{ji}^0}(\overline{\overline{z}}_{ji}^*) = 1, \; \gamma_{C_{jk}^0}(\overline{z}_{jk}^*) > 0, \; jk \in \overline{I}, \; \gamma_{\widetilde{C}_{ji}^0}(\overline{\overline{z}}_{ji}^*) > 0, \; ji \in \widetilde{\overline{I}}, \; and$

$\overline{z}_{ef}^* = 0_{X^*}, \; ef \in V \setminus \overline{I}, \; \overline{\overline{z}}_{ed}^* = 0_{X^*}, \; ed \in \widetilde{V} \setminus \widetilde{\overline{I}}.$

(b) If there exists $\overline{x} \in S$ such that for some $(\overline{z}^*, \overline{\overline{z}}^*)$ and the corresponding index sets $\overline{I}$ and $\widetilde{\overline{I}}$ the conditions (i)-(vii) are fulfilled, then $\overline{x}$ is an optimal solution to (P^M), $(\overline{z}^*, \overline{\overline{z}}^*)$ is an optimal solution to $(\widetilde{D}^M)$ and $v(P^M) = v(\widetilde{D}^M)$.

Proof. (a) Theorem 4.32 implies for an optimal solution $\overline{x} \in S$ of (P^M) the existence of $(\overline{z}^*, \overline{\overline{z}}^*) \in (X^*)^{|V|} \times (X^*)^{|\widetilde{V}|}$ and index sets $\overline{I}$ and $\widetilde{\overline{I}}$, an optimal solution to $(\widetilde{D}^M)$, such that $v(P^M) = v(\widetilde{D}^M)$, i.e.

$$\max\left\{ \left(\gamma_{C_{ef}}(\overline{x}_e - \overline{x}_f)\right)_{ef\in V}, \; \left(\gamma_{\widetilde{C}_{ed}}(\overline{x}_e - p_d)\right)_{ed\in\widetilde{V}} \right\}$$

$$= -\sigma_S \left(-\sum\limits_{jk\in\overline{I}} A_{jk}^* \overline{z}_{jk}^* - \sum\limits_{ji\in\widetilde{\overline{I}}} B_{ji}^* \overline{\overline{z}}_{ji}^* \right) - \sum\limits_{ji\in\widetilde{\overline{I}}} \langle \overline{\overline{z}}_{ji}^*, p_i \rangle$$

$$\Leftrightarrow \max\left\{ \left(\gamma_{C_{ef}}(\overline{x}_e - \overline{x}_f)\right)_{ef\in V}, \; \left(\gamma_{\widetilde{C}_{ed}}(\overline{x}_e - p_d)\right)_{ed\in\widetilde{V}} \right\}$$

$$+ \sigma_S \left(-\sum\limits_{jk\in\overline{I}} A_{jk}^* \overline{z}_{jk}^* - \sum\limits_{ji\in\widetilde{\overline{I}}} B_{ji}^* \overline{\overline{z}}_{ji}^* \right) + \sum\limits_{ji\in\widetilde{\overline{I}}} \langle \overline{\overline{z}}_{ji}^*, p_i \rangle = 0$$

$$\Leftrightarrow \max\left\{ \left(\gamma_{C_{ef}}(\overline{x}_e - \overline{x}_f)\right)_{ef\in V}, \ \left(\gamma_{\widetilde{C}_{ed}}(\overline{x}_e - p_d)\right)_{ed\in\widetilde{V}} \right\}$$

$$+ \sigma_S\left(-\sum_{jk\in\overline{I}} A_{jk}^*\overline{z}_{jk}^* - \sum_{ji\in\overline{\overline{I}}} B_{ji}^*\overline{\overline{z}}_{ji}^* \right) + \sum_{ji\in\overline{\overline{I}}} \langle \overline{\overline{z}}_{ji}^*, p_i \rangle$$

$$+ \sum_{jk\in\overline{I}} [\gamma_{C_{jk}^0}(\overline{z}_{jk}^*)\gamma_{C_{jk}}(\overline{x}_j - \overline{x}_k) - \langle \overline{z}_{jk}^*, \overline{x}_j - \overline{x}_k \rangle]$$

$$- \sum_{jk\in\overline{I}} [\gamma_{C_{jk}^0}(\overline{z}_{jk}^*)\gamma_{C_{jk}}(\overline{x}_j - \overline{x}_k) - \langle \overline{z}_{jk}^*, \overline{x}_j - \overline{x}_k \rangle]$$

$$+ \sum_{ji\in\overline{\overline{I}}} [\gamma_{\widetilde{C}_{ji}^0}(\overline{\overline{z}}_{ji}^*)\gamma_{\widetilde{C}_{ji}}(\overline{x}_j - p_i) - \langle \overline{\overline{z}}_{ji}^*, \overline{x}_j \rangle] - \sum_{ji\in\overline{\overline{I}}} [\gamma_{\widetilde{C}_{ji}^0}(\overline{\overline{z}}_{ji}^*)\gamma_{\widetilde{C}_{ji}}(\overline{x}_j - p_i) - \langle \overline{\overline{z}}_{ji}^*, \overline{x}_j \rangle] = 0$$

$$\Leftrightarrow \left[\max\left\{ \left(\gamma_{C_{ef}}(\overline{x}_e - \overline{x}_f)\right)_{ef\in V}, \ \left(\gamma_{\widetilde{C}_{ed}}(\overline{x}_e - p_d)\right)_{ed\in\widetilde{V}} \right\} \right.$$

$$\left. - \sum_{jk\in\overline{I}} \gamma_{C_{jk}^0}(\overline{z}_{jk}^*)\gamma_{C_{jk}}(\overline{x}_j - \overline{x}_k) - \sum_{ji\in\overline{\overline{I}}} \gamma_{\widetilde{C}_{ji}^0}(\overline{\overline{z}}_{ji}^*)\gamma_{\widetilde{C}_{ji}}(\overline{x}_j - p_i) \right]$$

$$+ \sum_{jk\in\overline{I}} [\gamma_{C_{jk}^0}(\overline{z}_{jk}^*)\gamma_{C_{jk}}(\overline{x}_j - \overline{x}_k) - \langle \overline{z}_{jk}^*, \overline{x}_j - \overline{x}_k \rangle]$$

$$+ \sum_{ji\in\overline{\overline{I}}} [\gamma_{\widetilde{C}_{ji}^0}(\overline{\overline{z}}_{ji}^*)\gamma_{\widetilde{C}_{ji}}(\overline{x}_j - p_i) - \langle \overline{\overline{z}}_{ji}^*, \overline{x}_j - p_i \rangle]$$

$$+ \left[\sigma_S\left(-\sum_{jk\in\overline{I}} A_{jk}^*\overline{z}_{jk}^* - \sum_{ji\in\overline{\overline{I}}} B_{ji}^*\overline{\overline{z}}_{ji}^* \right) + \langle \overline{z}_{jk}^*, \overline{x}_j - \overline{x}_k \rangle + \langle \overline{\overline{z}}_{ji}^*, \overline{x}_j \rangle \right] = 0$$

$$\Leftrightarrow \left[\max\left\{ \left(\gamma_{C_{ef}}(\overline{x}_e - \overline{x}_f)\right)_{ef\in V}, \ \left(\gamma_{\widetilde{C}_{ed}}(\overline{x}_e - p_d)\right)_{ed\in\widetilde{V}} \right\} \right.$$

$$\left. - \sum_{jk\in\overline{I}} \gamma_{C_{jk}^0}(\overline{z}_{jk}^*)w_{jk}\gamma_{C_{jk}}(\overline{x}_j - \overline{x}_k) - \sum_{ji\in\overline{\overline{I}}} \gamma_{\widetilde{C}_{ji}^0}(\overline{\overline{z}}_{ji}^*)\widetilde{w}_{ji}\gamma_{\widetilde{C}_{ji}}(\overline{x}_j - p_i) \right]$$

$$+ \sum_{jk\in\overline{I}} [\gamma_{C_{jk}^0}(\overline{z}_{jk}^*)\gamma_{C_{jk}}(\overline{x}_j - \overline{x}_k) - \langle \overline{z}_{jk}^*, \overline{x}_j - \overline{x}_k \rangle]$$

$$+ \sum_{ji\in\overline{\overline{I}}} [\gamma_{\widetilde{C}_{ji}^0}(\overline{\overline{z}}_{ji}^*)\gamma_{\widetilde{C}_{ji}}(\overline{x}_j - p_i) - \langle \overline{\overline{z}}_{ji}^*, \overline{x}_j - p_i \rangle]$$

$$+ \left[\sigma_S\left(-\sum_{jk\in\overline{I}} A_{jk}^*\overline{z}_{jk}^* - \sum_{ji\in\overline{\overline{I}}} B_{ji}^*\overline{\overline{z}}_{ji}^* \right) + \langle A_{jk}^*\overline{z}_{jk}^*, \overline{x} \rangle + \langle B_{ji}^*\overline{\overline{z}}_{ji}^*, \overline{x} \rangle \right] = 0.$$

Lemma 4.2 implies that the first bracket is non-negative, from the generalized Cauchy-Schwarz inequality (see Lemma 4.3) follows that the brackets in the two sums are non-negative and from the Young-Fenchel inequality we get that the last bracket is also non-negative. Hence, the statements (i)-(iv) are proved. Now, we take a careful look at the first bracket

$$\max\left\{\left(\gamma_{C_{ef}}(\overline{x}_e - \overline{x}_f)\right)_{ef\in V}, \left(\gamma_{\widetilde{C}_{ed}}(\overline{x}_e - p_d)\right)_{ed\in\widetilde{V}}\right\}$$

$$= \sum_{jk\in\overline{I}}\gamma_{C_{jk}^0}(\overline{z}_{jk}^*)w_{jk}\gamma_{C_{jk}}(\overline{x}_j - \overline{x}_k) + \sum_{ji\in\widetilde{\overline{I}}}\gamma_{\widetilde{C}_{ji}^0}(\overline{\overline{z}}_{ji}^*)\widetilde{w}_{ji}\gamma_{\widetilde{C}_{ji}}(\overline{x}_j - p_i)$$

$$\leq \sum_{jk\in\overline{I}}\gamma_{C_{jk}^0}(\overline{z}_{jk}^*)\max\left\{\left(\gamma_{C_{ef}}(\overline{x}_e - \overline{x}_f)\right)_{ef\in V}, \left(\gamma_{\widetilde{C}_{ed}}(\overline{x}_e - p_d)\right)_{ed\in\widetilde{V}}\right\}$$

$$+ \sum_{ji\in\widetilde{\overline{I}}}\gamma_{\widetilde{C}_{ji}^0}(\overline{\overline{z}}_{ji}^*)\max\left\{\left(\gamma_{C_{ef}}(\overline{x}_e - \overline{x}_f)\right)_{ef\in V}, \left(\gamma_{\widetilde{C}_{ed}}(\overline{x}_e - p_d)\right)_{ed\in\widetilde{V}}\right\}$$

$$\leq \max\left\{\left(\gamma_{C_{ef}}(\overline{x}_e - \overline{x}_f)\right)_{ef\in V}, \left(\gamma_{\widetilde{C}_{ed}}(\overline{x}_e - p_d)\right)_{ed\in\widetilde{V}}\right\},$$

from which follows on the one hand that

$$\sum_{jk\in I}\gamma_{C_{jk}^0}(z_{jk}^*) + \sum_{ji\in\widetilde{I}}\gamma_{\widetilde{C}_{ji}^0}(\widetilde{z}_{ji}^*) = 1,$$

i.e. condition (vii), and on the other hand that

$$\sum_{jk\in\overline{I}}\gamma_{C_{jk}^0}(\overline{z}_{jk}^*)\max\left\{\left(\gamma_{C_{ef}}(\overline{x}_e - \overline{x}_f)\right)_{ef\in V}, \left(\gamma_{\widetilde{C}_{ed}}(\overline{x}_e - p_d)\right)_{ed\in\widetilde{V}}\right\}$$

$$+ \sum_{ji\in\widetilde{\overline{I}}}\gamma_{\widetilde{C}_{ji}^0}(\overline{\overline{z}}_{ji}^*)\max\left\{\left(\gamma_{C_{ef}}(\overline{x}_e - \overline{x}_f)\right)_{ef\in V}, \left(\gamma_{\widetilde{C}_{ed}}(\overline{x}_e - p_d)\right)_{ed\in\widetilde{V}}\right\}$$

$$= \sum_{jk\in\overline{I}}\gamma_{C_{jk}^0}(\overline{z}_{jk}^*)\gamma_{C_{jk}}(\overline{x}_j - \overline{x}_k) + \sum_{ji\in\widetilde{\overline{I}}}\gamma_{\widetilde{C}_{ji}^0}(\overline{\overline{z}}_{ji}^*)\gamma_{\widetilde{C}_{ji}}(\overline{x}_j - p_i)$$

$$\Leftrightarrow \sum_{jk\in\overline{I}}\gamma_{C_{jk}^0}(\overline{z}_{jk}^*)\left[\max\left\{\left(\gamma_{C_{ef}}(\overline{x}_e - \overline{x}_f)\right)_{ef\in V}, \left(\gamma_{\widetilde{C}_{ed}}(\overline{x}_e - p_d)\right)_{ed\in\widetilde{V}}\right\} - \gamma_{C_{jk}}(\overline{x}_j - \overline{x}_k)\right]$$

$$+ \sum_{ji\in\widetilde{\overline{I}}}\gamma_{\widetilde{C}_{ji}^0}(\overline{\overline{z}}_{ji}^*)\left[\max\left\{\left(\gamma_{C_{ef}}(\overline{x}_e - \overline{x}_f)\right)_{ef\in V}, \left(\gamma_{\widetilde{C}_{ed}}(\overline{x}_e - p_d)\right)_{ed\in\widetilde{V}}\right\} - \gamma_{\widetilde{C}_{ji}}(\overline{x}_j - p_i)\right]$$

$$= 0.$$

As $\gamma_{C_{jk}^0}(\overline{z}_{jk}^*) > 0$, $jk \in \overline{I}$, as well as $\gamma_{\widetilde{C}_{ji}^0}(\overline{\overline{z}}_{ji}^*) > 0$, $ji \in \widetilde{\overline{I}}$, we obtain that the brackets are non-negative and must therefore be equal to zero, which finally yields the conditions (v) and (vi).

(b) All calculations done within part (a) can also be made in the reverse direction. $\qquad\square$

Remark 4.45. *Analogously to Remark 4.44 one can determine equivalent formulations of the optimality conditions given in Theorem 4.33 by using the subdifferential, which look like as follows*

(i) $\left(\left(\gamma_{C_{jk}^0}(\overline{z}_{jk}^*) \right)_{jk\in V}, \left(\gamma_{\widetilde{C}_{ji}^0}(\overline{\widetilde{z}}_{ji}^*) \right)_{ji\in\widetilde{V}} \right) \in \partial f \left((\gamma_{C_{ef}}(\overline{x}_e - \overline{x}_f))_{ef\in V}, (\gamma_{\widetilde{C}_{ed}}(\overline{x}_e - p_d))_{ed\in\widetilde{V}} \right),$

(ii) $-\sum\limits_{jk\in\overline{I}} A_{jk}^* \overline{z}_{jk}^* - \sum\limits_{ji\in\overline{\widetilde{I}}} B_{ji}^* \overline{\widetilde{z}}_{ji}^* \in \partial\delta_S(\overline{x}) = N_S(\overline{x}),$

(iii) $\overline{z}_{jk}^* \in \partial\left(\gamma_{C_{jk}^0}(\overline{z}_{jk}^*)\gamma_{C_{jk}} \right)(\overline{x}_j - \overline{x}_k) = \partial\left(\gamma_{C_{jk}^0}(\overline{z}_{jk}^*)\gamma_{C_{jk}} \right)(A_{jk}\overline{x})$

$\Leftrightarrow A_{jk}^* \overline{z}_{jk}^* \in A_{jk}^* \partial\left(\gamma_{C_{jk}^0}(\overline{z}_{jk}^*)\gamma_{C_{jk}} \circ A_{jk} \right)(\overline{x}), \ jk\in\overline{I},$

(iv) $\overline{\widetilde{z}}_{ji}^* \in \partial\left(\gamma_{\widetilde{C}_{ji}^0}(\overline{\widetilde{z}}_{ji}^*)\gamma_{\widetilde{C}_{ji}} \right)(\overline{x}_j - p_i) = \partial\left(\gamma_{\widetilde{C}_{ji}^0}(\overline{\widetilde{z}}_{ji}^*)\gamma_{\widetilde{C}_{ji}} \right)(B_{ji}\overline{x} - p_i)$

$\Leftrightarrow B_{ji}^* \overline{\widetilde{z}}_{ji}^* \in B_{ji}^* \partial\left(\left(\gamma_{\widetilde{C}_{ji}^0}(\overline{\widetilde{z}}_{ji}^*)\gamma_{\widetilde{C}_{ji}} \circ B_{ji} \right)(\cdot - p_i) \right)(\overline{x}), \ ji\in\overline{\widetilde{I}}.$

Combining (ii) − (iv) yields

$$\sum\limits_{jk\in\overline{I}} A_{jk}^* \overline{z}_{jk}^* + \sum\limits_{ji\in\overline{\widetilde{I}}} B_{ji}^* \overline{\widetilde{z}}_{ji}^* \in$$

$$\left(\sum\limits_{jk\in\overline{I}} A_{jk}^* \partial\left(\gamma_{C_{jk}^0}(\overline{z}_{jk}^*)\gamma_{C_{jk}} \circ A_{jk} \right)(\overline{x}) + \sum\limits_{ji\in\overline{\widetilde{I}}} B_{ji}^* \partial\left(\left(\gamma_{\widetilde{C}_{ji}^0}(\overline{\widetilde{z}}_{ji}^*)\gamma_{\widetilde{C}_{ji}} \circ B_{ji} \right)(\cdot - p_i) \right)(\overline{x}) \right)$$

$$\bigcap (-N_S(\overline{x})), \qquad\qquad (4.114)$$

where the optimality conditions (iii) and (iv) of Theorem 4.33 give a detailed characterization of the subdifferentials of the gauges involved in 4.114.

4.5.2 Unconstrained multifacility minmax location problem in the Euclidean space

In this section we are interested in a detailed analysis of the situation when $S = X^m$ and $X = \mathbb{R}^d$ and the gauges are defined by the weighted Euclidean norm, while the sets V and $\widetilde{V}$ are given as in Remark 6.3, i.e. $V = \{jk \in J : 1 \leq j < k \leq m, \ w_{jk} > 0\}$ and $\widetilde{V} = \{ji \in J : w_{ji} > 0\}$. In other words, we will explore in the following the location problem

$$(P_N^M) \quad \inf_{x_i\in\mathbb{R}^d, \ i=1,...,m} \max\left\{ (w_{jk}\|x_j - x_k\|)_{jk\in V}, \ (\widetilde{w}_{ji}\|x_j - p_i\|)_{ji\in\widetilde{V}} \right\}. \qquad (4.115)$$

For the dual of the location problem (P_N^M) we get by (4.113)

$$(\widetilde{D}_N^M) \quad \sup_{(z^*,\widetilde{z}^*)\in\widetilde{\mathcal{B}}_N} \left\{ -\sum\limits_{ji\in\widetilde{I}} \langle\overline{\widetilde{z}}_{ji}^*, p_i\rangle \right\}, \qquad (4.116)$$

where

$$\widetilde{\mathcal{B}}_N = \left\{ (z^*, \widetilde{z}^*) = \left((z^*_{jk})_{jk \in V}, (\widetilde{z}^*_{ji})_{ji \in \widetilde{V}} \right) \in (\mathbb{R}^d)^{|V|} \times (\mathbb{R}^d)^{|\widetilde{V}|} : I \subseteq V, \ \widetilde{I} \subseteq \widetilde{V}, \right.$$

$$z^*_{jk} \in \mathbb{R}^d, \ jk \in I, \ \widetilde{z}^*_{ji} \in \mathbb{R}^d, \ ji \in \widetilde{I}, \ \sum_{jk \in I} \frac{1}{w_{jk}} \|z^*_{jk}\| + \sum_{ji \in \widetilde{I}} \frac{1}{\widetilde{w}_{ji}} \|\widetilde{z}^*_{ji}\| \le 1,$$

$$\left. \sum_{jk \in I} A^*_{jk} z^*_{jk} + \sum_{ji \in \widetilde{I}} B^*_{ji} \widetilde{z}^*_{ji} = 0_{\underbrace{\mathbb{R}^d \times \ldots \times \mathbb{R}^d}_{m-\text{times}}}, \ z^*_{ef} = 0_{\mathbb{R}^d}, \ ef \in V \setminus I, \ \widetilde{z}^*_{ed} = 0_{\mathbb{R}^d}, \ ed \in \widetilde{V} \setminus \widetilde{I} \right\}.$$

The next theorems are direct consequences of the results of the previous section.

Theorem 4.34. *(strong duality) Between* (P^M_N) *and* $(\widetilde{D}^M_N)$ *strong duality holds, i.e.* $v(P^M_N) = v(\widetilde{D}^M_N)$ *and the dual problem has an optimal solution.*

Theorem 4.35. *(optimality conditions) (a) Let* $(\overline{x}_1, ..., \overline{x}_m)$ *be an optimal solution to the problem* (P^M_N). *Then there exists an optimal solution to* $(\widetilde{D}^M_N)$ $(\overline{z}^*, \overline{\widetilde{z}}^*)$ *with the corresponding index sets* $\overline{I} \subseteq V$ *and* $\overline{\widetilde{I}} \subseteq \widetilde{V}$ *such that*

(i) $\max \left\{ (w_{jk} \|x_j - x_k\|)_{jk \in V}, \ (\widetilde{w}_{ji} \|x_j - p_i\|)_{ji \in \widetilde{V}} \right\}$
$= \sum_{jk \in \overline{I}} \|\overline{z}^*_{jk}\| \|\overline{x}_j - \overline{x}_k\| + \sum_{ji \in \overline{\widetilde{I}}} \|\overline{\widetilde{z}}^*_{ji}\| \|\overline{x}_j - p_i\|,$

(ii) $\sum_{jk \in I} A^*_{jk} z^*_{jk} + \sum_{ji \in \widetilde{I}} B^*_{ji} \widetilde{z}^*_{ji} = 0_{\mathbb{R}^d \times \ldots \times \mathbb{R}^d},$

(iii) $\|\overline{z}^*_{jk}\| \|\overline{x}_j - \overline{x}_k\| = \langle \overline{z}^*_{jk}, \overline{x}_j - \overline{x}_k \rangle, \ jk \in \overline{I},$

(iv) $\|\overline{\widetilde{z}}^*_{ji}\| \|\overline{x}_j - p_i\| = \langle \overline{\widetilde{z}}^*_{ji}, \overline{x}_j - p_i \rangle, \ ji \in \overline{\widetilde{I}},$

(v) $\max \left\{ (w_{jk} \|x_j - x_k\|)_{jk \in V}, \ (\widetilde{w}_{ji} \|x_j - p_i\|)_{ji \in \widetilde{V}} \right\} = w_{jk} \|\overline{x}_j - \overline{x}_k\|, \ jk \in \overline{I},$

(vi) $\max \left\{ (w_{jk} \|x_j - x_k\|)_{jk \in V}, \ (\widetilde{w}_{ji} \|x_j - p_i\|)_{ji \in \widetilde{V}} \right\} = \widetilde{w}_{ji} \|\overline{x}_j - p_i\|, \ ji \in \overline{\widetilde{I}},$

(vii) $\sum_{jk \in I} \frac{1}{w_{jk}} \|\overline{z}^*_{jk}\| + \sum_{ji \in \widetilde{I}} \frac{1}{\widetilde{w}_{ji}} \|\overline{\widetilde{z}}^*_{ji}\| = 1, \ \overline{z}^*_{jk} \in \mathbb{R}^d \setminus \{0_{\mathbb{R}^d}\}$ *for* $jk \in \overline{I}, \overline{\widetilde{z}}^*_{ji} \in \mathbb{R}^d \setminus \{0_{\mathbb{R}^d}\}$ *for*

$ji \in \overline{\widetilde{I}}$ *and* $\overline{z}^*_{jk} = 0_{\mathbb{R}^d}$ *for* $jk \in V \setminus \overline{I}, \ \overline{\widetilde{z}}^*_{ji} = 0_{\mathbb{R}^d}$ *for* $ji \in \widetilde{V} \setminus \overline{\widetilde{I}}.$

(b) If there exists $(\overline{x}_1, ..., \overline{x}_m)$ *such that for some* $(\overline{z}^*, \overline{\widetilde{z}}^*)$ *and the corresponding index sets* $\overline{I}$ *and* $\overline{\widetilde{I}}$ *the conditions (i)-(vii) are fulfilled, then* $\overline{x}$ *is an optimal solution to* (P^M_N), $(\overline{z}^*, \overline{\widetilde{z}}^*)$ *is an optimal solution to* $(\widetilde{D}^M_N)$ *and* $v(P^M_N) = v(\widetilde{D}^M_N).$

Remark 4.46. *The dual problem* $(\widetilde{D}_N^M)$ *can equivalently be written in the form (see Remark 4.43)*

$$(\widetilde{D}_N^M) \qquad \sup_{(z^*,\widetilde{z}^*)\in\widetilde{\mathcal{B}}_N} \left\{ -\sum_{ji\in\widetilde{V}} \langle \widetilde{z}_{ji}^*, p_i \rangle \right\},$$

where

$$\widetilde{\mathcal{B}}_N = \left\{ (z^*,\widetilde{z}^*) = \left((z_{jk}^*)_{jk\in V}, (\widetilde{z}_{ji}^*)_{ji\in\widetilde{V}} \right) \in (\mathbb{R}^d)^{|V|} \times (\mathbb{R}^d)^{|\widetilde{V}|} : \right.$$

$$\left. \sum_{jk\in V} \frac{1}{w_{jk}}\|z_{jk}^*\| + \sum_{ji\in\widetilde{V}} \frac{1}{\widetilde{w}_{ji}}\|\widetilde{z}_{ji}^*\| \leq 1, \ \sum_{jk\in V} A_{jk}^* z_{jk}^* + \sum_{ji\in\widetilde{V}} B_{ji}^* \widetilde{z}_{ji}^* = 0_{\underbrace{\mathbb{R}^d \times \dots \times \mathbb{R}^d}_{m-times}} \right\}.$$

For its corresponding Lagrange dual problem we obtain

$$(D\widetilde{D}_N^M) \qquad \inf_{\substack{\lambda\geq 0, \\ x=(x_1,\dots,x_m)\in\mathbb{R}^d\times\dots\times\mathbb{R}^d}} \sup_{(z^*,\widetilde{z}^*)\in\widetilde{\mathcal{B}}_N} \left\{ -\sum_{ji\in\widetilde{V}} \langle \widetilde{z}_{ji}^*, p_i \rangle + \right.$$

$$\left. \left\langle x, \sum_{jk\in V} A_{jk}^T z_{jk}^* + \sum_{ji\in\widetilde{V}} B_{ji}^T \widetilde{z}_{ji}^* \right\rangle - \lambda \left(\sum_{jk\in V} \frac{1}{w_{jk}}\|z_{jk}^*\| + \sum_{ji\in\widetilde{V}} \frac{1}{\widetilde{w}_{ji}}\|\widetilde{z}_{ji}^*\| - 1 \right) \right\}$$

$$= \inf_{\substack{\lambda\geq 0, \\ x_i\in\mathbb{R}^d,\ i=1,\dots,m}} \left\{ \lambda + \sup_{(z^*,\widetilde{z}^*)\in\widetilde{\mathcal{B}}_N} \left\{ -\sum_{ji\in\widetilde{V}} \langle \widetilde{z}_{ji}^*, p_i \rangle \right. \right.$$

$$\left. \left. + \sum_{jk\in V} \langle x, A_{jk}^T z_{jk}^* \rangle + \sum_{ji\in\widetilde{V}} \langle x, B_{ji}^T \widetilde{z}_{ji}^* \rangle - \sum_{jk\in V} \frac{\lambda}{w_{jk}}\|z_{jk}^*\| - \sum_{ji\in\widetilde{V}} \frac{\lambda}{\widetilde{w}_{ji}}\|\widetilde{z}_{ji}^*\| \right\} \right\}$$

$$= \inf_{\substack{\lambda\geq 0, \\ x_i\in\mathbb{R}^d,\ i=1,\dots,m}} \left\{ \lambda + \sum_{jk\in V} \sup_{z_{jk}^*\in\mathbb{R}^d} \left\{ \langle A_{jk}x, z_{jk}^* \rangle - \frac{\lambda}{w_{jk}}\|z_{jk}^*\| \right\} \right.$$

$$\left. + \sum_{ji\in\widetilde{V}} \sup_{\widetilde{z}_{ji}^*\in\mathbb{R}^d} \left\{ \langle B_{ji}x, \widetilde{z}_{ji}^* \rangle - \langle p_i, \widetilde{z}_{ji}^* \rangle \} - \frac{\lambda}{w_{ji}}\|\widetilde{z}_{ji}^*\| \right\} \right\}$$

$$= \inf_{\substack{\lambda\geq 0, \\ x_i\in\mathbb{R}^d,\ i=1,\dots,m}} \left\{ \lambda + \sum_{jk\in V} \sup_{z_{jk}^*\in\mathbb{R}^d} \left\{ \langle x_j - x_k, z_{jk}^* \rangle - \frac{\lambda}{w_{jk}}\|z_{jk}^*\| \right\} \right.$$

$$\left. + \sum_{ji\in\widetilde{V}} \sup_{\widetilde{z}_{ji}^*\in\mathbb{R}^d} \left\{ \langle x_j - p_i, \widetilde{z}_{ji}^* \rangle - \frac{\lambda}{\widetilde{w}_{ji}}\|\widetilde{z}_{ji}^*\| \right\} \right\}.$$

The case $\lambda = 0$ *leads to* $x_j - p_i = 0$, $ji \in \widetilde{V}$, *and* $x_j - x_k = 0$, $jk \in V$, *which contradicts our assumption that the given points* p_i, $i = 1,\dots,n$, *are distinct, such that we can assume* $\lambda > 0$.

For this reason we can write for the Lagrange dual problem, or rather, the bidual of the location problem (P_N^M),

$$(D\widetilde{D}_N^M) \quad \inf_{\substack{\lambda>0, \\ (x_1,\ldots,x_m)\in\mathbb{R}^d\times\ldots\times\mathbb{R}^d}} \left\{ \lambda + \sum_{jk\in V} \frac{\lambda}{w_{jk}} \sup_{z_{jk}^*\in\mathbb{R}^d} \left\{ \left\langle \frac{w_{jk}}{\lambda}(x_j-x_k), z_{jk}^* \right\rangle - \|z_{jk}^*\| \right\} \right.$$

$$\left. + \sum_{ji\in\widetilde{V}} \frac{\lambda}{\widetilde{w}_{ji}} \sup_{\widetilde{z}_{ji}^*\in\mathbb{R}^d} \left\{ \left\langle \frac{\widetilde{w}_{ji}}{\lambda}(x_j-p_i), \widetilde{z}_{ji}^* \right\rangle - \|\widetilde{z}_{ji}^*\| \right\} \right\}$$

$$= \inf_{\substack{\lambda>0, (x_1,\ldots,x_m)\in\mathbb{R}^d\times\ldots\times\mathbb{R}^d, \\ w_{jk}\|x_j-x_k\|\leq\lambda, jk\in V, \widetilde{w}_{ji}\|x_j-p_i\|\leq\lambda, ji\in\widetilde{V}}} \lambda$$

$$= \inf_{(x_1,\ldots,x_m)\in\mathbb{R}^d\times\ldots\times\mathbb{R}^d} \max\left\{ (w_{jk}\|x_j-x_k\|)_{jk\in V}, (\widetilde{w}_{ji}\|x_j-p_i\|)_{ji\in\widetilde{V}} \right\}.$$

By using the Lagrange dual concept we transformed the dual problem $(\widetilde{D}_N^M)$ *back into the multifacility minmax location problem* (P_N^M), *showing that one has a full symmetry between the location problem* (P_N^M), *its dual problem* $(\widetilde{D}_N^M)$ *and the Lagrange dual problem* $(D\widetilde{D}_N^M)$. *In addition, we see that the Lagrange multiplier associated to the equality constraint can be identified as the optimal solution to the multifacility minmax location problem* (P_N^M) *and the Lagrange multiplier associated to the inequality constraint as the optimal objective value. A similar fact was stated in [69] for the case of a multifacility minsum location problem (see also Remark 4.37).*

The next corollary gives an estimation of the length of the vectors z_{jk}^*, $jk\in V$, and $\widetilde{z}_{ji}^*$, $ji\in\widetilde{V}$, feasible to the dual problem $(\widetilde{D}_N^M)$.

Corollary 4.5. *Let* $\overline{w}_s := \max\{(w_{jk})_{jk\in V}, (w_{ji})_{ji\in\widetilde{V}}\}$, *then for any feasible solution* $(z^*, \widetilde{z}^*)$ *of the problem* $(\widetilde{D}_N^M)$ *it holds*

$$\|z_{jk}^*\| \leq \frac{\overline{w}_s w_{jk}}{\overline{w}_s + w_{jk}} \text{ for } jk\in V \text{ and } \|\widetilde{z}_{ji}^*\| \leq \frac{\overline{w}_s w_{ji}}{\overline{w}_s + w_{ji}} \text{ for } ji\in\widetilde{V}.$$

Proof. As $(z^*, \widetilde{z}^*)$ is a feasible solution to $(\widetilde{D}_N^M)$, it holds

$$\sum_{jk\in V} A_{jk}^* z_{jk}^* + \sum_{ji\in\widetilde{V}} B_{ji}^* \widetilde{z}_{ji}^* = 0_{\mathbb{R}^d\times\ldots\times\mathbb{R}^d} \Leftrightarrow -A_{uv}^* z_{uv}^* = \sum_{\substack{jk\in V, \\ jk\neq uv}} A_{jk}^* z_{jk}^* + \sum_{ji\in\widetilde{V}} B_{ji}^* \widetilde{z}_{ji}^*$$

$$\Rightarrow \|A_{uv}^* z_{uv}^*\| = \|\sum_{\substack{jk\in V, \\ jk\neq uv}} A_{jk}^* z_{jk}^* + \sum_{ji\in\widetilde{V}} B_{ji}^* \widetilde{z}_{ji}^*\| \Rightarrow \|A_{uv}^* z_{uv}^*\| \leq \sum_{\substack{jk\in V, \\ jk\neq uv}} \|A_{jk}^* z_{jk}^*\| + \sum_{ji\in\widetilde{V}} \|B_{ji}^* \widetilde{z}_{ji}^*\|$$

$$\Leftrightarrow \sqrt{2}\|z_{uv}^*\| \leq \sum_{\substack{jk\in V, \\ jk\neq uv}} \sqrt{2}\|z_{jk}^*\| + \sum_{ji\in\widetilde{V}} \|\widetilde{z}_{ji}^*\| \Leftrightarrow \|z_{uv}^*\| \leq \sum_{\substack{jk\in V, \\ jk\neq uv}} \|z_{jk}^*\| + \frac{1}{\sqrt{2}} \sum_{ji\in\widetilde{V}} \|\widetilde{z}_{ji}^*\|$$

$$\Rightarrow \|z_{uv}^*\| \leq \sum_{\substack{jk\in V, \\ jk\neq uv}} \|z_{jk}^*\| + \sum_{ji\in\widetilde{V}} \|\widetilde{z}_{ji}^*\|, \quad uv\in V,$$

and more than that, it holds

$$1 \geq \sum_{jk \in V} \frac{1}{w_{jk}} \|z_{jk}^*\| + \sum_{ji \in \widetilde{V}} \frac{1}{\widetilde{w}_{ji}} \|\widetilde{z}_{ji}^*\| = \frac{1}{w_{uv}} \|z_{uv}^*\| + \sum_{\substack{jk \in V, \\ jk \neq uv}} \frac{1}{w_{jk}} \|z_{jk}^*\| + \sum_{ji \in \widetilde{V}} \frac{1}{\widetilde{w}_{ji}} \|\widetilde{z}_{ji}^*\|$$

$$\geq \frac{1}{w_{uv}} \|z_{uv}^*\| + \frac{1}{\overline{w}_s} \left(\sum_{\substack{jk \in V, \\ jk \neq uv}} \|z_{jk}^*\| + \sum_{ji \in \widetilde{V}} \|\widetilde{z}_{ji}^*\| \right) \geq \frac{1}{w_{uv}} \|z_{uv}^*\| + \frac{1}{\overline{w}_s} \|z_{uv}^*\|$$

$$= \frac{\overline{w}_s + w_{uv}}{\overline{w}_s w_{uv}} \|z_{uv}^*\|,$$

which means that

$$\|z_{jk}^*\| \leq \frac{\overline{w}_s w_{jk}}{\overline{w}_s + w_{jk}}, \ jk \in V.$$

In the same way, we get

$$\|\widetilde{z}_{ji}^*\| \leq \frac{\overline{w}_s w_{ji}}{\overline{w}_s + w_{ji}}, \ ji \in \widetilde{V}.$$

$\square$

Example 4.5. *For the existing facilities* $p_1 = (0,0)^T$, $p_2 = (-2,3)^T$ *and* $p_3 = (5,8)^T$ *(t=3) we want to locate two new facilities (m=2) in the plane (d = 2) . The weights are given by* $w_{12} = \widetilde{w}_{11} = \widetilde{w}_{13} = \widetilde{w}_{21} = \widetilde{w}_{22} = 1$ *and* $\widetilde{w}_{12} = \widetilde{w}_{23} = 0$ *and define the following multifacility minmax location problem*

$$(P_N^M) \qquad \inf_{(x_1, x_2) \in \mathbb{R}^2 \times \mathbb{R}^2} \max \{ \|x_1 - x_2\|, \|x_1 - p_1\|, \|x_1 - p_3\|, \|x_2 - p_1\|, \|x_2 - p_2\| \},$$

i.e. $V = \{12\}$, $|V| = 1$, $\widetilde{V} = \{11, 13, 21, 22\}$ *and* $|\widetilde{V}| = 4$. *From the* MATLAB *Optimization Toolbox we obtained the following solution* $\overline{x}_1 = (2.5, 4)$ *and* $\overline{x}_2 = (0, 0)^T$. *The corresponding objective value was* $v(P_N^M) = 4.72$.
The dual problem (see Remark 4.43)

$$(\widetilde{D}_N^M) \qquad \max_{(z_{12}^*, \widetilde{z}_{11}^*, \widetilde{z}_{13}^*, \widetilde{z}_{21}^*, \widetilde{z}_{22}^*) \in \widetilde{\mathcal{B}}_N} \{ \langle \widetilde{z}_{11}^* + \widetilde{z}_{21}^*, p_1 \rangle + \langle \widetilde{z}_{22}^*, p_2 \rangle + \langle \widetilde{z}_{13}^*, p_3 \rangle \},$$

where

$$\widetilde{\mathcal{B}}_N = \{ (z_{12}^*, \widetilde{z}_{11}^*, \widetilde{z}_{13}^*, \widetilde{z}_{21}^*, \widetilde{z}_{22}^*) \in \mathbb{R}^2 \times \mathbb{R}^2 \times \mathbb{R}^2 \times \mathbb{R}^2 \times \mathbb{R}^2 : z_{12}^* + \widetilde{z}_{11}^* + \widetilde{z}_{13}^* = 0_{\mathbb{R}^2},$$
$$\widetilde{z}_{21}^* + \widetilde{z}_{22}^* = 0_{\mathbb{R}^2}, \ \|z_{12}^*\| + \|\widetilde{z}_{11}^*\| + \|\widetilde{z}_{21}^*\| + \|\widetilde{z}_{22}^*\| + \|\widetilde{z}_{13}^*\| \leq 1 \},$$

was also solved by the MATLAB *Optimization Toolbox. The following solution was obtained*

$$\overline{z}_{12}^* = \overline{\widetilde{z}}_{11}^* = (0.13, 0.21)^T, \ \overline{\widetilde{z}}_{13}^* = (-0.26, -0.42)^T, \ \overline{\widetilde{z}}_{21}^* = \overline{\widetilde{z}}_{22}^* = (0, 0)^T,$$

with the corresponding objective value $v(\widetilde{D}_N^M) = 4.72 = v(P_N^M)$, i.e. $\overline{I} = \{12\} \subseteq V$ and $\widetilde{\overline{I}} = \{11, 13\} \subseteq \widetilde{V}$.

In the situation when we have only the solution of the dual problem one can reconstruct the optimal solution of the primal problem in a recursive way by using the necessary and sufficient optimality conditions given in Theorem 4.35. By condition (iv) we know that there exists $\widetilde{\alpha}_{11} > 0$ such that

$$\widetilde{\overline{z}}_{11}^* = \widetilde{\alpha}_{11}(\overline{x}_1 - p_1), \quad i.e. \ \|\widetilde{\overline{z}}_{11}^*\| = \widetilde{\alpha}_{11}\|\overline{x}_1 - p_1\|, \tag{4.117}$$

and as, by condition (vi) it holds

$$v(\widetilde{D}_N^M) = v(P_N^M) = \|\overline{x}_1 - p_1\| = \frac{\|\widetilde{\overline{z}}_{11}^*\|}{\widetilde{\alpha}_{11}}, \tag{4.118}$$

we get by combining (4.117) and (4.118) that

$$\widetilde{\overline{z}}_{11}^* = \frac{\|\widetilde{\overline{z}}_{11}^*\|}{v(\widetilde{D}_N^M)}(\overline{x}_1 - p_1) \Leftrightarrow \overline{x}_1 = \frac{v(\widetilde{D}_N^M)}{\|\widetilde{\overline{z}}_{11}^*\|}\widetilde{\overline{z}}_{11}^* + p_1 = \frac{4.72}{0.25}(0.13, 0.21)^T = (2.5, 4)^T.$$

More than that, by condition (iii) there exists $\alpha_{12} > 0$ such that

$$\overline{z}_{12}^* = \alpha_{12}(\overline{x}_1 - \overline{x}_2), \quad i.e. \ \|\overline{z}_{12}^*\| = \alpha_{12}\|\overline{x}_1 - \overline{x}_2\|, \tag{4.119}$$

and therefore, we derive from condition (v) that

$$v(\widetilde{D}_N^M) = v(P_N^M) = \|\overline{x}_1 - \overline{x}_2\| = \frac{\|\overline{z}_{12}^*\|}{\alpha_{12}}. \tag{4.120}$$

Finally, taking (4.119) and (4.120) together yields

$$\overline{z}_{12}^* = \frac{\|\overline{z}_{12}^*\|}{v(\widetilde{D}_N^M)}(\overline{x}_1 - \overline{x}_2) \Leftrightarrow \overline{x}_2 = \overline{x}_1 - \frac{v(\widetilde{D}_N^M)}{\|\overline{z}_{12}^*\|}\overline{z}_{12}^* = (2.5, 4)^T - \frac{4.72}{0.25}(0.13, 0.21)^T = (0, 0)^T.$$

For a geometrical illustration see Figure 4.4.

Geometrical interpretation.
In the following we provide a geometrical characterization of the set of optimal solutions of the dual problem by Theorem 4.35. By the conditions (iii) and (iv) it is clear that for $jk \in \overline{I}$ and $ji \in \widetilde{\overline{I}}$ the vectors $\overline{z}_{jk}^*$ and $\widetilde{\overline{z}}_{ji}^*$ are parallel to the vectors $\overline{x}_j - \overline{x}_k$ and $\overline{x}_j - p_i$ directed to $\overline{x}_j$, respectively. In addition, if we take into account the conditions (v), (vi) and (vii), then it is also evident that $jk \in \overline{I}$ and $ji \in \widetilde{\overline{I}}$, i.e. $\overline{z}_{jk}^* \neq 0_{\mathbb{R}^d}$ and $\widetilde{\overline{z}}_{ji}^* \neq 0_{\mathbb{R}^d}$, if the points $\overline{x}_k$ and p_i are lying on the border of the minimum covering ball with radius $v(P_N^M)$ centered in $\overline{x}_j$, respectively. Vice versa, if $jk \in V \setminus \overline{I}$ and $ji \in \widetilde{V} \setminus \widetilde{\overline{I}}$, then $\overline{z}_{jk}^* = 0_{\mathbb{R}^d}$ and $\widetilde{\overline{z}}_{ji}^* = 0_{\mathbb{R}^d}$, which is exactly the case when the corresponding weights are zero or the

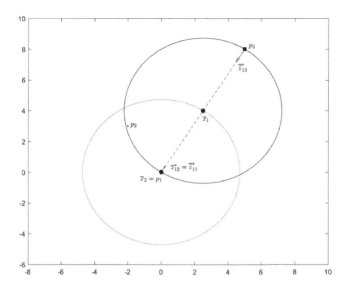

Figure 4.4: Illustration of the Example 4.5.

points $\overline{x}_k$ and p_i are lying inside the minimum covering ball centered in $\overline{x}_j$, respectively. Therefore, analogously to the geometrical interpretation presented in Section 4.3.4 for single minmax location problems, one can identify the vectors $\overline{z}^*_{jk}$, $jk \in \overline{I}$, and $\overline{\overline{z}}^*_{ji}$, $ji \in \overline{\overline{I}}$, as force vectors, which pull the points lying on the borders of the minimum covering balls inside the balls in direction to the their corresponding centers, the gravity points $\overline{x}_j$ (see Figure 4.4).

5 Solving minmax location problems via epigraphical projection

For solving minmax location problems in Hilbert spaces $\mathcal{H}_i$, $i = 1, ..., n$, numerically by proximal methods we present in this chapter first a general formula of the projection onto the epigraph of the function $h : \mathcal{H}_1 \times ... \times \mathcal{H}_n \to \mathbb{R}$, defined by $h(x_1, ..., x_n) := \sum_{i=1}^{n} w_i \|x_i\|_{\mathcal{H}_i}^{\beta_i}$. We consider the situations when $\beta_i = 1$, $i = 1, ..., n$, and $w_i = 1$, $\beta_i = 2$, $i = 1, ..., n$, where the formulae given for instance in [3, 33, 34] turn out to be special cases for $n = 1$ of our considerations.

Moreover, we develop a formula for the projection onto the epigraph of a gauge function $\gamma_C : X \to \overline{\mathbb{R}}$ of a closed and convex set $C \subseteq \mathcal{H}$ with $0_{\mathcal{H}} \in C$. As a consequence, we derive a formula for the projection onto the epigraph of the gauge of a closed and convex cone as well as the sum of gauges. Finally, two examples are considered to demonstrate how the latter formula can be used to determine the projector.

This chapter is based on the author's article [103].

5.1 Motivation

As argued in a large number of papers, the proximal method is an excellent tool for solving in an efficient way optimization problems of the form

$$\min_{x \in \mathcal{H}} \left\{ \sum_{i=1}^{n} f_i(x) \right\}, \tag{5.1}$$

where $\mathcal{H}$ is a real Hilbert space and $f_i : \mathcal{H} \to \overline{\mathbb{R}}$ is a proper, lower semicontinuous and convex function, $i = 1, ..., n$. This kind of problems occur for instance in areas like image processing [9, 18, 19, 33], portfolio optimization [12, 88], cluster analysis [11, 31], statistical learning theory [21], machine learning [16] and location theory [12, 17, 35, 64]. In the main step of this method it is necessary to determine the proximity operators of the functions involved in the formulation of the associated optimization problem. The *proximity operator* (a.k.a. *proximal mapping*) of a proper, lower semicontinuous and convex function $f : \mathcal{H} \to \overline{\mathbb{R}}$ denoted by prox_f is defined by

$$\mathrm{prox}_f x : \mathcal{H} \to \mathcal{H}, \ \mathrm{prox}_f x := \arg\min_{y \in \mathcal{H}} \left\{ f(y) + \frac{1}{2} \|x - y\|_{\mathcal{H}}^2 \right\} \ \forall x \in \mathcal{H}. \tag{5.2}$$

© Springer Fachmedien Wiesbaden GmbH, part of Springer Nature 2020
O. Wilfer, *Multi-Composed Programming with Applications to Facility Location*,
Mathematische Optimierung und Wirtschaftsmathematik I Mathematical
Optimization and Economathematics, https://doi.org/10.1007/978-3-658-30580-2_5

The proximity operator can be understood as a generalization of the projection onto a convex set, as for a non-empty, closed and convex set $A \subseteq \mathcal{H}$, i.e. δ_A is proper, convex and lower semicontinuous, we have

$$\text{prox}_{\delta_A} x = P_A x \; \forall x \in \mathcal{H}, \tag{5.3}$$

where P_A is the projection operator which maps every point x in $\mathcal{H}$ to its unique projection onto the set A (see [3]).

From (5.2) follows that the determination of the proximity operators of the functions f_i, $i = 1, ..., n$, of (5.1) requires the solving of n subproblems, where a favorable situation exists, when a closed formula of a proximity operator can be given. This in turn has a positive effect on the solving of optimization problems from the numerical point of view.

Motivated by this background, our aim is to solve numerically extended multifacility minmax location problems given by

$$(EP_N^{M,\beta}) \qquad \min_{(x_1,...,x_m)\in\mathbb{R}^d\times...\times\mathbb{R}^d} \max_{1\leq i\leq n} \left\{ \sum_{j=1}^m w_{ij}\|x_j - p_i\|^{\beta_i} \right\}, \tag{5.4}$$

where $w_{ij} > 0$, $\beta_i \geq 1$ and $p_i \in \mathbb{R}^d$ are distinct points, $j = 1, ..., m$, $i = 1, ..., n$. In this framework we first need to rewrite this kind of location problems into the form of (5.1) where the objective function is a sum of proper, lower semicontinuous and convex functions. For this purpose we introduce an additional variable and obtain for $(EP_N^{M,\beta})$ the following formulation

$$(EP_N^{M,\beta}) \qquad \min_{\substack{(x_1,...,x_m,t)\in\mathbb{R}^d\times...\times\mathbb{R}^d\times\mathbb{R}, \\ \sum_{j=1}^m w_{ij}\|x_j-p_i\|^{\beta_i}\leq t, \; i=1,...,n}} t = \min_{\substack{(x_1,...,x_m,t)\in\mathbb{R}^d\times...\times\mathbb{R}^d\times\mathbb{R}, \\ (x_1,...,x_m,t)\in\text{epi}\left(\sum_{j=1}^m w_{ij}\|\cdot-p_i\|^{\beta_i}\right), \; i=1,...,n}} t$$

$$= \min_{(x_1,...,x_m,t)\in\mathbb{R}^d\times...\times\mathbb{R}^d\times\mathbb{R}} \left\{ t + \sum_{i=1}^n \delta_{\text{epi}\left(\sum_{j=1}^m w_{ij}\|\cdot-p_i\|^{\beta_i}\right)}(x_1,...,x_m,t) \right\}. \tag{5.5}$$

Now, to apply the proximal method to $(EP_N^{M,\beta})$ one needs to calculate the proximity operators of the functions involved in the objective function of (5.5). For this reason and especially in the context of (5.3), we give in Section 5.2 formulae for the projections onto the epigraphs of several sums of powers of weighted norms. As the power of norm in (5.5) can be replaced by a gauge function, we present also formulae of projections onto the epigraphs of gauges.

To point out the benefits of the presented formulae we consider then examples of location problems in different settings and compare the numerical results with a method proposed by Cornejo and Michelot in [35]. The difference between these two methods is that the one given by Cornejo and Michelot splits the sum of powers of weighted norms by introducing $n \cdot m$ additional variables.

In this situation one gets the following presentation of the extended multifacility minmax location problem

$$(EP_N^{M,\beta}) \quad \min_{\substack{t,\ t_{ij} \in \mathbb{R},\ x_j \in \mathbb{R}^d, \\ j=1,\ldots,m, i=1,\ldots,n}} \left\{ t + \sum_{j=1}^{m}\sum_{i=1}^{n} \delta_{\mathrm{epi}\left(w_{ij}\|\cdot - p_i\|^{\beta_i}\right)}(x_j, t_{ij}) + \sum_{i=1}^{n} \delta_{\mathrm{epi}\,\tau_i}(t_{i1},\ldots,t_{im},t) \right\},$$

(5.6)

where $\tau_i(t_{i1},\ldots,t_{im}) := \sum_{j=1}^{m} t_{ij}$, $i = 1,\ldots,n$. In Section 5.1 we show that this concept makes the solving process for the considered examples of location problems very slow and the advantage of our approach more clear. The numerical tests are based on the parallel splitting algorithm, which can be found for instance in [3].

Finally, we collect some properties of Hilbert spaces, which can be found with proofs for instance in [3] and [34].

If f is Gâteaux-differentiable at $x \in \mathcal{H}$, then $\partial f(x) = \{\nabla f(x)\}$. The set of global minimizers of a function $f : \mathcal{H} \to \mathbb{R}$ is denoted by $\mathrm{Argmin}\, f$ and if f has a unique minimizer, it is denoted by $\arg\min_{x \in \mathcal{H}} f(x)$. It holds

$$x \in \mathrm{Argmin}\, f \Leftrightarrow 0_{\mathcal{H}} \in \partial f(x)\ \forall x \in \mathcal{H}.$$

(5.7)

It holds

$$y = \mathrm{prox}_f x \Leftrightarrow x - y \in \partial f(y)\ \forall x \in \mathcal{H},\ \forall y \in \mathcal{H}.$$

(5.8)

In addition, we make for the rest of this chapter the convention that $\frac{0}{0} = 0$ and $\frac{1}{0}\cdot 0_{\mathcal{H}} = 0_{\mathcal{H}}$.

In the following let $\mathcal{H}_1 \times \ldots \times \mathcal{H}_n$ be real Hilbert space endowed with inner product and norm, respectively defined by

$$\langle (x_1,\ldots,x_n),(y_1,\ldots,y_n)\rangle_{\mathcal{H}_1\times\ldots\times\mathcal{H}_n} = \sum_{i=1}^{n}\langle x_i, y_i\rangle_{\mathcal{H}_i}\ \text{and}\ \|(x_1,\ldots,x_n)\|_{\mathcal{H}_1\times\ldots\times\mathcal{H}_n} = \sqrt{\sum_{i=1}^{n}\|x_i\|_{\mathcal{H}_i}^2},$$

where $(x_1,\ldots,x_n) \in \mathcal{H}_1 \times \ldots \times \mathcal{H}_n$ and $(y_1,\ldots,y_n) \in \mathcal{H}_1 \times \ldots \times \mathcal{H}_n$.

We close this section with a lemma, which presents a formula for the projection onto a unit ball generated by the weighted sum of norms and generalizes the results given in [96] to real Hilbert spaces $\mathcal{H}_i$, $i = 1,\ldots,n$. Let $w_i > 0$, $i = 1,\ldots,n$, and $C := \{(x_1,\ldots,x_n) \in \mathcal{H}_1 \times \ldots \times \mathcal{H}_n : \sum_{i=1}^{n} w_i\|x_i\|_{\mathcal{H}_i} \leq 1\}$, then the following statement holds.

Lemma 5.1. *For all* $(x_1,\ldots,x_n) \in \mathcal{H}_1 \times \ldots \times \mathcal{H}_n$ *it holds*

$$\mathrm{P}_C(x_1,\ldots,x_n) = \begin{cases} (x_1,\ldots,x_n), & \text{if } \sum_{i=1}^{n} w_i\|x_i\|_{\mathcal{H}_i} \leq 1, \\ (\overline{y}_1,\ldots,\overline{y}_n), & \text{otherwise,} \end{cases}$$

where

$$\overline{y}_i = \frac{\max\{\|x_i\|_{\mathcal{H}_i} - \overline{\lambda} w_i, 0\}}{\|x_i\|_{\mathcal{H}_i}} x_i, \ i = 1, ..., n,$$

with

$$\overline{\lambda} = \frac{\sum\limits_{i=k+1}^{n} w_i^2 \tau_i - 1}{\sum\limits_{i=k+1}^{n} w_i^2}$$

and $k \in \{0, 1, ..., n-1\}$ is the unique integer such that $\tau_k \leq \overline{\lambda} \leq \tau_{k+1}$, where the values $\tau_0, ..., \tau_n$ are defined by $\tau_0 := 0$ and $\tau_i := \|x_i\|_{\mathcal{H}_i}/w_i$, $i = 1, ..., n$, and in ascending order.

Proof. In order to determine the projection onto the set C, we consider for fixed $(x_1, ..., x_n) \in \mathcal{H}_1 \times ... \times \mathcal{H}_n$ the following optimization problem

$$\min_{\substack{(y_1,...,y_n) \in \mathcal{H}_1 \times ... \times \mathcal{H}_n, \\ \sum_{i=1}^{n} w_i \|y_i\|_{\mathcal{H}_i} \leq 1}} \left\{ \sum_{i=1}^{n} \frac{1}{2} \|y_i - x_i\|_{\mathcal{H}_i}^2 \right\}. \tag{5.9}$$

Obviously, if $\sum_{i=1}^{n} w_i \|x_i\|_{\mathcal{H}_i} \leq 1$, i.e. $(x_1, ..., x_n) \in C$, then the unique solution is $\overline{y}_i = x_i$, $i = 1, ..., n$. In the following we consider the non-trivial situation where $\sum_{i=1}^{n} w_i \|x_i\|_{\mathcal{H}_i} > 1$, i.e. $(x_1, ..., x_n) \notin C$ and define the function $f : \mathcal{H}_1 \times ... \times \mathcal{H}_n \to \mathbb{R}$ by $f(y_1, ..., y_n) := \sum_{i=1}^{n} (1/2) \|y_i - x_i\|_{\mathcal{H}_i}^2$ and the function $g : \mathcal{H}_1 \times ... \times \mathcal{H}_n \to \mathbb{R}$ by $g(y_1, ..., y_n) := \sum_{i=1}^{n} w_i \|y_i\|_{\mathcal{H}_i} - 1$. Hence, by [3, Proposition 26.18] it holds for the unique solution $(\overline{y}_1, ..., \overline{y}_n)$ of (5.9) that

$$\nabla f(\overline{y}_1, ..., \overline{y}_n) \in -\overline{\lambda} \partial g(\overline{y}_1, ..., \overline{y}_n) \Leftrightarrow \overline{y}_i - x_i \in -\overline{\lambda} \partial \left(w_i \| \cdot \|_{\mathcal{H}_i} \right) (\overline{y}_i), \ i = 1, ..., n,$$

as well as

$$\overline{\lambda} \left(\sum_{i=1}^{n} w_i \|\overline{y}_i\|_{\mathcal{H}_i} - 1 \right) = 0 \text{ and } \sum_{i=1}^{n} w_i \|\overline{y}_i\|_{\mathcal{H}_i} \leq 1,$$

where $\overline{\lambda} \geq 0$ is the associated Lagrange multiplier of $(\overline{y}_1, ..., \overline{y}_n)$. If $\overline{\lambda} = 0$, then $\overline{y}_i = x_i$, $i = 1, ..., n$, and by the feasibility condition we obtain $\sum_{i=1}^{n} w_i \|x_i\|_{\mathcal{H}_i} \leq 1$, which contradicts our assumption. Therefore, $\overline{\lambda} > 0$ and we get by (5.8) that

$$\overline{y}_i - x_i \in -\overline{\lambda} \partial \left(w_i \| \cdot \|_{\mathcal{H}_i} \right) (\overline{y}_i) \Leftrightarrow x_i - \overline{y}_i \in \partial (\overline{\lambda} w_i \| \cdot \|_{\mathcal{H}_i})(\overline{y}_i) \Leftrightarrow \overline{y}_i = \text{prox}_{\overline{\lambda} w_i \| \cdot \|_{\mathcal{H}_i}} x_i,$$

$i = 1, ..., n$. Using [34, Proposition 2.8] reveals that

$$\overline{y}_i = \begin{cases} x_i - \frac{\overline{\lambda} w_i}{\|x_i\|_{\mathcal{H}_i}} x_i, & \text{if } \|x_i\|_{\mathcal{H}_i} > \overline{\lambda} w_i, \\ 0_{\mathcal{H}_i}, & \text{if } \|x_i\|_{\mathcal{H}_i} \leq \overline{\lambda} w_i \end{cases} = \frac{\max\{\|x_i\|_{\mathcal{H}_i} - \overline{\lambda} w_i, 0\}}{\|x_i\|_{\mathcal{H}_i}} x_i, \ i = 1, ..., n,$$

and as $\sum_{i=1}^{n} w_i \|\overline{y}_i\|_{\mathcal{H}_i} = 1$, we conclude that

$$\sum_{i=1}^{n} w_i \max\left\{\|x_i\|_{\mathcal{H}_i} - \overline{\lambda} w_i, 0\right\} = 1. \tag{5.10}$$

Now, we define the function $\kappa : \mathbb{R} \to \mathbb{R}$ by $\kappa(\lambda) = \sum_{i=1}^{n} w_i^2 \max\{\tau_i - \lambda, 0\} - 1$. Note, that there exists $\widetilde{\lambda} \geq \tau_i$ for all $i = 1, ..., n$, such that $\kappa(\widetilde{\lambda}) = -1 < 0$. Moreover, κ is a piecewise linear function with $\kappa(0) = w_i^2 \tau_i - 1$ and its slope changes at $\lambda = \tau_i$, $i = 1, ..., n$. To be more precise, at $\lambda = 0$ the slope of κ is $-\sum_{i=1}^{n} w_i^2$ and increases by w_1^2 when $\lambda = \tau_1$. If we continue in this matter for $i = 2, ..., n$, the slope keeps increasing and when $\lambda \geq \tau_n$, $\kappa(\lambda) = -1$ such that the slope is 0. In summary, to find the zero of κ one needs to determine the unique integer $k \in \{0, 1, ..., n - 1\}$ such that $\kappa(\tau_k) \geq 0$ and $\kappa(\tau_{k+1}) \leq 0$. In the light of the above, it holds

$$\kappa(\lambda) = \sum_{i=k+1}^{n} w_i^2 \tau_i - \lambda \sum_{i=k+1}^{n} w_i^2 - 1,$$

where $\tau_k \leq \lambda \leq \tau_{k+1}$, and hence, one gets for $\overline{\lambda}$ such that $\kappa(\overline{\lambda}) = 0$,

$$\overline{\lambda} = \frac{\sum\limits_{i=k+1}^{n} w_i^2 \tau_i - 1}{\sum\limits_{i=k+1}^{n} w_i^2}.$$

$\square$

5.2 Formulae of epigraphical projection

The first aim of this section is to give formulae for the projection operators onto the epigraphs of several sums of powers of weighted norms. For this purpose, we give a general formula in our central theorem, from which we deduce special cases used in our numerical tests.

The second aim is to present formulae of the projection operators onto the epigraphs of gauges. In this part of this section we use the properties of gauge functions listed in Section 4.1.1. Especially, by using the fact that the sum of gauges is again a gauge, we also present a formula of the projector onto the epigraph of the sum of gauges. Two examples in the cases of norms close this section.

5.2.1 Sum of weighted norms

Let us consider the following function $h : \mathcal{H}_1 \times ... \times \mathcal{H}_n \to \mathbb{R}$ defined as

$$h(x_1, ..., x_n) := \sum_{i=1}^{n} w_i \|x_i\|_{\mathcal{H}_i}^{\beta_i}, \tag{5.11}$$

where $w_i > 0$ and $\beta_i \geq 1$, $i = 1, ..., n$. By defining the sets

$$L := \{l \in \{1, ..., n\} : \beta_l > 1\} \text{ and } R := \{r \in \{1, ..., n\} : \beta_r = 1\},$$

we can state the following formula for the projection onto the epigraph of the sum of powers of weighted norms, which generalizes the results given for instance in [3, 33, 34, 88].

Theorem 5.1. *Assume that h is given by (5.11). Then, for every $(x_1, ..., x_n, \xi) \in \mathcal{H}_1 \times ... \times \mathcal{H}_n \times \mathbb{R}$ one has*

$$P_{\text{epi} h}(x_1, ..., x_n, \xi) = \begin{cases} (x_1, ..., x_n, \xi), & \text{if } \sum_{i=1}^{n} w_i \|x_i\|_{\mathcal{H}_i}^{\beta_i} \leq \xi, \\ (\overline{y}_1, ..., \overline{y}_n, \overline{\theta}), & \text{otherwise,} \end{cases} \tag{5.12}$$

with

$$\overline{y}_r = \frac{\max\{\|x_r\|_{\mathcal{H}_r} - \overline{\lambda} w_r, 0\}}{\|x_r\|_{\mathcal{H}_r}} x_r, \ r \in R,$$

$$\overline{y}_l = \frac{\|x_l\|_{\mathcal{H}_l} - \eta_l(\overline{\lambda})}{\|x_l\|_{\mathcal{H}_l}} x_l, \ l \in L,$$

$$\overline{\theta} = \xi + \overline{\lambda},$$

where $\eta_l(\overline{\lambda})$ is the unique non-negative real number that solves the equation

$$\eta_l(\overline{\lambda}) + \left(\frac{\eta_l(\overline{\lambda})}{\overline{\lambda} w_l \beta_l}\right)^{\frac{1}{\beta_l - 1}} = \|x_l\|_{\mathcal{H}_l}, \ l \in L, \tag{5.13}$$

and $\overline{\lambda} > 0$ is a solution of the equation

$$\sum_{r \in R} w_r \max\{\|x_r\|_{\mathcal{H}_r} - \lambda w_r, 0\} + \sum_{l \in L} w_l(\|x_l\|_{\mathcal{H}_l} - \eta_l(\lambda))^{\beta_l} = \lambda + \xi. \tag{5.14}$$

Proof. For given $\xi \in \mathbb{R}$ and $(x_1, ..., x_n) \in \mathcal{H}_1 \times ... \times \mathcal{H}_n$, let us consider the following optimization problem

$$\min_{\substack{(y_1, ..., y_n, \theta) \in \mathcal{H}_1 \times ... \times \mathcal{H}_n \times \mathbb{R} \\ \sum_{i=1}^{n} w_i \|y_i\|_{\mathcal{H}_i}^{\beta_i} \leq \theta}} \left\{ \frac{1}{2}(\theta - \xi)^2 + \sum_{i=1}^{n} \frac{1}{2} \|y_i - x_i\|_{\mathcal{H}_i}^2 \right\}. \tag{5.15}$$

It is clear that in the situation when $\sum_{i=1}^{n} w_i \|x_i\|_{\mathcal{H}_i}^{\beta_i} \leq \xi$, i.e. $(x_1, ..., x_n, \xi) \in \text{epi} h$, the unique solution of (5.15) is $\overline{y}_i = x_i$, $i = 1, ..., n$, and $\overline{\theta} = \xi$. Therefore, we consider in the following the non-trivial case where $\sum_{i=1}^{n} w_i \|x_i\|_{\mathcal{H}_i}^{\beta_i} > \xi$, i.e. $(x_1, ..., x_n, \xi) \notin \text{epi} h$.

Let us now define the function $f : \mathcal{H}_1 \times ... \times \mathcal{H}_n \times \mathbb{R} \to \mathbb{R}$ by $f(y_1, ..., y_n, \theta) := (1/2)(\theta - \xi)^2 + \sum_{i=1}^{n}(1/2)\|y_i - x_i\|_{\mathcal{H}_i}^2$ and the function $g : \mathcal{H}_1 \times ... \times \mathcal{H}_n \times \mathbb{R} \to \mathbb{R}$ by $g(y_1, ..., y_n, \theta) :=$

$\sum_{i=1}^{n} w_i \|y_i\|_{\mathcal{H}_i}^{\beta_i} - \theta$, then by [3, Proposition 26.18] there exists $\overline{\lambda} \geq 0$, such that for the unique solution $(\overline{y}_1, ..., \overline{y}_n, \overline{\theta})$ of (5.15) it holds

$$\nabla f(\overline{y}_1, ..., \overline{y}_n, \overline{\theta}) \in -\overline{\lambda}\partial g(\overline{y}_1, ..., \overline{y}_n, \overline{\theta}) \Leftrightarrow \begin{cases} \overline{y}_i - x_i \in -\overline{\lambda}\partial(w_i\| \cdot \|_{\mathcal{H}_i}^{\beta_i})(\overline{y}_i), \ i = 1, ...n, \\ \overline{\theta} - \xi = \overline{\lambda}, \end{cases}$$

(5.16)

where $\overline{\lambda}$ is the associated Lagrange multiplier of $(\overline{y}_1, ..., \overline{y}_n, \overline{\theta})$. If $\overline{\lambda} = 0$, then one gets by (5.16) that $\overline{y}_i = x_i$, $i = 1, ..., n$, and $\overline{\theta} = \xi$ and by the feasibility of the solution it follows that $\sum_{i=1}^{n} w_i \|x_i\|_{\mathcal{H}_i}^{\beta_i} \leq \xi$, which contradicts our assumption. Hence, it holds $\overline{\lambda} > 0$ and by (5.8) and (5.16) we have

$$\begin{cases} x_i - \overline{y}_i \in \partial(\overline{\lambda}w_i\| \cdot \|_{\mathcal{H}_i}^{\beta_i})(\overline{y}_i), \ i = 1, ...n, \\ \overline{\theta} = \overline{\lambda} + \xi, \end{cases} \Leftrightarrow \begin{cases} \overline{y}_i = \text{prox}_{\overline{\lambda}w_i\| \cdot \|_{\mathcal{H}_i}^{\beta_i}} x_i, \ i = 1, ...n, \\ \overline{\theta} = \overline{\lambda} + \xi. \end{cases}$$

Further, from [34, Proposition 2.8] it follows for the case $r \in R$, i.e. $\beta_r = 1$, that

$$\overline{y}_r = \begin{cases} x_r - \frac{\overline{\lambda}w_r}{\|x_r\|_{\mathcal{H}_r}}x_r, & \text{if } \|x_r\|_{\mathcal{H}_r} > \overline{\lambda}w_r, \\ 0_{\mathcal{H}_r}, & \text{if } \|x_r\|_{\mathcal{H}_r} \leq \overline{\lambda}w_r \end{cases} = \frac{\max\{\|x_r\|_{\mathcal{H}_r} - \overline{\lambda}w_r, 0\}}{\|x_r\|_{\mathcal{H}_r}}x_r,$$

(5.17)

and for the case $l \in L$, i.e. $\beta_l > 0$, that

$$\overline{y}_l = x_l - \frac{\eta_l(\overline{\lambda})}{\|x_l\|_{\mathcal{H}_l}}x_l = \frac{\|x_l\|_{\mathcal{H}_l} - \eta_l(\overline{\lambda})}{\|x_l\|_{\mathcal{H}_l}}x_l,$$

(5.18)

where $\eta_l(\overline{\lambda})$ is the unique non-negative real number that solves the following equation

$$\eta_l(\overline{\lambda}) + \left(\frac{\eta_l(\overline{\lambda})}{\overline{\lambda}w_l\beta_l}\right)^{\frac{1}{\beta_l-1}} = \|x_l\|_{\mathcal{H}_l}$$

(5.19)

(notice that by (5.19) follows that $\|x_l\|_{\mathcal{H}_l} - \eta_l(\overline{\lambda}) \geq 0$). Furthermore, the complementary slackness condition

$$\overline{\lambda}\left(\sum_{i=1}^{n} w_i \|\overline{y}_i\|_{\mathcal{H}_i}^{\beta_i} - \overline{\theta}\right) = 0$$

(5.20)

implies that

$$\sum_{i=1}^{n} w_i \|\overline{y}_i\|_{\mathcal{H}_i}^{\beta_i} = \overline{\theta},$$

(5.21)

and from here follows by (5.17) and (5.18) that

$$\sum_{i=1}^{n} w_i \|\overline{y}_i\|_{\mathcal{H}_i}^{\beta_i} = \sum_{r \in R} w_r \max\{\|x_r\|_{\mathcal{H}_r} - \overline{\lambda}w_r, 0\} + \sum_{l \in L} w_l(\|x_l\|_{\mathcal{H}_l} - \eta_l(\overline{\lambda}))^{\beta_l} = \overline{\lambda} + \xi. \quad (5.22)$$

$\square$

Remark 5.1. *In the situation when $\beta_i > 1$ for all $i=1,...,n$, we get by summarizing the formulae (5.13) and (5.14)*

$$\eta_i(\lambda) + \left(\frac{\eta_i(\lambda)}{w_i\beta_i \left(\sum_{j=1}^n w_j(\|x_j\|_{\mathcal{H}_j} - \eta_j(\lambda))^{\beta_j} \right) - w_i\beta_i\xi} \right)^{\frac{1}{\beta_i-1}} = \|x_i\|_{\mathcal{H}_i}$$

$$\Leftrightarrow \frac{\eta_i(\lambda)}{w_i\beta_i \left(\sum_{j=1}^n w_j(\|x_j\|_{\mathcal{H}_j} - \eta_j(\lambda))^{\beta_j} \right) - w_i\beta_i\xi} = (\|x_i\|_{\mathcal{H}_i} - \eta_i(\lambda))^{\beta_i-1}, \qquad (5.23)$$

$i = 1, ..., n$. *By setting $\chi_i = \|x_i\|_{\mathcal{H}_i} - \eta_i(\lambda) \geq 0$, $i = 1, ..., n$, formula (5.23) can be expressed by*

$$\frac{\|x_i\|_{\mathcal{H}_i} - \chi_i}{w_i\beta_i \left(\sum_{j=1}^n w_j\chi_j^{\beta_j} \right) - w_i\beta_i\xi} = \chi_i^{\beta_i-1}$$

$$\Leftrightarrow w_i\beta_i\chi_i^{\beta_i-1} \sum_{j=1}^n w_j\chi_j^{\beta_j} - \xi w_i\beta_i\chi_i^{\beta_i-1} + \chi_i = \|x_i\|_{\mathcal{H}_i}$$

$$\Leftrightarrow w_i^2\beta_i\chi_i^{2\beta_i-1} + w_i\beta_i\chi_i^{\beta_i-1} \sum_{\substack{j=1 \\ j\neq i}}^n w_j\chi_j^{\beta_j} - \xi w_i\beta_i\chi_i^{\beta_i-1} + \chi_i = \|x_i\|_{\mathcal{H}_i}, \ i = 1, ..., n.$$

Hence, it holds for every $(x_1, ..., x_n, \xi) \in \mathcal{H}_1 \times ... \times \mathcal{H}_n \times \mathbb{R}$

$$\mathrm{P_{epi\,h}}(x_1, ..., x_n, \xi) = \begin{cases} (x_1, ..., x_n, \xi), & \text{if } \sum_{i=1}^n w_i\|x_i\|_{\mathcal{H}_i}^{\beta_i} \leq \xi, \\ (\overline{y}_1, ..., \overline{y}_n, \overline{\theta}), & \text{otherwise,} \end{cases}$$

with

$$\overline{y}_i = \frac{\overline{\chi}_i}{\|x_i\|_{\mathcal{H}_i}} x_i, \ i = 1, ...n, \ and \ \overline{\theta} = \sum_{i=1}^n w_i\overline{\chi}_i^{\beta_i},$$

where $\overline{\chi}_i \geq 0$, $i = 1, ..., n$, are the unique real numbers that solve a polynomial equation system of the form

$$w_i^2\beta_i\chi_i^{2\beta_i-1} + w_i\beta_i\chi_i^{\beta_i-1} \sum_{\substack{j=1 \\ j\neq i}}^n w_j\chi_j^{\beta_j} - \xi w_i\beta_i\chi_i^{\beta_i-1} + \chi_i = \|x_i\|_{\mathcal{H}_i}, \ i = 1, ..., n.$$

Let us additionally mention that the case where $n = 1$ was considered for instance in [33].

An important consequence of Theorem 5.1 where $\beta_i = 1$ for all $i = 1, ..., n$, follows.

Corollary 5.1. *Let* h *be given by (5.11) where* $\beta_i = 1$ *for all* $i = 1, ..., n$. *Then for all* $(x_1, ..., x_n, \xi) \in \mathcal{H}_1 \times ... \times \mathcal{H}_n \times \mathbb{R}$ *it holds*

$$P_{\text{epi}\,h}(x_1, ..., x_n, \xi) = \begin{cases} (x_1, ..., x_n, \xi), & \text{if } \sum_{i=1}^{n} w_i \|x_i\|_{\mathcal{H}_i} \leq \xi, \\ (0_{\mathcal{H}_1}, ..., 0_{\mathcal{H}_n}, 0), & \text{if } \xi < 0 \text{ and } \|x_i\|_{\mathcal{H}_i} \leq -\xi w_i, \ i = 1, ..., n, \\ (\overline{y}_1, ..., \overline{y}_n, \overline{\theta}), & \text{otherwise,} \end{cases}$$

(5.24)

where

$$\overline{y}_i = \frac{\max\{\|x_i\|_{\mathcal{H}_i} - \overline{\lambda} w_i, 0\}}{\|x_i\|_{\mathcal{H}_i}} x_i, \ i = 1, ..., n, \text{ and } \overline{\theta} = \xi + \overline{\lambda},$$

with

$$\overline{\lambda} = \frac{\sum_{i=k+1}^{n} w_i^2 \tau_i - \xi}{\sum_{i=k+1}^{n} w_i^2 + 1}$$

(5.25)

and $k \in \{0, 1, ..., n-1\}$ *is the unique integer such that* $\tau_k \leq \overline{\lambda} \leq \tau_{k+1}$, *where the values* $\tau_0, ..., \tau_n$ *are defined by* $\tau_0 := 0$ *and* $\tau_i := \|x_i\|_{\mathcal{H}_i}/w_i$, $i = 1, ..., n$ *and in ascending order.*

Proof. As $\beta_i = 1$ for all $i = 1, ..., n$, Theorem 5.1 yields

$$P_{\text{epi}\,h}(x_1, ..., x_n, \xi) = \begin{cases} (x_1, ..., x_n, \xi), & \text{if } \sum_{i=1}^{n} w_i \|x_i\|_{\mathcal{H}_i} \leq \xi, \\ (\overline{y}_1, ..., \overline{y}_n, \overline{\theta}), & \text{otherwise,} \end{cases}$$

with

$$\overline{y}_i = \frac{\max\{\|x_i\|_{\mathcal{H}_i} - \overline{\lambda} w_i, 0\}}{\|x_i\|_{\mathcal{H}_i}} x_i, \ i = 1, ..., n, \text{ and } \overline{\theta} = \xi + \overline{\lambda},$$

where $\overline{\lambda} > 0$ is a solution of the equation

$$\sum_{i=1}^{n} w_i \max\{\|x_i\|_{\mathcal{H}_i} - \lambda w_i, 0\} = \lambda + \xi.$$

Now, we consider the case where $\sum_{i=1}^{n} w_i \|x_i\|_{\mathcal{H}_i} > \xi$ and distinguish two cases.

(a) Let $\xi < 0$. If $\|x_i\|_{\mathcal{H}_i} + \xi w_i \leq 0$ for all $i = 1, ..., n$, we have by $0 \leq \overline{\theta} = \xi + \overline{\lambda}$, i.e. $\xi \geq -\overline{\lambda}$, that

$$0 \geq \|x_i\|_{\mathcal{H}_i} + \xi w_i \geq \|x_i\|_{\mathcal{H}_i} - \overline{\lambda} w_i \ \forall i = 1, ..., n,$$

(5.26)

and from here follows that

$$\overline{\lambda} + \xi = \sum_{i=1}^{n} w_i \max\{\|x_i\|_{\mathcal{H}_i} - \overline{\lambda} w_i, 0\} = 0, \text{ i.e. } \overline{\lambda} = -\xi. \tag{5.27}$$

But this means that $(\overline{y}_1, ..., \overline{y}_n, \overline{\theta}) = (0_{\mathcal{H}_1}, ..., 0_{\mathcal{H}_n}, 0)$, which verifies the second case of (5.24).

If we now assume that there exists $j \in \{1, ..., n\}$ such that $\|x_j\|_{\mathcal{H}_j} + \xi w_j > 0$, then we define the function $g : \mathbb{R} \to \mathbb{R}$ by

$$g(\lambda) := \sum_{i=1}^{n} w_i^2 \max\{\tau_i - \lambda, 0\} - \lambda - \xi. \tag{5.28}$$

Moreover, this assumption yields

$$g(\lambda) = \sum_{i=1}^{n} w_i^2 \max\{\tau_i - \lambda, 0\} - \lambda - \xi < \sum_{i=1}^{n} w_i^2 \max\{\tau_i - \lambda, 0\} - \lambda + \frac{\|x_j\|_{\mathcal{H}_j}}{w_j}.$$

Now, we choose $\widetilde{\lambda} > 0$ such that $\|x_i\|_{\mathcal{H}_i} - w_i \widetilde{\lambda} < 0$ for all $i = 1, ..., n$, and get

$$g(\widetilde{\lambda}) < -\widetilde{\lambda} + \frac{\|x_j\|_{\mathcal{H}_j}}{w_j} < 0.$$

(b) Let $\xi \geq 0$. If there exists $j \in \{1, ..., n\}$ such that $\|x_j\|_{\mathcal{H}_j} + \xi w_j < 0$, we derive a contradiction. Therefore, it holds $\|x_i\|_{\mathcal{H}_i} + \xi w_i \geq 0$ for all $i = 1, ..., n$, and for the function g we have

$$g(\lambda) = \sum_{i=1}^{n} w_i^2 \max\{\tau_i - \lambda, 0\} - \lambda - \xi \leq \sum_{i=1}^{n} w_i^2 \max\{\tau_i - \lambda, 0\} - \lambda.$$

Now, we can take $\widetilde{\lambda} > 0$ such that $\|x_i\|_{\mathcal{H}_i} - w_i \widetilde{\lambda} < 0$ for all $i = 1, ..., n$, and derive that $g(\widetilde{\lambda}) \leq -\widetilde{\lambda} < 0$.

In summary, we can secure the existence of $\widetilde{\lambda} > 0$ such that $g(\widetilde{\lambda}) < 0$. Additionally, take note that, if $\lambda = 0$, then $g(0) = \sum_{i=1}^{n} w_i \|x_i\|_{\mathcal{H}_i} - \xi > 0$. The rest of the proof is oriented on the Algorithm I given in [96] to determine the projection onto an l_1-norm ball.

Since, the values $\tau_0, ..., \tau_n$ are in ascending order, g is a piecewise linear function in λ, where the slope of g changes at $\lambda = \tau_i$, $i = 0, ..., n$. More precisely, at $\lambda = 0$ the slope of g is $-(\sum_{i=1}^{n} w_i^2 + 1)$ and increases by w_1^2 when $\lambda = \tau_1$. If we proceed in this way, one may see that the slope keeps increasing when λ takes the values τ_k, $k = 2, ..., n$. In the case when $\lambda \geq \tau_n$ the slope of g is -1.

Hence, to determine λ such that $g(\lambda) = 0$, we have to locate the interval where g changes its sign from a positive to a negative value. In other words, we have to find the unique integer $k \in \{0, ..., n-1\}$ such that $g(\tau_k) \geq 0$ and $g(\tau_{k+1}) \leq 0$. Hence, we have

$$g(\lambda) = -\left(\sum_{i=k+1}^{n} w_i^2 + 1 \right) \lambda + \sum_{i=k+1}^{n} w_i^2 \tau_i - \xi,$$

where $\tau_k \leq \lambda \leq \tau_{k+1}$. Finally, we can determine $\overline{\lambda}$ such that $g(\overline{\lambda}) = 0$:

$$\overline{\lambda} = \frac{\sum\limits_{i=k+1}^{n} w_i^2 \tau_i - \xi}{\sum\limits_{i=k+1}^{n} w_i^2 + 1}.$$

$\square$

Remark 5.2. *From the ideas of the previous proof, we can now construct an algorithm to determine $\overline{\lambda}$ of Corollary 5.1.*
Algorithm:

1. *If $\sum_{i=1}^{n} w_i \|x_i\|_{\mathcal{H}_i} \leq \xi$, then $\overline{\lambda} = 0$.*

2. *If $\xi < 0$ and $\|x_i\|_{\mathcal{H}_i} \leq -\xi w_i$ for all $i = 1, ..., n$, then $\overline{\lambda} = -\xi$.*

3. *Otherwise, define $\tau_0 := 0$, $\tau_i := \|x_i\|_{\mathcal{H}_i}/w_i$, $i = 1, ..., n$, and sort $\tau_0, ..., \tau_n$ in ascending order.*

4. *Determine the values of g defined in (5.28) at $\lambda = \tau_i$, $i = 0, ..., n$.*

5. *Find the unique $k \in \{0, ..., n-1\}$ such that $g(\tau_k) \geq 0$ and $g(\tau_{k+1}) \leq 0$.*

6. *Calculate $\overline{\lambda}$ by (5.25).*

Corollary 5.2. *Let h be given by (5.11) where $\beta_i = 2$ and $w_i = 1$ for all $i = 1, ..., n$, then it holds*

$$P_{\mathrm{epi}\, h}(x_1, ..., x_n, \xi) = \begin{cases} (x_1, ..., x_n, \xi), & \text{if } \sum\limits_{i=1}^{n} \|x_i\|_{\mathcal{H}_i}^2 \leq \xi, \\ (\overline{y}_1, ..., \overline{y}_n, \overline{\theta}), & \text{otherwise,} \end{cases}$$

where

$$\overline{y}_i = \frac{1}{2\overline{\lambda} + 1} x_i, \ i = 1, ..., n, \ \text{and } \overline{\theta} = \xi + \overline{\lambda},$$

and $\overline{\lambda} > 0$ is a solution of a cubic equation of the form

$$\lambda^3 + (1 + \xi)\lambda^2 + \frac{1}{4}(1 + 4\xi)\lambda + \frac{1}{4}\left(\xi - \sum_{i=1}^{n} \|x_i\|_{\mathcal{H}_i}^2\right) = 0. \tag{5.29}$$

Proof. By Theorem 5.1 we get that

$$P_{\mathrm{epi}\, h}(x_1, ..., x_n, \xi) = \begin{cases} (x_1, ..., x_n, \xi), & \text{if } \sum\limits_{i=1}^{n} \|x_i\|_{\mathcal{H}_i}^2 \leq \xi, \\ (\overline{y}_1, ..., \overline{y}_n, \overline{\theta}), & \text{otherwise,} \end{cases}$$

with

$$\overline{y}_i = \frac{\|x_i\|_{\mathcal{H}_i} - \eta_i(\overline{\lambda})}{\|x_i\|_{\mathcal{H}_i}} x_i, \ i = 1, ..., n, \ \text{and} \ \overline{\theta} = \xi + \overline{\lambda}, \tag{5.30}$$

where $\eta_i(\overline{\lambda})$ is the unique non-negative real number that solves the equation

$$\eta_i(\overline{\lambda}) + \frac{\eta_i(\overline{\lambda})}{2\overline{\lambda}} = \|x_i\|_{\mathcal{H}_i}, \ i = 1, ..., n, \tag{5.31}$$

and $\overline{\lambda} > 0$ is a solution of the equation

$$\sum_{i=1}^{n} (\|x_i\|_{\mathcal{H}_i} - \eta_i(\lambda))^2 = \lambda + \xi. \tag{5.32}$$

From (5.31) we get immediately

$$\eta_i(\overline{\lambda})\left(1 + \frac{1}{2\overline{\lambda}}\right) = \|x_i\|_{\mathcal{H}_i} \Leftrightarrow \eta_i(\overline{\lambda}) = \frac{2\overline{\lambda}}{2\overline{\lambda} + 1}\|x_i\|_{\mathcal{H}_i}, \ i = 1, ..., n, \tag{5.33}$$

and in combination with (5.32) we derive

$$\sum_{i=1}^{n} \left(\|x_i\|_{\mathcal{H}_i} - \frac{2\overline{\lambda}}{2\overline{\lambda} + 1}\|x_i\|_{\mathcal{H}_i}\right)^2 = \overline{\lambda} + \xi \Leftrightarrow \frac{1}{(2\overline{\lambda} + 1)^2}\sum_{i=1}^{n}\|x_i\|_{\mathcal{H}_i}^2 = \overline{\lambda} + \xi$$

$$\Leftrightarrow (2\overline{\lambda} + 1)^2(\overline{\lambda} + \xi) - \sum_{i=1}^{n}\|x_i\|_{\mathcal{H}_i}^2 = 0$$

$$\Leftrightarrow 4\overline{\lambda}^3 + 4(1 + \xi)\overline{\lambda}^2 + (1 + 4\xi)\overline{\lambda} + \xi - \sum_{i=1}^{n}\|x_i\|_{\mathcal{H}_i}^2 = 0.$$

In the end, formula (5.33) implies that

$$\overline{y}_i = \frac{\|x_i\|_{\mathcal{H}_i} - \frac{2\overline{\lambda}}{2\overline{\lambda}+1}\|x_i\|_{\mathcal{H}_i}}{\|x_i\|_{\mathcal{H}_i}} x_i = \frac{1}{2\overline{\lambda} + 1} x_i, \ i = 1, ..., n, \tag{5.34}$$

which completes the proof. $\square$

The next remark discusses the question whether the solution $\overline{\lambda} > 0$ of Corollary 5.2 is unique.

Remark 5.3. Let $(x_1, ..., x_n, \xi) \in \mathcal{H}_1 \times ... \times \mathcal{H}_n \times \mathbb{R}$ be such that $\sum_{i=1}^{n}\|x_i\|_{\mathcal{H}_i}^2 > \xi$ and $g : \mathbb{R} \to \mathbb{R}$ be defined by $g(\lambda) := \lambda^3 + (1 + \xi)\lambda^2 + (1/4)(1 + 4\xi)\lambda + (1/4)(\xi - \sum_{i=1}^{n}\|x_i\|_{\mathcal{H}_i}^2)$, then $g'(\lambda) = 3\lambda^2 + 2(1 + \xi)\lambda + (1/4)(1 + 4\xi)$ as well as $g''(\lambda) = 6\lambda + 2(1 + \xi)$. From the zeros of g' we derive the local extrema of g as follows

$$\lambda_{1/2} = -\frac{1}{3}(1 + \xi) \pm \sqrt{\frac{(1 + \xi)^2}{9} - \frac{1 + 4\xi}{12}} = -\frac{1}{3}(1 + \xi) \pm \sqrt{\frac{4(1 + 2\xi + \xi^2) - 3(1 + 4\xi)}{36}}$$

$$= -\frac{1}{3}(1 + \xi) \pm \sqrt{\frac{1 - 4\xi + 4\xi^2}{36}} = -\frac{1}{3}(1 + \xi) \pm \frac{1}{6}(1 - 2\xi)$$

and hence, $\lambda_1 = -(1/6)(1 + 4\xi)$ and $\lambda_2 = -(1/2)$.

Further, if $\xi > 1/2 \Leftrightarrow -1 + 2\xi > 0$, then g is strongly monotone increasing on $\mathbb{R}_+$, $g''(\lambda_1) = 1 - 2\xi < 0$ and $g''(\lambda_2) = -1 + 2\xi > 0$, which means that g has in λ_1 a local maximum and in λ_2 a local minimum. As $\lambda_1 < \lambda_2 < 0$ and $g(0) = (1/4)(\xi - \sum_{i=1}^n \|x_i\|_{\mathcal{H}_i}^2) < 0$, the function g has exactly one positive zero in this situation.

If $\xi < 1/2 \Leftrightarrow 1 - 2\xi > 0$, then $g''(\lambda_1) = 1 - 2\xi > 0$ and $g''(\lambda_2) = -1 + 2\xi < 0$ and we derive a local minimum in λ_1 and a local maximum in λ_2. From $g(0) < 0$ and $\lambda_2 < \lambda_1$ we conclude that g has also in this situation exactly one positive zero.

Finally, let us consider the case where $\xi = 1/2$, then g is strongly monotone increasing on $\mathbb{R}_+$, $\lambda_1 = \lambda_2 = -1/2$ and $g''(\lambda_1) = 0$, i.e. g has at the point $-(1/2)$ a saddle point. From the fact that $g''(\lambda) \leq 0$ for all $\lambda \in (-\infty, -(1/2)]$ and $g''(\lambda) > 0$ for all $\lambda \in (-(1/2), +\infty)$, it is clear that g has again exactly one positive zero.

In conclusion, the function g has in all situations exactly one positive zero, i.e. $\overline{\lambda} > 0$ is unique.

Remark 5.4. In the framework of Corollary 5.2, let us consider the case where $n = 1$. Then, by Remark 5.1 we have to find a real number $\overline{\chi} \geq 0$ that solves the equation

$$2\chi^3 + (1 - 2\xi)\chi - \|x\|_{\mathcal{H}} = 0, \tag{5.35}$$

to get a formula of the projection onto the epigraph of h.

As one may see by (5.29), the arithmetic effort for the case $n > 1$ is not much higher compared to the case $n = 1$. In both situations we have to solve a cubic equation to derive a formula for the projection onto the epigraph of h.

As a direct consequence of Corollary 5.1 one gets the following well-known statement (see for instance [3] or [33]).

Corollary 5.3. Let h be given by (5.11) where $n = 1$, $w_1 = w \geq 1$ and $\beta_1 = 1$, i.e. $h(x) = w\|x\|_{\mathcal{H}}$. Then, for every $(x, \xi) \in \mathcal{H} \times \mathbb{R}$

$$P_{\text{epi } w\|\cdot\|_{\mathcal{H}}}(x, \xi) = \begin{cases} (x, \xi), & \text{if } w\|x\|_{\mathcal{H}} \leq \xi, \\ (0, 0), & \text{if } \|x\|_{\mathcal{H}} \leq -w\xi, \\ \left(\frac{\|x\|_{\mathcal{H}} + w\xi}{\|x\|_{\mathcal{H}}(w^2 + 1)}x, \frac{w\|x\|_{\mathcal{H}} + w^2\xi}{w^2 + 1}\right), & \text{otherwise.} \end{cases}$$

For our numerical tests we need two lemmas more.

Lemma 5.2. *For* $p_i \in \mathcal{H}$, $i = 1, ..., n$, *it holds*

$$\mathrm{P}_{\mathrm{epi}\left(\sum\limits_{i=1}^{n} w_i\|\cdot - p_i\|_{\mathcal{H}_i}^{\beta_i}\right)}(x_1, ..., x_n, \xi) = \mathrm{P}_{\mathrm{epi}\left(\sum\limits_{i=1}^{n} w_i\|\cdot\|_{\mathcal{H}_i}^{\beta_i}\right)}(x_1 - p_1, ..., x_n - p_n, \xi) + (p_1, ..., p_n, 0).$$

Proof. For $p_i \in \mathcal{H}_i$, $i = 1, ..., n$ one has

$$(x_1, ..., x_n, \xi) \in \mathrm{epi}\left(\sum_{i=1}^{n} w_i\|\cdot - p_i\|_{\mathcal{H}_i}^{\beta_i}\right) \Leftrightarrow \sum_{i=1}^{n} w_i\|x_i - p_i\|_{\mathcal{H}_i}^{\beta_i} \leq \xi$$

$$\Leftrightarrow (x_1 - p_1, ..., x_n - p_n, \xi) \in \mathrm{epi}\left(\sum_{i=1}^{n} w_i\|\cdot\|_{\mathcal{H}_i}^{\beta_i}\right)$$

$$\Leftrightarrow (x_1, ..., x_n, \xi) \in \mathrm{epi}\left(\sum_{i=1}^{n} w_i\|\cdot\|_{\mathcal{H}_i}^{\beta_i}\right) + (p_1, ..., p_n, 0).$$

Thus, by [3, Proposition 3.17] follows

$$\mathrm{P}_{\mathrm{epi}\left(\sum\limits_{i=1}^{n} w_i\|\cdot - p_i\|_{\mathcal{H}_i}^{\beta_i}\right)}(x_1, ..., x_n, \xi) = \mathrm{P}_{\mathrm{epi}\left(\sum\limits_{i=1}^{n} w_i\|\cdot\|_{\mathcal{H}_i}^{\beta_i}\right) + (p_1, ..., p_n, 0)}(x_1, ..., x_n, \xi)$$

$$= \mathrm{P}_{\mathrm{epi}\left(\sum\limits_{i=1}^{n} w_i\|\cdot\|_{\mathcal{H}_i}^{\beta_i}\right)}(x_1 - p_1, ..., x_n - p_n, \xi) + (p_1, ..., p_n, 0).$$

□

Lemma 5.3. *Let* $w > 0$ *and* $A : \mathcal{K} \to \mathcal{H}$ *be a linear operator with* $AA^* = \mu Id$, $\mu > 0$, *where* $\mathcal{K}$ *is a real Hilbert space. Then,*

$$\mathrm{P}_{\mathrm{epi}\, w\|A\cdot\|_{\mathcal{H}}}(x, \xi) = (x, \xi) + \left(\frac{1}{\sqrt{\mu}}A^* \times \mathrm{Id}\right)\left(\mathrm{P}_{\mathrm{epi}\, w\sqrt{\mu}\|\cdot\|_{\mathcal{H}}}\left(\frac{1}{\sqrt{\mu}}Ax, \xi\right) - \left(\frac{1}{\sqrt{\mu}}Ax, \xi\right)\right),$$

where $\frac{1}{\sqrt{\mu}}A^* \times \mathrm{Id} : \mathcal{H} \times \mathbb{R} \to \mathcal{K} \times \mathbb{R}$ *is defined as* $\left(\frac{1}{\sqrt{\mu}}A^* \times \mathrm{Id}\right)(y, \zeta) = \left(\frac{1}{\sqrt{\mu}}A^*y, \zeta\right)$.

Proof. We have

$$\delta_{\mathrm{epi}(w\|A\cdot\|_{\mathcal{H}})}(x, \xi) = \delta_{\mathrm{epi}(w\sqrt{\mu}\|\cdot\|_{\mathcal{H}})}\left(\frac{1}{\sqrt{\mu}}Ax, \xi\right) = \left(\delta_{\mathrm{epi}(w\sqrt{\mu}\|\cdot\|_{\mathcal{H}})} \circ \left(\frac{1}{\sqrt{\mu}}A \times \mathrm{Id}\right)\right)(x, \xi).$$

By [3, Proposition 23.32] (with $L = (1/\sqrt{\mu})A \times \mathrm{Id}$) it follows that

$$\mathrm{prox}_{\delta_{\mathrm{epi}\, w\|A\cdot\|_{\mathcal{H}}}}(x, \xi) = \mathrm{prox}_{\delta_{\mathrm{epi}(w\sqrt{\mu}\|\cdot\|_{\mathcal{H}})} \circ \left(\frac{1}{\sqrt{\mu}}A \times \mathrm{Id}\right)}(x, \xi)$$

$$= (x, \xi) + \left(\frac{1}{\sqrt{\mu}}A \times \mathrm{Id}\right)^*\left(\mathrm{prox}_{\delta_{\mathrm{epi}\, w\sqrt{\mu}\|\cdot\|_{\mathcal{H}}}}\left(\frac{1}{\sqrt{\mu}}Ax, \xi\right) - \left(\frac{1}{\sqrt{\mu}}Ax, \xi\right)\right)$$

$$\Leftrightarrow \mathrm{P}_{\mathrm{epi}\, w\|A\cdot\|_{\mathcal{H}}}(x, \xi)$$

$$= (x, \xi) + \left(\frac{1}{\sqrt{\mu}}A^* \times \mathrm{Id}\right)\left(\mathrm{P}_{\mathrm{epi}\, w\sqrt{\mu}\|\cdot\|_{\mathcal{H}}}\left(\frac{1}{\sqrt{\mu}}Ax, \xi\right) - \left(\frac{1}{\sqrt{\mu}}Ax, \xi\right)\right).$$

□

5.2.2 Gauges

The next considerations are devoted to gauge functions of closed convex sets defined on Hilbert spaces.

Theorem 5.2. *Let C be a closed convex subset of $\mathcal{H}$ such that $0_{\mathcal{H}} \in C$, then it holds for every $(x, \xi) \in \mathcal{H} \times \mathbb{R}$*

$$
\mathrm{P}_{\mathrm{epi}\,\gamma_C}(x, \xi) = \begin{cases} (x, \xi), & \text{if } \gamma_C(x) \leq \xi, \\ \left(\mathrm{P}_{\mathrm{cl}(\mathrm{dom}\,\gamma_C)}(x), \xi\right), & \text{if } x \notin \mathrm{dom}\,\gamma_C \text{ and } \gamma_C\left(\mathrm{P}_{\mathrm{cl}(\mathrm{dom}\,\gamma_C)}(x)\right) \leq \xi < \gamma_C(x), \\ (\overline{y}, \overline{\theta}), & \text{otherwise,} \end{cases}
$$

where

$$
\overline{y} = x - \overline{\lambda}\,\mathrm{P}_{C^0}\left(\frac{1}{\overline{\lambda}}x\right) \ \text{ and } \ \overline{\theta} = \overline{\lambda} + \xi
$$

and $\overline{\lambda} > 0$ is a solution of an equation of the form

$$
\lambda + \xi = \left\langle x, \mathrm{P}_{C^0}\left(\frac{1}{\lambda}x\right) \right\rangle_{\mathcal{H}} - \lambda \left\| \mathrm{P}_{C^0}\left(\frac{1}{\lambda}x\right) \right\|_{\mathcal{H}}^2.
$$

Proof. Let us consider for fixed $(x, \xi) \in \mathcal{H} \times \mathbb{R}$ the following optimization problem

$$
\min_{\substack{(y, \theta) \in \mathcal{H} \times \mathbb{R}, \\ \gamma_C(y) \leq \theta}} \left\{ \frac{1}{2}(\theta - \xi)^2 + \frac{1}{2}\|y - x\|_{\mathcal{H}}^2 \right\}. \tag{5.36}
$$

If $\gamma_C(x) \leq \xi$, i.e. $(x, \xi) \in \mathrm{epi}\,\gamma_C$, then it is obvious that $(\overline{y}, \overline{\theta}) = (x, \xi)$. In the following we consider the non-trivial situation where $\gamma_C(x) > \xi$.

We define the function $f : \mathcal{H} \times \mathbb{R} \to \mathbb{R}$ by $f(y, \theta) := (1/2)(\theta - \xi)^2 + (1/2)\|y - x\|_{\mathcal{H}}^2$ and the function $g : \mathcal{H} \times \mathbb{R} \to \overline{\mathbb{R}}$ by $g(y, \theta) = \gamma_C(y) - \theta$, then it is clear that f is continuous and strongly convex and g is proper, lower semicontinuous and convex by Theorem 4.1. As $\gamma_C(0) < 1$, it follows by [14, Theorem 3.3.16] (see also [14, Remark 3.3.8]) that

$$
0 \in \partial(f + (\overline{\lambda}g))(\overline{y}, \overline{\theta}) \tag{5.37}
$$

and

$$
\begin{cases} (\overline{\lambda}g)(\overline{y}, \overline{\theta}) = 0, \\ g(\overline{y}, \overline{\theta}) \leq 0, \end{cases} \Leftrightarrow \begin{cases} \overline{\lambda}(\gamma_C(\overline{y}) - \overline{\theta}) = 0, \\ \gamma_C(\overline{y}) \leq \overline{\theta}, \end{cases} \tag{5.38}
$$

where $(\overline{y}, \overline{\theta})$ is the unique solution of (5.36) and $\overline{\lambda} \geq 0$ the associated Lagrange multiplier. Furthermore, from [14, Theorem 3.5.13] one gets that

$$
0 \in \partial(f + (\overline{\lambda}g))(\overline{y}, \overline{\theta}) \Leftrightarrow 0 \in \partial f(\overline{y}, \overline{\theta}) + \partial(\overline{\lambda}g)(\overline{y}, \overline{\theta}). \tag{5.39}
$$

If $\overline{\lambda} = 0$, then it follows by (5.8) and (5.3)

$$0 \in \partial f(\overline{y}, \overline{\theta}) + \partial\delta_{\mathrm{dom}\,g}(\overline{y}, \overline{\theta}) \Leftrightarrow 0 \in (\overline{y} - x, \overline{\theta} - \xi) + \partial\delta_{\mathrm{dom}\,\gamma_C \times \mathbb{R}}(\overline{y}, \overline{\theta})$$

$$\Leftrightarrow 0 \in (\overline{y} - x, \overline{\theta} - \xi) + \partial\delta_{\mathrm{cl}(\mathrm{dom}\,\gamma_C) \times \mathbb{R}}(\overline{y}, \overline{\theta}) \Leftrightarrow (x - \overline{y}, \xi - \overline{\theta}) \in \partial\delta_{\mathrm{cl}(\mathrm{dom}\,\gamma_C) \times \mathbb{R}}(\overline{y}, \overline{\theta})$$

$$\Leftrightarrow (\overline{y}, \overline{\theta}) = \mathrm{P}_{\mathrm{cl}(\mathrm{dom}\,\gamma_C) \times \mathbb{R}}(x, \xi) \Leftrightarrow \begin{cases} \overline{y} = \mathrm{P}_{\mathrm{cl}(\mathrm{dom}\,\gamma_C)}(x), \\ \overline{\theta} = \xi, \end{cases}$$

and thus, it holds by the feasibility condition (5.38) that $\gamma_C(\mathrm{P}_{\mathrm{cl}(\mathrm{dom}\,\gamma_C)}(x)) \leq \xi$, from which follows that $\mathrm{P}_{\mathrm{cl}(\mathrm{dom}\,\gamma_C)}(x) \in \mathrm{dom}\,\gamma_C$. If $x \in \mathrm{dom}\,\gamma_C$, this means that $\mathrm{P}_{\mathrm{cl}(\mathrm{dom}\,\gamma_C)}(x) = x$ and again by the feasibility condition (5.38) that $\gamma_C(x) \leq \xi$, which contradicts our assumption. Therefore, if $x \notin \mathrm{dom}\,\gamma_C$ and the inequalities $\gamma_C(\mathrm{P}_{\mathrm{cl}(\mathrm{dom}\,\gamma_C)(x)}) \leq \xi < \gamma_C(x)$ hold, then $(\overline{y}, \overline{\theta}) = (\mathrm{P}_{\mathrm{cl}(\mathrm{dom}\,\gamma_C)}(x), \xi)$.

Now, let $\overline{\lambda} > 0$, then it follows from (5.39) and (5.8)

$$0 \in \partial(f + (\overline{\lambda}g))(\overline{y}, \overline{\theta}) \Leftrightarrow 0 \in \partial f(\overline{y}, \overline{\theta}) + \overline{\lambda}\partial g(\overline{y}, \overline{\theta})$$

$$\Leftrightarrow \nabla f(\overline{y}, \overline{\theta}) \in -\overline{\lambda}\partial g(\overline{y}, \overline{\theta}) \Leftrightarrow \begin{cases} \overline{y} - x \in -\overline{\lambda}\partial\gamma_C(\overline{y}), \\ \overline{\theta} - \xi = \overline{\lambda}, \end{cases} \Leftrightarrow \begin{cases} \overline{y} = \mathrm{prox}_{\overline{\lambda}\gamma_C}x, \\ \overline{\theta} = \xi - \overline{\lambda}, \end{cases} \tag{5.40}$$

by combining (5.40) and (5.38) we derive that $\gamma_C(\overline{y}) = \xi + \overline{\lambda}$. Finally, as by Lemma 4.4 and Remark 4.5 it holds that $\gamma_C^* = \delta_{C^\circ}$, one gets by [3, Theorem 14.3(iii)] the following equivalences

$$\gamma_C(\overline{y}) = \xi + \overline{\lambda} \tag{5.41}$$

$$\Leftrightarrow \xi + \overline{\lambda} = \gamma_C\left(\mathrm{prox}_{\overline{\lambda}\gamma_C}x\right) + \delta_{C^\circ}\left(\mathrm{P}_{C^\circ}\left(\frac{1}{\overline{\lambda}}x\right)\right) = \left\langle \mathrm{prox}_{\overline{\lambda}\gamma_C}x, \mathrm{P}_{C^\circ}\left(\frac{1}{\overline{\lambda}}x\right)\right\rangle_{\mathcal{H}}$$

$$\Leftrightarrow \xi + \overline{\lambda} = \left\langle x - \overline{\lambda}\mathrm{P}_{C^\circ}\left(\frac{1}{\overline{\lambda}}x\right), \mathrm{P}_{C^\circ}\left(\frac{1}{\overline{\lambda}}x\right)\right\rangle_{\mathcal{H}}.$$

$$\square$$

Corollary 5.4. *Let $C \subseteq \mathcal{H}$ be a closed convex cone, then $\gamma_C = \delta_C$ and*

$$\mathrm{P}_{\mathrm{epi}\,\gamma_C}(x, \xi) = \mathrm{P}_{C \times \mathbb{R}_+}(x, \xi) = \begin{cases} (x, \xi), & \text{if } x \in C \text{ and } \xi \geq 0, \\ (\mathrm{P}_C\,x, \max\{0, \xi\}), & \text{otherwise.} \end{cases}$$

Proof. We use Theorem 5.2. Let $x \in \mathrm{dom}\,\gamma_C$ such that $\gamma_C(x) > \xi$, then one has from [3, Proposition 28.22] and [3, Theorem 6.29] that

$$\overline{y} = x - \overline{\lambda}\mathrm{P}_{C^\circ}\left(\frac{1}{\overline{\lambda}}x\right) = x - \mathrm{P}_{C^\circ}\,x = \mathrm{P}_C\,x. \tag{5.42}$$

Moreover, as $\gamma_C = \delta_C$ it holds that $\mathrm{dom}\,\gamma_C = C$ and by (5.38) we have $\gamma_C(\overline{y}) = \overline{\theta}$, which yields $\mathrm{P}_{\mathrm{epi}\,\gamma_C}(x, \xi) = (\mathrm{P}_C\,x, \gamma_C(\mathrm{P}_C\,x)) = (\mathrm{P}_C\,x, 0)$.

If $x \notin \mathrm{dom}\,\gamma_C = C$, then $0 = \gamma_C(\mathrm{P}_{\mathrm{cl}(\mathrm{dom}\,\gamma_C)}(x)) \leq \xi < \gamma_C(x) = +\infty$ and so, $\mathrm{P}_{\mathrm{epi}\,\gamma_C}(x, \xi) = (\mathrm{P}_{\mathrm{cl}(\mathrm{dom}\,\gamma_C)}(x), \xi) = (\mathrm{P}_C\,x, \xi)$, which implies the statement. $\square$

Corollary 5.5. *Let C_i be a closed convex subset of $\mathcal{H}_i$ such that $0_{\mathcal{H}_i} \in \text{int } C_i$, $i = 1, ..., n$, and the gauge $\gamma_C : \mathcal{H}_1 \times ... \times \mathcal{H}_n \to \overline{\mathbb{R}}$ be defined by $\gamma_C(x_1, ..., x_n) = \sum_{i=1}^{n} \gamma_{C_i}(x_i)$. Then it holds for every $(x_1, ..., x_n, \xi) \in \mathcal{H}_1 \times ... \times \mathcal{H}_n \times \mathbb{R}$*

$$
P_{\text{epi}\,\gamma_C}(x_1, ..., x_n, \xi) = \begin{cases} (x_1, ..., x_n, \xi), & \text{if } \sum_{i=1}^{n} \gamma_{C_i}(x_i) \leq \xi, \\ (\overline{y}_1, ..., \overline{y}_n, \overline{\theta}), & \text{otherwise,} \end{cases}
$$

where

$$
\overline{y}_i = x_i - \overline{\lambda} P_{C_i^0}\left(\frac{1}{\overline{\lambda}} x_i\right), \ i = 1, ..., n, \ \text{and } \overline{\theta} = \overline{\lambda} + \xi \tag{5.43}
$$

and $\overline{\lambda} > 0$ is a solution of an equation of the form

$$
\lambda + \xi = \sum_{i=1}^{n} \left[\left\langle x_i, P_{C_i^0}\left(\frac{1}{\lambda} x_i\right) \right\rangle_{\mathcal{H}_i} - \lambda \left\| P_{C_i^0}\left(\frac{1}{\lambda} x_i\right) \right\|^2_{\mathcal{H}_i} \right]. \tag{5.44}
$$

Proof. As $0_{\mathcal{H}_i} \in \text{int } C_i$, $i = 1, ..., n$, it is clear that the gauges are well-defined, i.e. $\text{dom } \gamma_{C_i} = \mathcal{H}_i$, $i = 1, ..., n$, and so, $\text{dom } \gamma_C = \mathcal{H}_1 \times ... \times \mathcal{H}_n$. Further, let us recall that the polar set C^0 of the set C can be characterized by the dual gauge γ_{C^0} as

$$
C^0 = \{x = (x_1, ..., x_n) \in \mathcal{H}_1 \times ... \times \mathcal{H}_n : \gamma_{C^0}(x) = \gamma_{C^0}(x_1, ..., x_n) \leq 1\}. \tag{5.45}
$$

This relation holds also for the polar set C_i^0 and its associated dual gauge $\gamma_{C_i^0}$, $i = 1, ..., n$. Moreover, in Lemma 4.5 it was shown that $\gamma_{C^0}(x) = \max_{1 \leq i \leq n}\{\gamma_{C_i^0}(x_i)\}$ and hence, the polar set in (5.45) can be written as

$$
\begin{aligned}
C^0 &= \left\{ (x_1, ..., x_n) \in \mathcal{H}_1 \times ... \times \mathcal{H}_n : \max_{1 \leq i \leq n}\{\gamma_{C_i^0}(x_i)\} \leq 1 \right\} \\
&= \left\{ (x_1, ..., x_n) \in \mathcal{H}_1 \times ... \times \mathcal{H}_n : \gamma_{C_i^0}(x_i) \leq 1, \ i = 1, ..., n \right\} \\
&= \{x_1 \in \mathcal{H}_1 : \gamma_{C_1^0}(x_1) \leq 1\} \times ... \times \{x_n \in \mathcal{H}_n : \gamma_{C_n^0}(x_n) \leq 1\} = C_1^0 \times ... \times C_n^0.
\end{aligned}
$$

From here follows that

$$
P_{C^0}(x) = P_{C_1^0 \times ... \times C_n^0}(x_1, ..., x_n) = P_{C_1^0}(x_1) \times ... \times P_{C_n^0}(x_n),
$$

which by using Theorem 5.2 directly implies (5.43) and (5.44). □

As one may see, the equation (5.44) of the previous corollary can be very hard to solve and hence, it can be very complicated to find a projection formula. The next two corollaries are examples, which demonstrate how one can determine the formula of the projector by using Corollary 5.5.

Corollary 5.6. *Let* $\gamma_C : \mathcal{H}_1 \times ... \times \mathcal{H}_n \to \mathbb{R}$ *be defined by*

$$\gamma_C(x_1, ..., x_n) := \max_{1 \leq i \leq n} \{\|x_i\|_{\mathcal{H}_i}\} + \|x_{n+1}\|_{\mathcal{H}_{n+1}},$$

then for every $(x_1, ..., x_n, \xi) \in \mathcal{H}_1 \times ... \times \mathcal{H}_n \times \mathbb{R}$

$$P_{\operatorname{epi}\gamma_C}(x_1, ..., x_{n+1}, \xi) = \begin{cases} (x_1, ..., x_{n+1}, \xi), & \text{if } \max_{1 \leq i \leq n}\{\|x_i\|_{\mathcal{H}_i}\} + \|x_{n+1}\|_{\mathcal{H}_{n+1}} \leq \xi, \\ (\overline{y}_1, ..., \overline{y}_{n+1}, \overline{\theta}), & \text{otherwise,} \end{cases}$$

where exactly one of the following four cases holds:

(i) $\sum_{i=1}^{n} \max\{\|x_i\|_{\mathcal{H}_i} - \|x_{n+1}\|_{\mathcal{H}_{n+1}} - \xi, 0\} < \|x_{n+1}\|_{\mathcal{H}_{n+1}}$,
$\sum_{i=1}^{n} \max\{\|x_i\|_{\mathcal{H}_i} - \|x_{n+1}\|_{\mathcal{H}_{n+1}} - \xi, 0\} < \sum_{i=1}^{n} \|x_i\|_{\mathcal{H}_i}$,
$\sum_{i=1}^{n} \|x_i\|_{\mathcal{H}_i} > (\|x_{n+1}\|_{\mathcal{H}_{n+1}} - \xi)/2 > -\xi$ *and* $\|x_{n+1}\|_{\mathcal{H}_{n+1}} > -\xi$, *then*

$$\overline{y}_i = x_i - \frac{\max\{\|x_i\|_{\mathcal{H}_i} - 2\overline{\lambda} - \xi + \|x_{n+1}\|_{\mathcal{H}_{n+1}}, 0\}}{\|x_i\|_{\mathcal{H}_i}} x_i, \ i = 1, ..., n,$$

$$\overline{y}_{n+1} = \frac{\|x_{n+1}\|_{\mathcal{H}_{n+1}} - \overline{\lambda}}{\|x_{n+1}\|_{\mathcal{H}_{n+1}}} x_{n+1} \ \text{and} \ \overline{\theta} = \overline{\lambda} + \xi,$$

where $\overline{\lambda} > 0$ *is a solution of the equation*

$$\sum_{i=1}^{n} \max\{\|x_i\|_{\mathcal{H}_i} - 2\overline{\lambda} - \xi + \|x_{n+1}\|_{\mathcal{H}_{n+1}}, 0\} = \overline{\lambda}.$$

(ii) $\|x_{n+1}\|_{\mathcal{H}_{n+1}} \leq \sum_{i=1}^{n} \max\{\|x_i\|_{\mathcal{H}_i} - \|x_{n+1}\|_{\mathcal{H}_{n+1}} - \xi, 0\} < \sum_{i=1}^{n} \|x_i\|_{\mathcal{H}_i}$ *and* $\sum_{i=1}^{n} \|x_i\|_{\mathcal{H}_i} > -\xi$, *then*

$$\overline{y}_i = x_i - \frac{\max\{\|x_i\|_{\mathcal{H}_i} - \overline{\lambda} - \xi, 0\}}{\|x_i\|_{\mathcal{H}_i}} x_i, \ i = 1, ..., n, \ \overline{y}_{n+1} = 0_{\mathcal{H}_{n+1}} \ \text{and} \ \overline{\theta} = \overline{\lambda} + \xi,$$

where $\overline{\lambda} > 0$ *is a solution of the equation*

$$\sum_{i=1}^{n} \max\{\|x_i\|_{\mathcal{H}_i} - \overline{\lambda} - \xi, 0\} = \overline{\lambda}.$$

(iii) $\sum_{i=1}^{n} \|x_i\|_{\mathcal{H}_i} \leq (\|x_{n+1}\|_{\mathcal{H}_{n+1}} - \xi)/2 < \|x_{n+1}\|_{\mathcal{H}_{n+1}}$ *and* $\|x_{n+1}\|_{\mathcal{H}_{n+1}} > -\xi$, *then*

$$\overline{y}_i = 0_{\mathcal{H}_i}, \ i = 1, ..., n, \ \overline{y}_{n+1} = \frac{\|x_{n+1}\|_{\mathcal{H}_{n+1}} + \xi}{2\|x_{n+1}\|_{\mathcal{H}_{n+1}}} x_{n+1} \ \text{and} \ \overline{\theta} = \frac{\|x_{n+1}\|_{\mathcal{H}_{n+1}} + \xi}{2}.$$

(iv) $\sum_{i=1}^{n} \|x_i\|_{\mathcal{H}_i} \leq -\xi$ *and* $\|x_{n+1}\|_{\mathcal{H}_{n+1}} \leq -\xi$, *then* $\overline{y}_i = 0_{\mathcal{H}_i}, i = 1, ..., n+1$, *and* $\overline{\theta} = 0$.

Proof. By Corollary 5.5 we have

$$P_{\text{epi}\,\gamma_C}(x_1,...,x_{n+1},\xi) = \begin{cases} (x_1,...,x_{n+1},\xi), & \text{if } \max_{1\leq i\leq n}\{\|x_i\|_{\mathcal{H}_i}\} + \|x_{n+1}\|_{\mathcal{H}_{n+1}} \leq \xi, \\ (\overline{y}_1,...,\overline{y}_{n+1},\overline{\theta}), & \text{otherwise,} \end{cases}$$

where

$$(\overline{y}_1,...,\overline{y}_n) = (x_1,...,x_n) - \overline{\lambda}\,\mathrm{P}_{C_1^0}\left(\frac{1}{\overline{\lambda}}(x_1,...,x_n)\right),\ \overline{y}_{n+1} = x_{n+1} - \overline{\lambda}\,\mathrm{P}_{C_2^0}\left(\frac{1}{\overline{\lambda}}x_{n+1}\right),$$

$\overline{\theta} = \overline{\lambda} + \xi$ and $\overline{\lambda} > 0$.

From Lemma 4.6 follows that dual gauge of $\gamma_{C_1}(x_1,...,x_n) = \max_{1\leq i\leq n}\{\|x_i\|_{\mathcal{H}_i}\}$ is given by $\gamma_{C_1^0}(x_1,...,x_n) = \sum_{i=1}^n \|x_i\|_{\mathcal{H}_i}$ and hence, the polar set of $C_1 = \{(x_1,...,x_n) \in \mathcal{H}_1 \times ... \times \mathcal{H}_n : \max_{1\leq i\leq n}\{\|x_i\|_{\mathcal{H}_i}\} \leq 1\}$ is $C_1^0 = \{(x_1,...,x_n) \in \mathcal{H}_1 \times ... \times \mathcal{H}_n : \sum_{i=1}^n \|x_i\|_{\mathcal{H}_i} \leq 1\}$. Thus, we derive for the case $(1/\overline{\lambda})(x_1,...,x_n) \notin C_1^0$, i.e. $\sum_{i=1}^n \|x_i\|_{\mathcal{H}_i} > \overline{\lambda}$, from Lemma 5.1 that

$$\mathrm{P}_{C_1^0}\left(\frac{1}{\overline{\lambda}}(x_1,...,x_n)\right) = (\overline{z}_1,...,\overline{z}_n) \in \mathcal{H} \times ... \times \mathcal{H} \text{ where } \overline{z}_i = \frac{\max\{\|x_i\|_{\mathcal{H}_i} - \overline{\lambda}\overline{\mu}, 0\}}{\overline{\lambda}\|x_i\|_{\mathcal{H}_i}}x_i,$$

$$(5.46)$$

$i = 1,...,n$, and $\overline{\mu} > 0$ is a solution of the equation (see (5.10) of the proof of Lemma 5.1)

$$\sum_{i=1}^n \max\left\{\frac{1}{\overline{\lambda}}\|x_i\|_{\mathcal{H}_i} - \mu, 0\right\} = 1 \Leftrightarrow \sum_{i=1}^n \max\left\{\|x_i\|_{\mathcal{H}_i} - \overline{\lambda}\mu, 0\right\} = \overline{\lambda}. \quad (5.47)$$

Furthermore, as $C_2 = C_2^0 = \{x_{n+1} \in \mathcal{H}_{n+1} : \|x_{n+1}\|_{\mathcal{H}_{n+1}} \leq 1\}$, it holds by [3, Example 3.16] (or also by Lemma 5.1 for $n = 1$)

$$\mathrm{P}_{C_2^0}\left(\frac{1}{\overline{\lambda}}x_{n+1}\right) = \frac{1}{\overline{\lambda}\max\left\{\frac{1}{\overline{\lambda}}\|x_{n+1}\|_{\mathcal{H}_{n+1}}, 1\right\}}x_{n+1}. \quad (5.48)$$

Now, we need to consider the following four conceivable cases.

(a) $(1/\overline{\lambda})(x_1,...,x_n) \notin C_1^0$, i.e. $\sum_{i=1}^n \|x_i\|_{\mathcal{H}_i} > \overline{\lambda}$, and $(1/\overline{\lambda})x_2 \notin C_2^0$, i.e. $\|x_{n+1}\|_{\mathcal{H}_{n+1}} > \overline{\lambda}$: Then one has $\mathrm{P}_{C_2^0}((1/\overline{\lambda})x_{n+1}) = (1/\|x_{n+1}\|_{\mathcal{H}_{n+1}})x_{n+1}$ and therefore, it follows together with (5.46), (5.47) and (5.48)

$$\overline{y}_i = x_i - \overline{\lambda}\overline{z}_i,\ i = 1,...,n,\ \overline{y}_{n+1} = x_{n+1} - \frac{\overline{\lambda}}{\|x_{n+1}\|_{\mathcal{H}_{n+1}}}x_{n+1} = \frac{\|x_{n+1}\|_{\mathcal{H}_{n+1}} - \overline{\lambda}}{\|x_{n+1}\|_{\mathcal{H}_{n+1}}}x_{n+1}$$

and $\overline{\theta} = \gamma_C(\overline{y}_1,...,\overline{y}_{n+1}) = \overline{\lambda} + \xi$.

As for $\|x_i\|_{\mathcal{H}_i} - \overline{\lambda}\overline{\mu} > 0$ we have that

$$\overline{y}_i = x_i - \overline{\lambda}\overline{z}_i = x_i - \frac{\|x_i\|_{\mathcal{H}_i} - \overline{\lambda}\overline{\mu}}{\|x_i\|_{\mathcal{H}_i}}x_i = \frac{\overline{\lambda}\overline{\mu}}{\|x_i\|_{\mathcal{H}_i}}x_i$$

and for $\|x_i\|_{\mathcal{H}_i} - \overline{\lambda}\overline{\mu} \leq 0$ that $\overline{y}_i = x_i$, $i = 1, ..., n$, it follows that $\max_{1 \leq i \leq n}\{\|\overline{y}_i\|_{\mathcal{H}_i}\} = \overline{\lambda}\overline{\mu}$ and so, $\overline{\lambda} + \xi = \gamma_C(\overline{y}_1, ..., \overline{y}_{n+1}) = \max_{1 \leq i \leq n}\{\|\overline{y}_i\|_{\mathcal{H}_i}\} + \|\overline{y}_{n+1}\|_{\mathcal{H}_{n+1}} = \overline{\lambda}\overline{\mu} + \|x_{n+1}\|_{\mathcal{H}_{n+1}} - \overline{\lambda}$, which means that $\overline{\lambda}\overline{\mu} = 2\overline{\lambda} + \xi - \|x_{n+1}\|_{\mathcal{H}_{n+1}} \geq 0$. For this reason, we can write for (5.47)

$$\sum_{i=1}^n \max\{\|x_i\|_{\mathcal{H}_i} - 2\overline{\lambda} - \xi + \|x_{n+1}\|_{\mathcal{H}_{n+1}}, 0\} = \overline{\lambda}.$$

Bringing the inequalities $\|x_{n+1}\|_{\mathcal{H}_{n+1}} > \overline{\lambda}$ and $\sum_{i=1}^n \|x_i\|_{\mathcal{H}_i} > \overline{\lambda}$ together with the last equality implies

$$\sum_{i=1}^n \max\{\|x_i\|_{\mathcal{H}_i} - \|x_{n+1}\|_{\mathcal{H}_{n+1}} - \xi, 0\} < \|x_{n+1}\|_{\mathcal{H}_{n+1}}$$

and

$$\sum_{i=1}^n \max\{\|x_i\|_{\mathcal{H}_i} - \|x_{n+1}\|_{\mathcal{H}_{n+1}} - \xi, 0\} < \sum_{i=1}^n \|x_i\|_{\mathcal{H}_i}.$$

Moreover, as $\overline{\lambda}\overline{\mu} > 0$, we have $\overline{\lambda} + \xi > \|x_{n+1}\|_{\mathcal{H}_{n+1}} - \overline{\lambda}$, which means that $\overline{\lambda} > (\|x_{n+1}\|_{\mathcal{H}_{n+1}} - \xi)/2$. From the assumption $\|x_{n+1}\|_{\mathcal{H}_{n+1}} > \overline{\lambda}$ and $\sum_{i=1}^n \|x_i\|_{\mathcal{H}_i} > \overline{\lambda}$ follows that $\|x_{n+1}\|_{\mathcal{H}_{n+1}} > -\xi$ and $\sum_{i=1}^n \|x_i\|_{\mathcal{H}_i} > (\|x_{n+1}\|_{\mathcal{H}_{n+1}} - \xi)/2$. This yields (i).

(b) $(1/\overline{\lambda})(x_1, ..., x_n) \notin C_1^0$, i.e. $\sum_{i=1}^n \|x_i\|_{\mathcal{H}_i} > \overline{\lambda}$, and $(1/\overline{\lambda})x_{n+1} \in C_2^0$, i.e. $\|x_{n+1}\|_{\mathcal{H}_{n+1}} \leq \overline{\lambda}$: Then one has that $P_{C_2^0}((1/\overline{\lambda})x_{n+1}) = (1/\overline{\lambda})x_{n+1}$, which means that $\overline{y}_{n+1} = 0_{\mathcal{H}_{n+1}}$ and as $(1/\overline{\lambda})(y_1, ..., y_n) \notin C_1^0$ it follows, as shown in the previous case, that $\max_{1 \leq i \leq n}\{\|\overline{y}_i\|_{\mathcal{H}_i}\} = \overline{\lambda}\overline{\mu}$. This means that $\overline{\theta} = \gamma_C(\overline{y}_1, ..., \overline{y}_{n+1}) = \max_{1 \leq i \leq n}\{\|\overline{y}_i\|_{\mathcal{H}_i}\} + \|\overline{y}_{n+1}\|_{\mathcal{H}_{n+1}} = \overline{\lambda}\overline{\mu} = \overline{\lambda} + \xi$ and for (5.47) we can write

$$\sum_{i=1}^n \max\{\|x_i\|_{\mathcal{H}_i} - \overline{\lambda} - \xi, 0\} = \overline{\lambda}.$$

As $\|x_{n+1}\|_{\mathcal{H}_{n+1}} \leq \overline{\lambda}$, it holds that

$$\|x_{n+1}\|_{\mathcal{H}_{n+1}} \leq \sum_{i=1}^n \max\{\|x_i\|_{\mathcal{H}_i} - \|x_{n+1}\|_{\mathcal{H}_{n+1}} - \xi, 0\} < \sum_{i=1}^n \|x_i\|_{\mathcal{H}_i}.$$

This verifies the case (ii).

(c) $(1/\overline{\lambda})(x_1, ..., x_n) \in C_1^0$, i.e. $\sum_{i=1}^n \|x_i\|_{\mathcal{H}_i} \leq \overline{\lambda}$, and $(1/\overline{\lambda})x_{n+1} \notin C_2^0$, which means that $\|x_{n+1}\|_{\mathcal{H}_{n+1}} > \overline{\lambda}$: Then

$$P_{C_1^0}((1/\overline{\lambda})(x_1, ..., x_n)) = (1/\overline{\lambda})(x_1, ..., x_n) \text{ and } P_{C_2^0}((1/\overline{\lambda})x_{n+1}) = (1/\|x_{n+1}\|_{\mathcal{H}_{n+1}})x_{n+1}$$

implies

$$\overline{y}_i = 0_{\mathcal{H}_i}, \ i = 1, ..., n, \ \overline{y}_{n+1} = \frac{\|x_{n+1}\|_{\mathcal{H}_{n+1}} - \overline{\lambda}}{\|x_{n+1}\|_{\mathcal{H}_{n+1}}}x_{n+1} \text{ and}$$

$$\overline{\lambda} + \xi = \overline{\theta} = \gamma_C(\overline{y}_1, ..., \overline{y}_{n+1}) = \max_{1 \leq i \leq n}\{\|\overline{y}_i\|_{\mathcal{H}_i}\} + \|\overline{y}_{n+1}\|_{\mathcal{H}_{n+1}} = \|x_{n+1}\|_{\mathcal{H}_{n+1}} - \overline{\lambda},$$

where from the last equality one gets $\overline{\lambda} = (\|x_{n+1}\|_{\mathcal{H}_{n+1}} - \xi)/2$. But this yields

$$\overline{y}_{n+1} = \frac{\|x_{n+1}\|_{\mathcal{H}_{n+1}} - \frac{\|x_{n+1}\|_{\mathcal{H}_{n+1}} - \xi}{2}}{\|x_{n+1}\|_{\mathcal{H}_{n+1}}} x_{n+1} = \frac{\|x_{n+1}\|_{\mathcal{H}_{n+1}} + \xi}{2\|x_{n+1}\|_{\mathcal{H}_{n+1}}} x_{n+1}$$

and

$$\overline{\theta} = \|x_{n+1}\|_{\mathcal{H}_{n+1}} - \frac{\|x_{n+1}\|_{\mathcal{H}_{n+1}} - \xi}{2} = \frac{\|x_{n+1}\|_{\mathcal{H}_{n+1}} + \xi}{2}.$$

Further, one has

$$\sum_{i=1}^{n} \|x_i\|_{\mathcal{H}_i} \leq \frac{\|x_{n+1}\|_{\mathcal{H}_{n+1}} - \xi}{2} < \|x_{n+1}\|_{\mathcal{H}_{n+1}} \text{ as well as } \|x_{n+1}\|_{\mathcal{H}_{n+1}} > -\xi$$

and this yields *(iii)*.

(d) $(1/\overline{\lambda})(x_1, ..., x_n) \in C_1^0$, i.e. $\sum_{i=1}^{n} \|x_i\|_{\mathcal{H}_i} \leq \overline{\lambda}$, and $(1/\overline{\lambda})x_{n+1} \in C_2^0$, i.e. $\|x_{n+1}\|_{\mathcal{H}_{n+1}} \leq \overline{\lambda}$: Then $P_{C_1^0}((1/\overline{\lambda})(x_1, ..., x_n)) = (1/\overline{\lambda})(x_1, ..., x_n)$ and $P_{C_2^0}((1/\overline{\lambda})x_{n+1}) = (1/\overline{\lambda})x_{n+1}$ implies

$$\overline{y}_i = 0, \ i = 1, ..., n+1, \ \text{and} \ \overline{\lambda} + \xi = \overline{\theta} = \gamma_C(\overline{y}_1, ..., \overline{y}_{n+1}) = 0,$$

which means that $\overline{\lambda} = -\xi$. Hence, one gets that $\xi < 0$ and $\sum_{i=1}^{n} \|x_i\|_{\mathcal{H}_i} \leq -\xi$ as well as $\|x_{n+1}\|_{\mathcal{H}_{n+1}} \leq -\xi$. This verifies *(iv)*.

As only these four cases are possible and exclude each other, we derive the statement of the corollary. $\qquad\square$

Corollary 5.7. *Let* $\gamma_C : \mathcal{H}_1 \times ... \times \mathcal{H}_n \to \mathbb{R}$ *be defined by*

$$\gamma_C(x_1, ..., x_n) := \max_{1 \leq i \leq n} \left\{ (\frac{1}{w_i} \|x_i\|_{\mathcal{H}_i}) \right\},$$

then it holds

$$P_{\text{epi}\,\gamma_C}(x_1, ..., x_n, \xi) = \begin{cases} (x_1, ..., x_n), \ if \ \max\limits_{1 \leq i \leq n} \left\{ \frac{1}{w_i} \|x_i\|_{\mathcal{H}_i} \right\} \leq \xi, \\ (0_{\mathcal{H}}, ..., 0_{\mathcal{H}}, 0), \ \ if \ \xi < 0 \ and \ \sum\limits_{i=1}^{n} w_i \|x_i\|_{\mathcal{H}_i} \leq -\xi, \\ (\overline{y}_1, ..., \overline{y}_n, \overline{\theta}), \ otherwise, \end{cases}$$

where

$$\overline{y}_i = x_i - \frac{\max\{\|x_i\|_{\mathcal{H}_i} - (\overline{\lambda} + \xi)w_i, 0\}}{\|x_i\|_{\mathcal{H}_i}} x_i, \ i = 1, ..., n, \ and \ \overline{\theta} = \frac{\sum\limits_{i=k+1}^{n} w_i^2 \tau_i + \xi}{\sum\limits_{i=k+1}^{n} w_i^2 + 1}$$

with

$$\overline{\lambda} = \frac{\sum\limits_{i=k+1}^{n} w_i^2 \tau_i - \xi \sum\limits_{i=k+1}^{n} w_i^2}{\sum\limits_{i=k+1}^{n} w_i^2 + 1} \tag{5.49}$$

and $k \in \{0, 1, \ldots, n-1\}$ *is the unique integer such that* $\tau_k + \xi \leq \overline{\lambda} \leq \tau_{k+1} + \xi$, *where the values* $\tau_0, \ldots, \tau_n$ *are defined by* $\tau_0 := 0$ *and* $\tau_i := (\|x_i\|_{\mathcal{H}_i})/w_i$, $i = 1, \ldots, n$, *and in ascending order.*

Proof. As $C = \{(x_1, \ldots, x_n) : \mathcal{H}_1 \times \ldots \times \mathcal{H}_n : \max_{1 \leq i \leq n}\{(1/w_i)\|x_i\|_{\mathcal{H}_i}\}\} \leq 1$ (see Remark 4.3), Corollary 5.5 reveals that

$$\mathrm{P}_{\mathrm{epi}\,\gamma_C}(x_1, \ldots, x_n, \xi) = \begin{cases} (x_1, \ldots, x_n, \xi), & \text{if } \max\limits_{1 \leq i \leq n}\{\frac{1}{w_i}\|x_i\|_{\mathcal{H}_i}\} \leq \xi, \\ (\overline{y}_1, \ldots, \overline{y}_n, \overline{\theta}), & \text{otherwise,} \end{cases}$$

where

$$(\overline{y}_1, \ldots, \overline{y}_n) = (x_1, \ldots, x_n) - \overline{\lambda}\mathrm{P}_{C^0}\left(\frac{1}{\overline{\lambda}}(x_1, \ldots, x_n)\right), \ \overline{\theta} = \overline{\lambda} + \xi \text{ and } \overline{\lambda} > 0.$$

By Lemma 4.6 the polar set of C looks like $C^0 = \{(x_1, \ldots, x_n) \in \mathcal{H}_1 \times \ldots \times \mathcal{H}_n : \sum_{i=1}^{n} w_i\|x_i\|_{\mathcal{H}_i} \leq 1\}$ and from Lemma 5.1 we derive that $\mathrm{P}_{C^0}\left(\frac{1}{\overline{\lambda}}(x_1, \ldots, x_n)\right) = (x_1, \ldots, x_n)$ if $\sum_{i=1}^{n} w_i\|x_i\|_{\mathcal{H}_i} \leq \overline{\lambda}$, i.e. $(\overline{y}_1, \ldots, \overline{y}_n) = (0_{\mathcal{H}}, \ldots, 0_{\mathcal{H}})$, which implies that $\max_{1 \leq i \leq n}\{\overline{y}_i\} = 0 = \overline{\theta} = \overline{\lambda} + \xi$ and hence, $\overline{\lambda} = -\xi$.

Otherwise, one has by Lemma 5.1

$$\mathrm{P}_{C^0}\left(\frac{1}{\overline{\lambda}}(x_1, \ldots, x_n)\right) = (\overline{z}_1, \ldots, \overline{z}_n) \in \mathcal{H}_1 \times \ldots \times \mathcal{H}_n,$$

where

$$\overline{z}_i = \frac{\max\{\|x_i\|_{\mathcal{H}_i} - \overline{\lambda}\overline{\mu}w_i, 0\}}{\overline{\lambda}\|x_i\|_{\mathcal{H}_i}}x_i, \ i = 1, \ldots, n,$$

and $\overline{\mu} > 0$ is a solution of the equation (see (5.10) of the proof of Lemma 5.1)

$$\sum_{i=1}^{n} w_i \max\{\|x_i\|_{\mathcal{H}_i} - \overline{\lambda}\overline{\mu}w_i, 0\} = \overline{\lambda}. \tag{5.50}$$

Therefore, it follows

$$\overline{y}_i = x_i - \frac{\max\{\|x_i\|_{\mathcal{H}_i} - \overline{\lambda}\overline{\mu}w_i, 0\}}{\|x_i\|_{\mathcal{H}_i}}x_i = \frac{\|x_i\|_{\mathcal{H}_i} - \max\{\|x_i\|_{\mathcal{H}_i} - \overline{\lambda}\overline{\mu}w_i, 0\}}{\|x_i\|_{\mathcal{H}_i}}x_i, \ i = 1, \ldots, n,$$

and as for $\|x_i\|_{\mathcal{H}_i} - \overline{\lambda}\overline{\mu}w_i \leq 0$ one gets $\overline{y}_i = x_i$, i.e. $\|\overline{y}_i\|_{\mathcal{H}_i} = \|x_i\|_{\mathcal{H}_i}$ and for $\|x_i\|_{\mathcal{H}_i} - \overline{\lambda}\overline{\mu}w_i > 0$, $\overline{y}_i = (\overline{\lambda}\overline{\mu}w_i/\|x_i\|_{\mathcal{H}_i})x_i$, i.e. $\|\overline{y}_i\|_{\mathcal{H}_i} = \overline{\lambda}\overline{\mu}w_i$, $i = 1, \ldots, n$, we obtain

$$\gamma_C(\overline{y}_1, \ldots, \overline{y}_n) = \max_{1 \leq i \leq n} \left\{ \frac{1}{w_i} \|\overline{y}_i\|_{\mathcal{H}_i} \right\} = \overline{\lambda}\overline{\mu} = \overline{\lambda} + \xi. \tag{5.51}$$

Bringing (5.50) and (5.51) together yields

$$\sum_{i=1}^{n} w_i \max \left\{ \|x_i\|_{\mathcal{H}_i} - (\overline{\lambda} + \xi)w_i, 0 \right\} = \overline{\lambda}. \tag{5.52}$$

Clearly, if $\|x_i\|_{\mathcal{H}_i} - \xi w_i \leq 0$ for all $i = 1, \ldots, n$, i.e. $\max_{1 \leq i \leq n}\{\|x_i\|_{\mathcal{H}_i}/w_i\} \leq \xi$, then $\|x_i\|_{\mathcal{H}_i} - \xi w_i - \overline{\lambda}w_i \leq 0$ for all $i = 1, \ldots, n$, and one gets by (5.52) that

$$\overline{\lambda} = \sum_{i=1}^{n} w_i \max \left\{ \|x_i\|_{\mathcal{H}_i} - (\overline{\lambda} + \xi)w_i, 0 \right\} = 0,$$

which means that $\overline{y}_i = x_i$ for all $i = 1, \ldots, n$, and $\overline{\theta} = \xi$.

Now, we define the function $g : \mathbb{R} \to \mathbb{R}$ by

$$g(\lambda) = \sum_{i=1}^{n} w_i^2 \max \left\{ \tau_i - (\lambda + \xi), 0 \right\} - \overline{\lambda} \tag{5.53}$$

Take note, that there exists $i \in \{1, \ldots, n\}$ such that $\tau_i > 0$ and so,

$$g(\tau_n - \xi) = \sum_{i=1}^{n} w_i^2 \max \left\{ \tau_i - \tau_n, 0 \right\} - \overline{\lambda} = -\overline{\lambda} < 0.$$

Moreover, as g is a piecewise linear function, one has, similarly to Corollary 5.1, to find the unique integer $k \in \{0, 1, \ldots, n-1\}$ such that $g(\tau_k - \xi) \geq 0$ and $g(\tau_{k+1} - \xi) \leq 0$. This leads to

$$\sum_{i=k+1}^{n} w_i^2\tau_i - \xi \sum_{i=k+1}^{n} w_i^2 - \overline{\lambda} \sum_{i=k+1}^{n} (w_i^2 + 1) = 0 \Leftrightarrow \overline{\lambda} = \frac{\sum\limits_{i=k+1}^{n} w_i^2\tau_i - \xi \sum\limits_{i=k+1}^{n} w_i^2}{\sum\limits_{i=k+1}^{n} w_i^2 + 1}$$

and hence, $\overline{\theta} = \overline{\lambda} + \xi = (\sum_{i=k+1}^{n} w_i^2\tau_i + \xi)/(\sum_{i=k+1}^{n} w_i^2 + 1)$. $\square$

Remark 5.5. *In [33] the formula in the previous corollary was given for the case where* $\mathcal{H}_i = \mathbb{R}$, $i = 1, ..., n$, *in other words, where* γ_C *is the weighted* l_∞-*norm.*

Remark 5.6. *The proof of the previous theorem allows us to construct an algorithm to determine* $\overline{\lambda}$ *of Corollary 5.7.*

Algorithm:

1. If $\max\limits_{1 \le i \le n} \left\{ \frac{1}{w_i} \|x_i\|_{\mathcal{H}_i} \right\} \le \xi$, then $\overline{\lambda} = 0$.

2. If $\xi < 0$ and $\sum\limits_{i=1}^{n} w_i \|x_i\|_{\mathcal{H}_i} \le -\xi$, then $\overline{\lambda} = -\xi$.

3. Otherwise, define $\tau_0 := 0$, $\tau_i := \|x_i\|_{\mathcal{H}_i}/w_i$, $i = 1, \ldots, n$, and sort $\tau_0, \ldots, \tau_n$ in ascending order.

4. Determine the values of g defined in (5.53) at $\lambda = \tau_i + \xi$, $i = 0, \ldots, n$.

5. Find the unique $k \in \{0, \ldots, n-1\}$ such that $g(\tau_k - \xi) \ge 0$ and $g(\tau_{k+1} - \xi) \le 0$.

6. Calculate $\overline{\lambda}$ by (5.49).

Remark 5.7. *Like in Lemma 5.3, one can give a formula for the projection onto the epigraph of a gauge composed with a linear operator $A : \mathcal{K} \to \mathcal{H}$ with $AA^* = \mu Id$, $\mu > 0$,*

$$P_{\mathrm{epi}\,\gamma_C(A\cdot)}(x, \xi) = (x, \xi) + \left(\frac{1}{\sqrt{\mu}} A^* \times \mathrm{Id} \right) \left(P_{\mathrm{epi}\,\sqrt{\mu}\gamma_C(\cdot)} \left(\frac{1}{\sqrt{\mu}} Ax, \xi \right) - \left(\frac{1}{\sqrt{\mu}} Ax, \xi \right) \right).$$

Moreover, it can easily be observed that for $p \in \mathcal{H}$ holds (similar to the proof of Lemma 5.2)

$$P_{\mathrm{epi}\,\gamma_C(\cdot - p)}(x, \xi) = P_{\mathrm{epi}\,\gamma_C}(x - p, \xi) + (p, 0).$$

We close this section with a characterization of the subdifferential of a gauge function by the projection operator.

Remark 5.8. *Let $C \subseteq \mathcal{H}$ be closed and convex such that $0_{\mathcal{H}} \in C$, then it holds by (5.3), (5.8), [102, Lemma 2.1], [102, Remark 2.2] and [3, Theorem 14.3(ii)] for all $x, y \in \mathcal{H}$ that*

$$x \in \partial\gamma_C(y) \Leftrightarrow x + y - y \in \partial\gamma_C(y) \Leftrightarrow y = \mathrm{prox}_{\gamma_C}(x + y)$$
$$\Leftrightarrow y = x + y - \mathrm{prox}_{\gamma_C^*}(x + y) \Leftrightarrow y = x + y - \mathrm{prox}_{\delta_{C^0}}(x + y)$$
$$\Leftrightarrow x = P_{C^0}(x + y).$$

From which follows that

$$\partial\gamma_C(y) = \{ x \in \mathcal{H} : x = P_{C^0}(x + y) \}.$$

In addition, if C is a closed convex cone, then it follows from [3, Theorem 6.29] that

$$\partial\gamma_C(y) = \{ x \in \mathcal{H} : x = x + y - P_C(x + y) \} = \{ x \in \mathcal{H} : y = P_C(x + y) \}.$$

6 Numerical experiments

In this chapter we solve minmax location problems via MATLAB by employing the formulae for the projection onto the epigraph of the weighted sum of powers of norms and onto the epigraphs of gauges and compare our method with the ones presented in [2, 35, 78, 82].

We begin our numerical investigations in Section 6.1 by solving numerically an extended multifacility minmax location problem from Section 4.4 and transforming it into an unconstrained optimization problem where its objective function is a sum of functions and apply the splitting algorithm (see [3, 33, 34]) as well as the projection formulae developed in Section 5.2. By some numerical examples we demonstrate that the parallel splitting algorithm together with the projection formulae presented in the previous chapter clearly outperforms the method proposed by Cornejo and Michelot in [35] from the viewpoints of accuracy, CPU speed and number of iterations.

In Section 6.2 we solve in MATLAB concrete location optimization problems considered in Subsection 4.2.1 and 4.3.2 as well as their conjugate duals, rewritten as unconstrained minimization problems. The computational results show that the primal optimal solutions are obtained faster when numerically solving the dual problems. One of these concrete examples was numerically solved in [78, 82] by means of a subgradient method and a comparison of the computational results is provided as well, stressing once again the superiority of the algorithm proposed in the present book. Another comparison is made with the log-exponential smoothing accelerated gradient method proposed in [2] in several examples, one of them including a large data set in high dimensions and our method turns out again to converge faster towards the optimal solution of the considered location problem.

This chapter is based on the author's publications [56, 103].

6.1 Extended multifacility minmax location problems

In this section we solve numerically extended multifacility minmax location problems considered in Section 4.4. Our numerical tests are implemented in MATLAB on a PC with an Intel Core i5-8400 CPU with 2.8GHz and 16 GB RAM. While the numerical tests in [35] were based on the partial inverse algorithm introduced by Spingarn in [97], we use here the *parallel splitting algorithm* from [3, Proposition 27.8].

© Springer Fachmedien Wiesbaden GmbH, part of Springer Nature 2020
O. Wilfer, *Multi-Composed Programming with Applications to Facility Location*,
Mathematische Optimierung und Wirtschaftsmathematik I Mathematical
Optimization and Economathematics, https://doi.org/10.1007/978-3-658-30580-2_6

Theorem 6.1. *(parallel splitting algorithm) Let n be an integer such that $n \geq 2$ and $f_i : \mathbb{R}^s \to \overline{\mathbb{R}}$ be a proper, lower semicontinuous and convex function for $i = 1, ..., n$. Suppose that the problem*

$$(P^{DR}) \qquad \min_{x \in \mathbb{R}^s} \left\{ \sum_{i=1}^n f_i(x) \right\}$$

has at least one solution and that $\operatorname{dom} f_1 \cap \bigcap_{i=2}^n \operatorname{int} \operatorname{dom} f_i \neq \emptyset$. *Let* $(\mu_k)_{k \in \mathbb{N}}$ *be a sequence in $[0, 2]$ such that $\sum_{k \in \mathbb{N}} \mu_k(2 - \mu_k) = +\infty$, let $\nu > 0$, and let $(x_{i,0})_{i=1}^n \in \mathbb{R}^s \times ... \times \mathbb{R}^s$. Set*

$$(\forall k \in \mathbb{N}) \quad \left| \begin{array}{l} r_k = \frac{1}{n} \sum_{i=1}^n x_{i,k}, \\[2mm] y_{i,k} = \operatorname{prox}_{\nu f_i} x_{i,k}, \ i = 1, ..., n, \\[2mm] q_k = \frac{1}{n} \sum_{i=1}^n y_{i,k}, \\[2mm] x_{i,k+1} = x_{i,k} + \mu_k(2q_k - r_k - y_{i,k}), \ i = 1, ..., n. \end{array} \right.$$

Then $(r_k)_{k \in \mathbb{N}}$ converges to an optimal solution to (P^{DR}).

In order to use the parallel splitting algorithm given in the previous theorem, we need to rewrite the *extended multifacility location problem* $(EP_N^{M,\beta})$ in (5.4) into an optimization problem with an objective function, which is a sum of proper, convex and lower semicontinuous functions.

The first way to reformulate this location problem is based on the introduction of an additional variable as presented in (5.5):

$$(EP_N^{M,\beta}) \qquad \min_{(x_1,...,x_m,t) \in \mathbb{R}^d \times ... \times \mathbb{R}^d \times \mathbb{R}} \left\{ t + \sum_{i=1}^n \delta_{\operatorname{epi}\left(\sum_{j=1}^m w_{ij}\|\cdot - p_i\|^{\beta_i}\right)}(x_1, ..., x_m, t) \right\}. \qquad (6.1)$$

We define the functions

$$f_1 : \mathbb{R}^d \times ... \times \mathbb{R}^d \times \mathbb{R} \to \mathbb{R}, \ f_1(x_1, ..., x_m, t) = t \text{ and}$$
$$f_i : \mathbb{R}^d \times ... \times \mathbb{R}^d \times \mathbb{R} \to \overline{\mathbb{R}}, \ f_i(x_1, ..., x_m, t) = \delta_{\operatorname{epi}\left(\sum_{j=1}^m w_{ij}\|\cdot - p_i\|^{\beta_i}\right)}(x_1, ..., x_m, t),$$

$i = 2, ..., n+1$, then $\operatorname{dom} f_1 = \mathbb{R}^d \times ... \times \mathbb{R}^d \times \mathbb{R}$ and

$$\left(0_{\mathbb{R}^d}, ..., 0_{\mathbb{R}^d}, \max_{1 \leq i \leq n} \left\{ \sum_{j=1}^m w_{ij}\|p_i\|^{\beta_i} \right\} + 1 \right) \in \operatorname{int} \operatorname{dom} f_i = \operatorname{int} \operatorname{epi}\left(\sum_{j=1}^m w_{ij}\| \cdot - p_i\|^{\beta_i} \right)$$

for all $i = 2, ..., n+1$, i.e., it holds that $\operatorname{dom} f_1 \cap \bigcap_{i=2}^{n+1} \operatorname{int} \operatorname{dom} f_i \neq \emptyset$. Therefore, the sequences generated by the algorithm from Theorem 6.1 converges to a solution of the location problem $(EP_N^{M,\beta})$ and the following formulae for the proximal points associated to the functions $f_1, ..., f_{n+1}$ can be formulated by using (5.8) and Lemma 5.2

$$(\overline{y}_1, ..., \overline{y}_m, \overline{\theta}) = \mathrm{prox}_{\nu f_1}(x_1, ..., x_m, t)$$

$$\Leftrightarrow (x_1, ..., x_m, t) - (\overline{y}_1, ..., \overline{y}_m, \overline{\theta}) \in \partial(\nu f_1)(\overline{y}_1, ..., \overline{y}_m, \overline{\theta}) = (0_{\mathbb{R}^d}, ..., 0_{\mathbb{R}^d}, \nu)$$

$$\Leftrightarrow x_i = \overline{y}_i, \ i = 1, ..., m, \ \text{and} \ \overline{\theta} = t - \nu \Leftrightarrow (\overline{y}_1, ..., \overline{y}_m, \overline{\theta}) = (x_1, ..., x_n, t - \nu)$$

and

$$\begin{aligned}
(\overline{y}_1, ..., \overline{y}_m, \overline{\theta}) &= \mathrm{prox}_{\nu f_i}(x_1, ..., x_m, t) = \mathrm{prox}_{\nu \delta_{\mathrm{epi}\left(\sum_{j=1}^m w_{ij}\|\cdot - p_i\|^{\beta_i}\right)}}(x_1, ..., x_m, t) \\
&= \mathrm{P}_{\mathrm{epi}\left(\sum_{j=1}^m w_{ij}\|\cdot - p_i\|^{\beta_i}\right)}(x_1, ..., x_m, t) \\
&= \mathrm{P}_{\mathrm{epi}\left(\sum_{j=1}^m w_{ij}\|\cdot\|^{\beta_i}\right)}(x_1 - p_i, ..., x_m - p_i, t) + (p_i, ..., p_i, 0).
\end{aligned} \tag{6.2}$$

The second way to rewrite the extended multifacility location problem $(EP_N^{M,\beta})$ into an optimization problem of the form of (P^{DR}) makes use of the ideas of Cornejo and Michelot given in [35] and splits the sums of weighted norms by $n \cdot m$ additional variables (see also (5.6)):

$$(EP_N^{M,\beta}) \quad \min_{\substack{t, \ t_{ij} \in \mathbb{R}, \ x_j \in \mathbb{R}^d, \\ j=1,...,m, i=1,...,n}} \left\{ t + \sum_{j=1}^m \sum_{i=1}^n \delta_{\mathrm{epi}\left(w_{ij}\|\cdot - p_i\|^{\beta_i}\right)}(x_j, t_{ij}) + \sum_{i=1}^n \delta_{\mathrm{epi}\,\tau_i}(t_{i1}, ..., t_{im}, t) \right\},$$

$$\tag{6.3}$$

where $\tau_i(t_{i1}, ..., t_{im}) := \sum_{j=1}^m t_{ij}, \ i = 1, ..., n$. Now, let $\widetilde{x} := (x_1, ..., x_m) \in \mathbb{R}^d \times ... \times \mathbb{R}^d$, $\widetilde{t} := (t_{ij})_{i=1,...,n, j=1,...,m}$,

$$f_1 : \underbrace{\mathbb{R}^d \times ... \times \mathbb{R}^d}_{m-\text{times}} \times \mathbb{R}^{mn} \times \mathbb{R} \to \mathbb{R}, \ f_1(\widetilde{x}, \widetilde{t}, t) := t,$$

$$f_{ij} : \mathbb{R}^d \times ... \times \mathbb{R}^d \times \mathbb{R}^{mn} \times \mathbb{R} \to \overline{\mathbb{R}}, \ f_{ij}(\widetilde{x}, \widetilde{t}, t) := \delta_{\mathrm{epi}\left(w_{ij}\|\cdot - p_i\|^{\beta_i}\right)}(x_j, t_{ij}),$$

$j = 1, ..., m, \ i = 1, ..., n$, and

$$\widetilde{f}_i : \mathbb{R}^d \times ... \times \mathbb{R}^d \times \mathbb{R}^{mn} \times \mathbb{R} \to \overline{\mathbb{R}}, \ \widetilde{f}_i(\widetilde{x}, \widetilde{t}, t) := \delta_{\mathrm{epi}\,\tau_i}(t_{i1}, ..., t_{im}, t), \ i = 1, ..., n.$$

As

$$\mathrm{dom}\, f_1 = \mathbb{R}^d \times ... \times \mathbb{R}^d \times \mathbb{R}^{mn} \times \mathbb{R},$$

$$\mathrm{dom}\, f_{ij} = \left\{ (\widetilde{x}, \widetilde{t}, t) \in \mathbb{R}^d \times ... \times \mathbb{R}^d \times \mathbb{R}^{mn} \times \mathbb{R} : (x_j, t_{ij}) \in \mathrm{epi}(w_{ij}\|\cdot - p_i\|^{\beta_i}) \right\},$$

$$i = 1, ..., n, \ j = 1, ..., m,$$

$$\mathrm{dom}\, \widetilde{f}_i = \left\{ (\widetilde{x}, \widetilde{t}, t) \in \mathbb{R}^d \times ... \times \mathbb{R}^d \times \mathbb{R}^{mn} \times \mathbb{R} : (t_{i1}, ..., t_{im}, t) \in \mathrm{epi}\,\tau_i \right\},$$

$$i = 1, ..., n,$$

and

$$\left(0_{\mathbb{R}^d}, ..., 0_{\mathbb{R}^d}, \max_{\substack{1\le i\le n,\\1\le j\le m}}\{w_{ij}\|p_i\|^{\beta_i}\}+1, ..., \max_{\substack{1\le i\le n,\\1\le j\le m}}\{w_{ij}\|p_i\|^{\beta_i}\}+1, m\max_{\substack{1\le i\le n,\\1\le j\le m}}\{w_{ij}\|p_i\|^{\beta_i}\}+m+1\right)$$

$$\in \operatorname{dom} f_1 \cap \left(\bigcap_{\substack{1\le i\le n,\\1\le j\le m}}\operatorname{int}\operatorname{dom} f_{ij}\right)\cap\left(\bigcap_{1\le i\le n}\operatorname{int}\operatorname{dom}\widetilde{f}_i\right),$$

convergence in the sense of Theorem 6.1 can be guaranteed. Now, let $\overline{\overline{y}} := (\overline{y}_1, ..., \overline{y}_m)$ and $\overline{\overline{\theta}} := (\overline{\theta}_{ij})_{1\le i\le n,\ 1\le j\le m}$, then one has by (5.8) for the corresponding proximal points of the functions f_1, f_{ij}, $j = 1, ..., m$, $i = 1, ..., n$, and $\widetilde{f}_i$, $i = 1, ..., n$,

$$(\overline{\overline{y}}, \overline{\overline{\theta}}, \overline{\theta}) = \operatorname{prox}_{\nu f_1}(\widetilde{x}, \widetilde{t}, t) = (\underbrace{0_{\mathbb{R}^d}, ..., 0_{\mathbb{R}^d}}_{m-\text{times}}, \underbrace{0, ..., 0}_{mn-\text{times}}, t - \nu)$$

and by (5.8) and Lemma 5.2

$$(\overline{\overline{y}}, \overline{\overline{\theta}}, \overline{\theta}) = \operatorname{prox}_{\nu f_{ij}}(\widetilde{x}, \widetilde{t}, t) \Leftrightarrow (\widetilde{x}, \widetilde{t}, t) - (\overline{\overline{y}}, \overline{\overline{\theta}}, \overline{\theta}) \in \partial(\nu f_{ij})(\overline{\overline{y}}, \overline{\overline{\theta}}, \overline{\theta})$$

$$\Leftrightarrow (x_j, t_{ij}) - (\overline{y}_j, \overline{\theta}_{ij}) \in \partial(\nu\delta_{\operatorname{epi}(w_{ij}\|\cdot - p_i\|^{\beta_i})})(\overline{y}_j, \overline{\theta}_{ij}) \text{ and}$$

$$\overline{y}_l = x_l, \ \overline{\theta}_{sl} = t_{sl}, \ \overline{\theta} = t, \ s = 1, ..., n, \ l = 1, ..., m, \ sl \ne ij,$$

$$\Leftrightarrow (\overline{y}_j, \overline{\theta}_{ij}) = \operatorname{prox}_{\nu\delta_{\operatorname{epi}(w_{ij}\|\cdot - p_i\|^{\beta_i})}}(x_j, t_{ij}) = P_{\operatorname{epi}(w_{ij}\|\cdot - p_i\|^{\beta_i})}(x_j, t_{ij})$$

$$= P_{\operatorname{epi}(w_{ij}\|\cdot\|^{\beta_i})}(x_j - p_i, t_{ij}) + (p_i, 0) \text{ and}$$

$$\overline{y}_l = x_l, \ \overline{\theta}_{sl} = t_{sl}, \ \overline{\theta} = t, \ s = 1, ..., n, \ l = 1, ..., m, \ sl \ne ij, \qquad (6.4)$$

$j = 1, ..., m$, $i = 1, ..., n$. Moreover, by (5.8) and [3, Example 28.17] follows

$$(\overline{\overline{y}}, \overline{\overline{\theta}}, \overline{\theta}) = \operatorname{prox}_{\nu \widetilde{f}_i}(\widetilde{x}, \widetilde{t}, t) \Leftrightarrow (\widetilde{x}, \widetilde{t}, t) - (\overline{\overline{y}}, \overline{\overline{\theta}}, \overline{\theta}) \in \partial(\nu\widetilde{f}_i)(\overline{\overline{y}}, \overline{\overline{\theta}}, \overline{\theta})$$

$$\Leftrightarrow (t_{i1}, ..., t_{im}, t) - (\overline{\theta}_{i1}, ..., \overline{\theta}_{im}, \overline{\theta}) \in \partial(\nu\delta_{\operatorname{epi}\tau_i})(\overline{\theta}_{i1}, ..., \overline{\theta}_{im}, \overline{\theta}) \text{ and}$$

$$(t_{l1}, ..., t_{lm}, t) = (\overline{\theta}_{l1}, ..., \overline{\theta}_{lm}, \overline{\theta}), \ l = 1, ..., n, \ l \ne i, \ (x_1, ..., x_m) = (\overline{y}_1, ..., \overline{y}_m)$$

$$\Leftrightarrow (\overline{\theta}_{i1}, ..., \overline{\theta}_{im}, \overline{\theta}) = \operatorname{prox}_{\nu\delta_{\operatorname{epi}\tau_i}}(t_{i1}, ..., t_{im}, t) = P_{\operatorname{epi}\tau_i}(t_{i1}, ..., t_{im}, t)$$

$$= \begin{cases} (\overline{\theta}_{i1}, ..., \overline{\theta}_{im}, \overline{\theta})^T, & \text{if } \sum_{j=1}^m t_{ij} - t \le 0, \\ (\overline{\theta}_{i1}, ..., \overline{\theta}_{im}, \overline{\theta})^T - \dfrac{\sum_{j=1}^m t_{ij} - t}{m+1}(1, ..., 1, -1)^T, & \text{if } \sum_{j=1}^m t_{ij} - t > 0, \end{cases}$$

and $(t_{l1}, ..., t_{lm}, t) = (\overline{\theta}_{l1}, ..., \overline{\theta}_{lm}, \overline{\theta})$, $l = 1, ..., n$, $l \ne i$, $(x_1, ..., x_m) = (\overline{y}_1, ..., \overline{y}_m)$, $i = 1, ..., n$.

The tables below illustrate the performance of our method using the formulae from Corollary 5.1 and 5.2 for the projection onto the epigraph of the sum of powers of weighted norms (EpiSumNorms) compared with the concept proposed by Cornejo and Michelot in [35], where only the projection onto the epigraph of a weighted norm (EpiNorm) is needed (see Corollary 5.3). We solved the problem $(EP_N^{M,\beta})$ in $\mathbb{R}^2$ and $\mathbb{R}^3$ for different choices of given and new facilities. The performance results are visualized by the associated figures, where we use the following notations:

NumGivFac: Number of given facilities
NumNewFac: Number of new facilities
Numit: Number of Iterations of the algorithm
CPUtime: CPU time in seconds
ObjFunVal: Function Value.

We used the following parameters for initialization: $\mu_n = 1$ for all $n \in \mathbb{N}$. Moreover, let us point out that we tested the algorithm of Theorem 6.1 for different values of the parameter ν, where some results are printed in the tables below and selected ones are visualized in the corresponding figures (in the tables the best results of the methods EpiSumNorms and EpiNorm concerning the CPU time and number of iterations are marked in bold, respectively). Notice also that in the context of the problem (6.1) the iterate r_k of Theorem 6.1 is of the form $r_k = (x_1, ..., x_m, t)$ and in the framework of (6.3) of the form $r_k = (x_1, ..., x_m, \tilde{t}, t)$ with $\tilde{t} = (t_{ij})_{i=1,...,n,\ j=1,...,m}$, where $x_1, ..., x_m$ converge to the optimal locations and t to the optimal objective value.

To be more precise, for our numerical experiments we proceed as follows. The points $p_1, ..., p_n$ were generated by the MATLAB command RANDN and the corresponding weights $a_1, ..., a_n$ by RAND. Further, we used in all numerical tests as starting point the origin (i.e. $x_1 = ... = x_m = 0_{\mathbb{R}^d}$) and ran the algorithm of the method EpiSumNorms for all examples five hundred thousand iterations. Then, we saved the determined solutions as the optimal solutions $\overline{x} = (\overline{x}_1, ..., \overline{x}_m)$ to the associated optimization problems and set the maximal number of iterations to one hundred thousand (i.e. MaxNumIt=100000). Finally, we ran for all examples the algorithm of the method EpiSumNorms a second time as well as the algorithm of the method EpiNorm and noticed the number of iterations as well as the time needed to generate a solution which is within the maximum bound from the optimal location(s) $\overline{x} = (\overline{x}_1, ..., \overline{x}_m)$ of 0.001, i.e. $\|\overline{x} - x\| \leq 0.001$, respectively.

First, we consider the situation where $\beta_i = 1$ for all $i = 1, ..., n$. In Table 6.1 it is shown that the parallel splitting algorithm converges very slow when employed in connection with the method proposed in [35], while our method performs much better. The corresponding figure shows that our method EpiSumNorms generates after 185 iterations a solution which is within the maximum bound from the optimal solution, while the method EpiNorm needs 2180 iterations. Take also note that in this example the location problem has in the form of EpiNorm 125 additional variables, while the examples in the Table 6.2 and 6.3 have 300 and 1200 additional variables, respectively. For this reason our method by far outperforms the concept EpiNorm on such optimization problems regarding the accuracy as well as the CPU speed and number of iterations.

Table 6.1: Performance evaluation for NumGiFac 25 and NumNewFac 5 in $\mathbb{R}^2$

MaxNumIt	EpiSumNorms		EpiNorm	
$= 100000$	NumIt	CPUtime	NumIt	CPUtime
$\nu = 0.1$	50879	83.8313	>100000	-
$\nu = 1$	5076	7.7140	>100000	-
$\nu = 5$	989	1.4959	21182	184.7405
$\nu = 30$	**185**	**0.3054**	**2180**	**16.3661**
$\nu = 100$	688	1.1056	2216	16.0718
$\nu = 500$	3507	5.7172	15121	122.9190
$\nu = 1000$	7012	11.4579	30390	281.2456

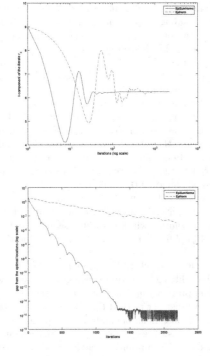

Figure 6.1: Comparison of the methods EpiSumNorms (blue solid line) and EpiNorm (red dashed line) in $\mathbb{R}^2$ for $\nu = 30$

Table 6.2: Performance evaluation for NumGiFac 30 and NumNewFac 10 in $\mathbb{R}^2$

MaxNumIt	EpiSumNorms		EpiNorm	
$= 100000$	NumIt	CPUtime	NumIt	CPUtime
$\nu = 0.1$	43058	106.5284	>100000	-
$\nu = 1$	4306	9.7233	>100000	-
$\nu = 10$	411	0.9267	26596	639.1180
$\nu = 18$	**269**	**0.6146**	14416	303.8991
$\nu = 50$	538	1.2406	**3478**	**62.2674**
$\nu = 100$	1122	2.6104	4762	88.3561
$\nu = 1000$	11324	26.8627	56615	1633.1276

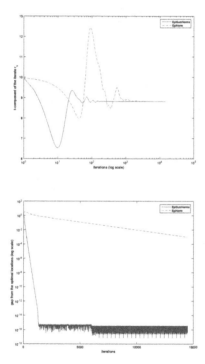

Figure 6.2: Comparison of the methods EpiSumNorms (blue solid line) and EpiNorm (red dashed line) in $\mathbb{R}^2$ for $\nu = 18$

Table 6.3: Performance evaluation for NumGiFac 60 and NumNewFac 20 in $\mathbb{R}^3$

MaxNumIt	EpiSumNorms		EpiNorm	
= 100000	NumIt	CPUtime	NumIt	CPUtime
$\nu = 1$	15415	89.1836	>100000	-
$\nu = 10$	1541	8.5561	>100000	-
$\nu = 98$	**592**	**3.3485**	28920	5332.9115
$\nu = 205$	1129	6.3206	**15697**	**2784.3976**
$\nu = 500$	2687	15.7556	16369	2831.1355
$\nu = 1000$	5346	29.7313	31429	5859.4418
$\nu = 5000$	26715	163.1581	>100000	-

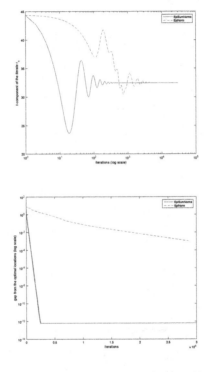

Figure 6.3: Comparison of the methods EpiSumNorms (blue solid line) and EpiNorm (red dashed line) in $\mathbb{R}^3$ for $\nu = 98$

Finally, we consider the situation where $w_i = 1$ and $\beta_i = 2$ for all $i = 1, ..., n$.

Table 6.4: Performance evaluation for NumGiFac 25 and NumNewFac 5 in $\mathbb{R}^2$

MaxNumIt	EpiSumNorms		EpiNorm	
= 100000	NumIt	CPUtime	NumIt	CPUtime
$\nu = 0.1$	4967	4.6351	>100000	-
$\nu = 1$	663	0.6409	53618	583.0896
$\nu = 5$	**306**	**0.3172**	9504	70.9192
$\nu = 39$	2645	2.8821	**2851**	**21.4682**
$\nu = 100$	6776	7.4103	7120	53.9206
$\nu = 500$	33904	39.3127	35740	340.2518
$\nu = 1000$	67806	84.4253	71456	881.7213

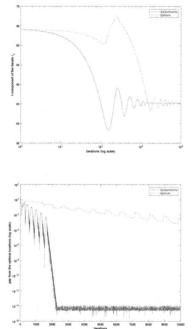

Figure 6.4: Comparison of the methods EpiSumNorms (blue solid line) and EpiNorm (red dashed line) in $\mathbb{R}^2$ for $\nu = 5$

Table 6.5: Performance evaluation for NumGiFac 60 and NumNewFac 10 in $\mathbb{R}^3$

MaxNumIt	EpiSumNorms		EpiNorm	
$= 100000$	NumIt	CPUtime	NumIt	CPUtime
$\nu = 0.1$	3391	8.2135	>100000	-
$\nu = 1$	**1042**	**2.5581**	>100000	-
$\nu = 10$	8270	20.7564	29000	1713.3125
$\nu = 50$	2714	6.8445	32168	2042.8242
$\nu = 110$	1691	4.3914	15167	821.7959
$\nu = 445$	6669	17.5405	**5224**	**273.9350**
$\nu = 1000$	15381	41.1793	14533	788.8773

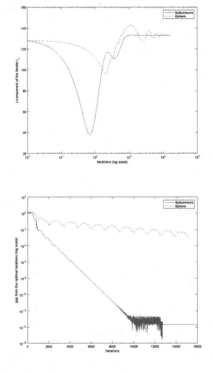

Figure 6.5: Comparison of the methods EpiSumNorm (blue solid line) and EpiNorm (red dashed line) in $\mathbb{R}^3$ for $\nu = 110$

The examples in the last two tables draw a similar picture as the examples in the previous ones. While the method EpiSumNorms generates a solution within the maximum bound from the optimal solution after few seconds, the method EpiNorm needs several minutes. This also points up the usefulness of our approach made in Section 5.2.

6.2 Minmax location problems with perturbed minimal time functions

The aim in this section is to solve numerically several types of concrete location problems and their associated duals discussed in Section 4.3.2 as well as to discuss the results generated via MATLAB. Here we set $X = \mathbb{R}^d$ and used for our numerical experiments a PC with an Intel Core i5 2400 CPU with 3.10 GHz and 8GB RAM. Note that in Section 4.2 we have considered for the theoretical investigations very general frameworks, however, in order to get closer to the real world applications, our numerical experiments are performed in finitely dimensional spaces.

First, we consider a location problem of the type analyzed in Subsection 4.3.2. To solve this kind of a location problem and its dual, rewritten as unconstrained optimization problems, we implemented in MATLAB the parallel splitting algorithm presented in Theorem 6.1 (see also [3, Proposition 27.8]). Note also that other recent proximal splitting methods could prove to be suitable for these problems, too, and comparisons of their performances on the problems we solve below could constitute an interesting follow-up of our investigations.

Take note, that for this purpose it is necessary to bring the location problem and its dual problem into the form of an unconstrained optimization problem where the objective function is a sum of proper, lower semicontinuous and convex functions. By following these ideas, we rewrite the location problem (see Remark 4.20)

$$(P_{\gamma_G}, \tau) \quad \inf_{x \in X} \max_{1 \leq i \leq n} \left\{ \mathcal{T}^{C_i}_{\Omega_i, \gamma_{G_i}}(x) \right\} = \min_{\substack{t>0,\ x \in \mathbb{R}^d,\ z_i \in \mathbb{R}^d, \\ \gamma_{C_i}(x-p_i-z_i)+\gamma_{G_i}(z_i) \leq t,\ i=1,\dots,n}} t$$

$$= \min_{\substack{t>0,\ x \in \mathbb{R}^d,\ z_i \in \mathbb{R}^d,\ \alpha_i \geq 0,\ \beta_i \geq 0, \\ \gamma_{C_i}(x-p_i-z_i) \leq \alpha_i,\ \gamma_{G_i}(z_i) \leq \beta_i, \\ \alpha_i+\beta_i=t,\ i=1,\dots,n}} t = \min_{\substack{t>0,\ x \in \mathbb{R}^d,\ z_i \in \mathbb{R}^d,\ \alpha_i \geq 0,\ \beta_i \geq 0, \\ (x-p_i-z_i,\alpha_i) \in \mathrm{epi}\,\gamma_{C_i},\ (z_i,\beta_i) \in \mathrm{epi}\,\gamma_{G_i}, \\ \alpha_i+\beta_i=t,\ i=1,\dots,n}} t,$$

where C_i and G_i are closed and convex subsets of $\mathbb{R}^d$ with $0_{\mathbb{R}^d} \in \mathrm{int}\, C_i$, $0_{\mathbb{R}^d} \in \mathrm{int}\, G_i$, and $\Omega_i = \{p_i\}$ with $p_i \in \mathbb{R}^d$, $i = 1, \dots, n$, as follows

$$(P_{\gamma_G}, \tau) \quad \min_{\substack{t>0,\ x \in \mathbb{R}^d,\ z_i \in \mathbb{R}^d, \\ \alpha_i \geq 0,\ \beta_i \geq 0,\ i=1,\dots,n}} \left\{ t + \sum_{i=1}^{n} \left[\delta_{\mathrm{epi}\,\gamma_{C_i}}(x - p_i - z_i, \alpha_i) + \delta_{\mathrm{epi}\,\gamma_{G_i}}(z_i, \beta_i) \right] + \delta_H(\alpha, \beta, t) \right\},$$

where $\alpha = (\alpha_1, \dots, \alpha_n)^\top$, $\beta = (\beta_1, \dots, \beta_n)^\top$ and $H = \{(\alpha, \beta, t)^\top : \alpha_i + \beta_i = t,\ i = 1, \dots, n\}$. Similarly, the dual problem

$$(D_{\gamma_G}, \tau) \quad \max_{\substack{z_i^* \geq 0,\ w_i^* \in \mathbb{R}^d,\ \gamma_{C_i^0}(w_i^*) \leq z_i^*, \\ \gamma_{G_i^0}(w_i^*) \leq z_i^*,\ i=1,\dots,n,\ \sum_{i=1}^{n} z_i^* \leq 1,\ \sum_{i=1}^{n} w_i^*=0_{\mathbb{R}^d}}} \left\{ -\sum_{i \in I} \sigma_{\Omega_i}(w_i^*) \right\},$$

can be equivalently rewritten as

$$- \min_{\substack{z_i^* \geq 0, \ w_i^* \in \mathbb{R}^d \\ i=1,\dots,n}} \left\{ \sum_{i=1}^{n} \left[p_i^\top w_i^* + \delta_{\text{epi}\,\gamma_{C_i}^0}(w_i^*, z_i^*) + \delta_{\text{epi}\,\gamma_{G_i}^0}(w_i^*, z_i^*) \right] + \delta_D(z^*) + \delta_E(w^*) \right\},$$

where $z^* = (z_1^*, \dots, z_n^*)^\top$, $w^* = (w_1^*, \dots, w_n^*)$, $D = \{z^* \in \mathbb{R}_+^n : \sum_{i=1}^n z_i^* \leq 1\}$ and $E = \{w^* \in \mathbb{R}^d \times \dots \times \mathbb{R}^d : \sum_{i=1}^n w_i^* = 0_{\mathbb{R}^d}\}$. For both these optimization problems the nonnegativity constraints can be omitted because they are implicitly contained in the indicator functions of the epigraphs of gauges from the objective functions and one can verify, in the same way as in the previous section, that the hypotheses of Theorem 6.1 are fulfilled. Moreover, for the full implementation of this algorithm for solving these problems numerically, one requires also formulae of the proximal mappings of the functions involved in the objective function of the primal and the dual problem.

Example 6.1. *Take $\gamma_{C_i} = \|\cdot\|$ and $\gamma_{G_i} = \|\cdot\|_\infty$, $i = 1, \dots, 5$, and set $d = 2$, $\Omega_i = \{p_i\}$, $i = 1, \dots, 5$, with $p_1 = (-8, -9)^T$, $p_2 = (10, 0)^T$, $p_3 = (11, 5)^T$, $p_4 = (-12, 10)^T$ and $p_5 = (4, 13)^T$. To compute the required proximal and projection points in the primal and dual problem we used Corollary 5.1, Lemma 5.2, Lemma 5.3, Theorem 5.7, [3, Proposition 23.32] and [3, Example 28.14 (iii)]. We ran our MATLAB programs for various step sizes ν and chose always the origin as the starting point and set the initialization parameters to the value 1. The best performance results of our tests are illustrated in Table 6.6. As stopping criterion for the iteration of both programs we used the values $\varepsilon_1 = 10^{-4}$ and $\varepsilon_2 = 10^{-8}$, which define the maximum bounds from the optimal solution, respectively. MATLAB computed for the location problem (primal) an optimal location at $\overline{x} = (-0.5, 2.2878)^T$ with*

$$\overline{z}_1 = (5.9851, 5.9851)^T, \ \overline{z}_2 = (-3.3359, 3.0337)^T, \ \overline{z}_3 = (-3.7516, -2.7122)^T,$$
$$\overline{z}_4 = (7.7122, -7.7122)^T, \ \overline{z}_5 = (-3.4694, -3.4402)^T,$$
$$(\overline{\alpha}_1, \dots, \overline{\alpha}_5)^T = (5.5149, 7.6765, 7.7484, 3.7878, 7.6526)^T$$
$$(\overline{\beta}_1, \dots, \overline{\beta}_5)^T = (5.9851, 3.8235, 3.7516, 7.7122, 3.8474)^T,$$

where the optimal objective value was $v(P_{\gamma_G}, \tau) = 11.5$. Note that the optimal solution of the location problem is not unique and may differ for each chosen step size and starting point. For the dual problem the following optimal solution was computed $\overline{w}_1^* = \overline{w}_2^* = \overline{w}_5^* = (0, 0)^T$, $\overline{w}_3^* = (-0.5, 0)^T$ and $\overline{w}_4^* = (0.5, 0)^T$ with the objective function value $v(D_{\gamma_G}, \tau) = 11.5$, i.e. $v(P_{\gamma_G}, \tau) = v(D_{\gamma_G}, \tau)$. Note that, similar to Remark 4.27, one can understand the vectors $\overline{w}_i^*$, $i = 1, \dots, n$, as force vectors fulfilling the optimality conditions of Theorem 4.6 and increasing the maximum norm balls centered at the given points and the Euclidean norm balls centered at $\overline{x}$ until their intersection is non-empty. Especially, it follows from the optimality conditions (iv) and (vi) that an index i belongs to the optimal index set $\overline{I}$, if the value of the associated extended minimal time function $\mathcal{T}_{\Omega_i, \gamma_{G_i}}^{C_i}$ at $\overline{x}$ is equal to $\overline{\lambda}$, which is exactly the case when the corresponding vector $\overline{w}_i^*$ is unequal to the zero vector (in our example $\overline{I} = \{3, 4\}$, see Figure 6.6). At this point it is also important to say, that for a better visualization we multiplied the vectors, characterizing the optimal solution of the dual problem, in all figures with the value 3. One can note in Table 6.6 that the primal method needs less iterations, while the dual method generates faster a solution which is within the maximum bound from the optimal solution.

Table 6.6: Performance evaluation for 5 points in $\mathbb{R}^2$

	$\varepsilon_1 = 10^{-4}$		$\varepsilon_2 = 10^{-8}$	
	primal ($\nu = 6$)	dual ($\nu = 0.08$)	primal ($\nu = 6$)	dual ($\nu = 0.08$)
CPUtime	0.2889	0.2049	0.7196	0.6352
NumIt	749	783	1949	2446

Table 6.7: Performance evaluation for 20 points in $\mathbb{R}^2$

	$\varepsilon_1 = 10^{-4}$		$\varepsilon_2 = 10^{-8}$	
	primal ($\nu = 6$)	dual ($\nu = 0.08$)	primal ($\nu = 6$)	dual ($\nu = 0.08$)
CPUtime	6.6447	12.6774	85.0740	68.3818
NumIt	3802	13155	29334	36604

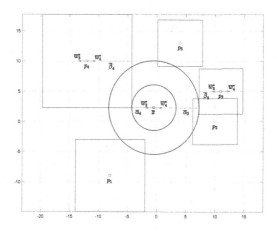

Figure 6.6: Visualization of the optimal solutions of the location problem (P_{γ_G}, τ) and its dual problem (D_{γ_G}, τ)

By using the formula from Remark 4.18, the optimal location can be determined immediately:

$$\overline{x} = \frac{1}{0.5 + 0.5} \left(0.5 \cdot (11,5)^T + 0.5 \cdot (-12,10)^T \right) = (-0.5, 7.5)^T.$$

We also considered primal and dual problems defined by 20 given points. The computational information can be seen in Table 6.7 and draws a similar picture as in the previous situation. If we increase the accuracy to $\varepsilon_2 = 10^{-8}$ the dual method is faster than the primal method, which could be especially a benefit for location problems with a large number of given points.

The second scenario of our numerical approach relies on the location problems discussed in Section 4.3.2. In this situation the location problem can be rewritten as follows

$$(P_T) \quad \min_{x \in \mathbb{R}^d} \max_{1 \le i \le n} \left\{ \mathcal{T}_{\Omega_i, \delta_{L_i}}^{C_i}(x) \right\} = \min_{\substack{t \ge 0, \, x \in \mathcal{H}, \\ \min_{y_i \in \Omega_i, \, z_i \in L_i} \{\gamma_{C_i}(x - y_i - z_i)\} \le t, \\ i=1,\dots,n}} t = \min_{\substack{t \ge 0, \, x \in \mathbb{R}^d, \, y_i \in \Omega_i, \, z_i \in L_i, \\ (x - y_i - z_i, t) \in \mathrm{epi}\,\gamma_{C_i}, \, i=1,\dots,n}} t$$

$$= \min_{\substack{t \ge 0, \, x, y_i, z_i \in \mathbb{R}^d, \\ i=1,\dots,n}} \left\{ t + \sum_{i=1}^{n} \left[\delta_{\mathrm{epi}\,\gamma_{C_i}}(x - y_i - z_i, t) + \delta_{\Omega_i}(y_i) + \delta_{L_i}(z_i) \right] \right\},$$

where C_i, $L_i \subseteq \mathbb{R}^d$ are closed and convex sets with $0_{\mathbb{R}^d} \in \mathrm{int}\, C_i$ and $\Omega_i \subseteq \mathbb{R}^d$ are convex and compact sets, $i = 1, \dots, n$, and likewise one gets for its dual problem

$$(\widetilde{D}_T) \quad \max_{\substack{y_i^* \in \mathbb{R}^d, \, i=1,\dots,n, \\ \sum_{i=1}^{n} \gamma_{C_i^0}(y_i^*) \le 1, \, \sum_{i=1}^{n} y_i^* = 0_{\mathbb{R}^d}}} \left\{ -\sum_{i=1}^{n} \left[\sigma_{L_i}(y_i^*) + \sigma_{\Omega_i}(y_i^*) \right] \right\}$$

$$= -\min_{y_i^* \in \mathbb{R}^d, \, i=1,\dots,n} \left\{ \sum_{i=1}^{n} \left[\sigma_{L_i}(y_i^*) + \sigma_{\Omega_i}(y_i^*) \right] + \delta_F(y^*) + \delta_E(y^*) \right\},$$

where $y^* = (y_1^*, \dots, y_n^*)$, $E = \{ y^* \in \mathbb{R}^d \times \dots \times \mathbb{R}^d : \sum_{i=1}^{n} y_i^* = 0_{\mathbb{R}^d} \}$ and $F = \{ y^* \in \mathbb{R}^d \times \dots \times \mathbb{R}^d : \sum_{i=1}^{n} \gamma_{C_i^0}(y_i^*) \le 1 \}$. The nonnegativity constraint of (P_T) can be omitted because it is implicitly contained in some indicator functions from the objective functions and one can then verify that the hypotheses of Theorem 6.1 are fulfilled.

Example 6.2. *Let $d = 2$, $p_1 = (-8, 8)^T$, $p_2 = (-7, 0)^T$, $p_3 = (-4, -1)^T$, $p_4 = (2, 0)^T$, $p_5 = (2, -6)^T$, $p_6 = (7, 1)^T$, $p_7 = (6, 5)^T$, $a_1 = 1$, $a_2 = 2$, $a_3 = 3$, $a_4 = 0.5$, $a_5 = 2$, $a_6 = 1$, $a_7 = 1$, $b_1 = 0.5$, $b_2 = 2$, $b_3 = 0.6$, $b_4 = 1$, $b_5 = 1.5$, $b_6 = 1$, $b_7 = 0.5$, $\Omega_i = \{x \in \mathbb{R}^2 : \|x - p_i\|_\infty \le a_i\}$, $L_i = \{x \in \mathbb{R}^2 : \|x\| \le b_i\}$ and $\gamma_{C_i} = \|\cdot\|$, $i = 1, \dots, 7$. Note that in this case $\sigma_{L_i} = \|\cdot\|$ and $\sigma_{\Omega_i} = \|\cdot\|_1$, $i = 1, \dots, 7$. Using the formulae given in Lemma 5.1, Corollary 5.1 and Corollary 5.3 to compute the proximal and projection points regarding the location problem and its dual, we tested various step sizes ν, where the starting point was always the origin and the initialization parameters were set to the value 1. The best performance results are presented in Table 6.8 and visualized in Figure 6.7 (note, that for a better visualization we multiplied the vectors, characterizing the optimal solution of the dual problem, with the value 3). The cancellation criterion for ending the iteration for both programs were the values $\varepsilon_1 = 10^{-4}$ and $\varepsilon_2 = 10^{-8}$, the maximum bounds from the optimal solution. The optimal location we obtained is $\overline{x} = (-1.0765, 3.7039)^T$ and the optimal objective value is $v(P_T) = 6.2788$. Let us remark that the optimal solution of the location problem is not unique and may differ for each chosen step size. The optimal solution of the dual problem was found at $\overline{y}_1^* = (0.4072, -0.2266)^T$, $\overline{y}_2^* = \overline{y}_3^* = \overline{y}_4^* = (0, 0)^T$, $\overline{y}_5^* = (-0.0186, 0.1330)^T$, $\overline{y}_6^* = (-0.3886, 0.0936)^T$ and $\overline{y}_7^* = (0, 0)^T$, while the objective function value was $v(\widetilde{D}_T) = 6.2788$,* *which means that $v(P_{\mathcal{T}}) = v(\widetilde{D}_{\mathcal{T}})$. In Table 6.8 one can note that the dual method needed less CPU time as well as fewer iterations to determine a solution which is within the maximum bound from the optimal solution compared to the method which solves the location problem directly. The optimal location can be reconstructed by using the formulae given in Remark 4.26.*

Table 6.8: Performance evaluation for 7 points in $\mathbb{R}^2$

	$\varepsilon_1 = 10^{-4}$		$\varepsilon_2 = 10^{-8}$	
	primal ($\nu = 39$)	dual ($\nu = 0.11$)	primal ($\nu = 39$)	dual ($\nu = 0.11$)
CPUtime	0.3786	0.1174	0.7640	0.2973
NumIt	541	330	1106	830

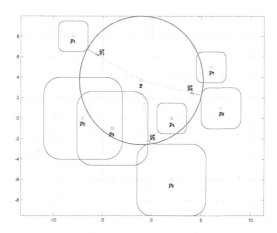

Figure 6.7: Visualization of the optimal solutions of the location problem ($P_{\mathcal{T}}$) and its dual problem ($\widetilde{D}_{\mathcal{T}}$)

Setting $L_i = \{0_{\mathbb{R}^d}\}$, $i = 1, \ldots, n$, we were able to compare our two methods with the one presented in [82, Theorem 4.1] (or [78, Theorem 4.69]) that employs the subgradient method for solving the following generalized Sylvester problem

$$(P_{\mathcal{T}}) \quad \min_{x \in \mathbb{R}^d} \max_{1 \le i \le n} \left\{ \mathcal{T}^{C_i}_{\Omega_i, \delta_{\{0_{\mathbb{R}^d}\}}}(x) \right\} = \min_{\substack{t \ge 0, \ x, v_i \in \mathbb{R}^d, \\ i=1,\ldots,n}} \left\{ t + \sum_{i=1}^{n} \left[\delta_{\text{epi} \, \gamma_{C_i}}(x - y_i, t) + \delta_{\Omega_i}(y_i) \right] \right\}.$$

The corresponding dual problem looks then as follows

$$(\widetilde{D}_T) \qquad \max_{\substack{v_i^* \in \mathbb{R}^d,\ i=1,\dots,n, \\ \sum_{i=1}^n \gamma_{C_i^0}(v_i^*)\leq 1,\ \sum_{i=1}^n v_i^*=0_{\mathbb{R}^d}}} \left\{ -\sum_{i=1}^n \sigma_{\Omega_i}(y_i^*) \right\} = -\min_{\substack{v_i^* \in \mathbb{R}^d, \\ i=1,\dots,n}} \left\{ \sum_{i=1}^n \sigma_{\Omega_i}(y^*) + \delta_F(y^*) + \delta_E(y^*) \right\}.$$

Theorem 6.2. *(cf. [78, Theorem 4.69]) Let $\mathcal{H} = \mathbb{R}^m$, fix $x_1 \in \mathbb{R}^m$ and define the sequences of iterates by*

$$x_{k+1} := x_k - \alpha_k v_k, \ k \in \mathbb{N},$$

where $\{\alpha_k\}$ are positive numbers, and where

$$v_k \in \begin{cases} \{0_{\mathbb{R}^m}\}, & \text{if } x_k \in \Omega_i + L_i, \\ [-\partial\|\cdot\|(w_k - x_k)] \cap N_{\Omega_i+L_i}(w_k), & \text{if } x_k \notin \Omega_i + L_i, \end{cases}$$

where $w_k = \mathrm{P}_{\Omega_i+L_i}(x_k)$ for some

$$i \in I = \left\{ j = 1,\dots,n : \mathcal{T}^{C_j}_{\Omega_j,\delta_{L_j}}(x) = \max_{1\leq l\leq n}\left\{ \mathcal{T}^{C_l}_{\Omega_l,\delta_{L_l}}(x) \right\} \right\}.$$

Define the value sequence

$$V_k := \min\left\{ \max_{1\leq l\leq n}\left\{ \mathcal{T}^{C_l}_{\Omega_l,\delta_{L_l}}(x) \right\} \right\}.$$

If the sequence $\{\alpha_k\}$ satisfies $\sum_{k=1}^\infty \alpha_k = \infty$ and $\sum_{k=1}^\infty \alpha_k^2 < \infty$, then $\{V_k\}$ converges to the optimal value $\overline{V}$ and $\{x_k\}$ converges to an optimal solution $\overline{x}$ to the problem (P_T).

Example 6.3. *(cf. [82, Example 4.3]) Let $d = 2$, $p_1 = (-8,8)^T$, $p_2 = (-7,0)^T$, $p_3 = (-4,-1)^T$, $p_4 = (2,0)^T$, $p_5 = (2,-6)^T$, $p_6 = (7,1)^T$, $p_7 = (6,5)^T$, $a_1 = 1$, $a_2 = 2$, $a_3 = 3$, $a_4 = 0.5$, $a_5 = 2$, $a_6 = 1$, $a_7 = 1$, $\Omega_i = \{x \in \mathbb{R}^2 : \|x-p_i\|_\infty \leq a_i\}$, $L_i = \{0_{\mathbb{R}^2}\}$, $\gamma_{C_i} = \|\cdot\|$, $i = 1,\dots,7$. We tested the subgradient method for the sequence $\alpha_{k_1} = 1/k$, which also was used in [82], as well as for the sequence $\alpha_{k_2} = 1/\sqrt{k+1}$ (see [86, Section 3.2.3]). Note that the subgradient method is simple to implement and can be employed for solving minmax location problems generated by various norms and generalized distances. The algorithms considered in this paper seem at the first look to be more complicated, also due to the necessity of determining some epigraphical projections, however, as seen below, they work faster and cheaper, and, taking for instance the results from Section 4.2, they can be employed for solving minmax location problems generated by various norms and generalized distances as well.*

For our numerical experiments we again used for all programs as starting point the origin and for the cancellation criterion for ending the iteration the values $\varepsilon_1 = 10^{-4}$ and $\varepsilon_2 = 10^{-8}$. Especially, we set again in the parallel splitting algorithms the initialization parameters to the value 1 and tested these methods for various step sizes ν. The best performance results were reached for the step sizes $\nu = 24$ and $\nu = 0.076$ for the location problem (primal) and its associated dual problem, respectively. The optimal location was found to be $\overline{x} = (-1.0556, 3.0556)^T$ and for the optimal objective value we got the value $v(P_T) = 7.1340$. The optimal solution of the dual problem was determined as $\overline{y}_1^* = (0.3755, -0.2491)^T$, $\overline{y}_2^* = \overline{y}_3^* = \overline{y}_4^* = \overline{y}_7^* = (0,0)^T$, $\overline{y}_5^* = (-0.0295, 0.1974)^T$,

$\overline{y}_6^* = (-0.3459, 0.0518)^T$ and the objective value $v(\widetilde{D}_\mathcal{T}) = 7.1340$, i.e. $v(P_\mathcal{T}) = v(\widetilde{D}_\mathcal{T})$ (see Figure 6.8). As the Tables 6.9 and 6.10 demonstrate, the dual method performs once again very well, especially for the accuracy $\varepsilon_2 = 10^{-8}$. While the subgradient method is the fastest one for the accuracy $\varepsilon_1 = 10^{-4}$ and the sequence $\alpha_{k_1} = 1/k$, it has not reached the precision $\varepsilon_2 = 10^{-8}$ after passing 500000 iterations.

In Tables 6.11 and 6.12 we present the computational results obtained while solving a location problem defined by 50 points, where the dual method performs once again very well and is faster as the primal method. For the sequence $\alpha_{k_1} = 1/k$ the subgradient method has not reached machine precision after passing 500000 iterations, whereas the sequence $\alpha_{k_2} = 1/\sqrt{k+1}$ performs surprisingly very well for $\varepsilon_2 = 10^{-8}$. Hence, for the accuracy $\varepsilon_2 = 10^{-8}$ the sequence $\alpha_{k_2} = 1/\sqrt{k+1}$ is the optimal strategy for the subgradient method, as it also was shown in [86, Section 3.2.3] under the additional assumption that the objective function is Lipschitz continuous.

Table 6.9: Performance evaluation for 7 points in $\mathbb{R}^2$ with $\varepsilon_1 = 10^{-4}$

	primal ($\nu = 24$)	dual ($\nu = 0.076$)	subgrad. (α_{k_1})	subgrad. (α_{k_2})
CPUtime	0.1904	0.0871	0.0416	1.2782
NumIt	399	181	918	70752

Table 6.10: Performance evaluation for 7 points in $\mathbb{R}^2$ with $\varepsilon_2 = 10^{-8}$

	primal ($\nu = 24$)	dual ($\nu = 0.076$)	subgrad. (α_{k_1})	subgrad. (α_{k_2})
CPUtime	0.3377	0.1608	0.7016	-
NumIt	730	453	37854	>500000

Table 6.11: Performance evaluation for 50 points in $\mathbb{R}^2$ with $\varepsilon_1 = 10^{-4}$

	primal ($\nu = 27$)	dual ($\nu = 0.025$)	subgrad. (α_{k_1})	subgrad. (α_{k_2})
CPUtime	5.6477	0.4292	-	27.1555
NumIt	2421	735	>500000	383782

Table 6.12: Performance evaluation for 50 points in $\mathbb{R}^2$ with $\varepsilon_2 = 10^{-8}$

	primal $(\nu = 27)$	dual $(\nu = 0.0046)$	subgrad. (α_{k_1})	subgrad. (α_{k_2})
CPUtime	16.1011	3.6020	-	32.2530
NumIt.	6983	7207	-	436138

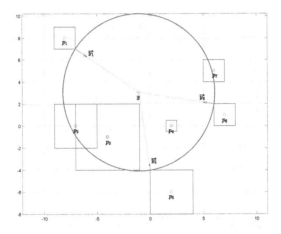

Figure 6.8: Visualization of the optimal solutions of the location problem $(P_{\mathcal{T}})$ and its dual problem $(\widetilde{D}_{\mathcal{T}})$

In the next, we present an example in the three-dimensional space, where we compared the two parallel splitting algorithms for the location problem and its dual, respectively.

Example 6.4. *Let* $d = 3$, $p_1 = (-8, 8, 8)^T$, $p_2 = (-7, 0, 0)^T$, $p_3 = (-4, -1, 1)^T$, $p_4 = (2, 0, 2)^T$, $p_5 = (2, -6, 2)^T$, $p_6 = (7, 1, 1)^T$, $p_7 = (6, 5, 4)^T$, $a_1 = \ldots = a_7 = 0.5$, $\Omega_i = \{x \in \mathbb{R}^3 : \|x - p_i\|_\infty \leq a_i\}$, $L_i = \{0_{\mathbb{R}^3}\}$, $\gamma_{C_i} = \|\cdot\|$, $i = 1, \ldots, 7$. For the numerical tests we used here the same values for the initialization parameters, starting point and stopping criterion for the iteration as in the previous example. The performance results were determined for the step sizes $\nu = 10$ and $\nu = 0.055$ for the location problem and its associated dual problem, respectively, and are presented in Table 6.13. The optimal location was identified at $\overline{x} = (-1.4350, 2.2492, 4.5693)^T$ and for the optimal objective value we got the value $v(P_{\mathcal{T}}) = 8.5408$. The optimal solution of the dual problem was determined as $\overline{y}_1^* = (0.3289, -0.2848, -0.1589)^T$, $\overline{y}_2^* = \overline{y}_3^* = \overline{y}_4^* = \overline{y}_7^* = (0, 0, 0)^T$, $\overline{y}_5^* = (-0.0997, 0.2632, 0.0703)^T$ and $\overline{y}_6^* = (-0.2292, 0.0216, 0.0887)^T$ and the objective value $v(D_{\mathcal{T}}) = 8.5408$, i.e. $v(P_{\mathcal{T}}) = v(D_{\mathcal{T}})$ (see Figure 6.9).*

Table 6.13: Performance evaluation for 7 points in $\mathbb{R}^3$

	$\varepsilon_1 = 10^{-4}$		$\varepsilon_2 = 10^{-8}$	
	primal ($\nu = 10$)	dual ($\nu = 0.055$)	primal ($\nu = 10$)	dual ($\nu = 0.055$)
CPUtime	0.1871	0.0992	0.4234	0.2042
NumIt	357	192	955	523

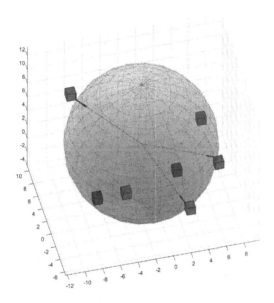

Figure 6.9: Visualization of the optimal solutions of the location problem $(P_\mathcal{T})$ and its dual problem $(\tilde{D}_\mathcal{T})$

As one can notice in Table 6.13, the dual algorithm needs roughly half of the time and the number of iterations needed by the primal one to solve the problem.

We close this section by a comparison of the fastest method of the above considered solving strategies, i.e. the dual method, and the one introduced in [2] regarding speed and especially precision in high dimensions.

Example 6.5. *We consider the location problem* (P_T) *as well as its associated dual one* $(\widetilde{D}_T)$ *in the setting where* $\Omega_i = \{x \in \mathbb{R}^d : \|x - p_i\|_\infty \leq a_i\}$, $L_i = \{0_{\mathbb{R}^d}\}$ *and* $\gamma_{C_i} = \|\cdot\|$, $i = 1, \dots, n$, *and compare the dual method with the numerical algorithm build on the log-exponential smoothing technique and Nesterov's accelerated gradient method (log-exp), which was developed in [2] for solving generalized Sylvester problems of the kind of* (P_T).

We implemented this algorithm in MATLAB *and used for our numerical tests the same settings as in [2, Remark 5.1] and [2, Example 6.4] (i.e.* $\varepsilon = 10^{-6}$, $\widetilde{\varepsilon} = 10^{-5}$, $p_0 = 5$ *and* $\gamma_0 = 0.5$). *We considered four situations where the test results are printed in the Tables 6.14 - 6.17. Notice, that in all situations the starting point was the origin and the points* $p_1, \dots, p_n$ *were generated by the command* RANDN, *while the corresponding radii* $a_1, \dots, a_n$ *were given by* RAND. *As mentioned, we are interested in an analysis regarding the precision of these two methods in such a way that the calculated objective values are exact up to six decimal places, which is especially important in the situation when these calculations are a part of a larger problem where the aim is to reduce rounding errors.*

So, to guarantee this desired precision in all four situations we tested the log-exponential smoothing algorithm for various numbers of iterations N. *For the scenario in Table 6.14 we received a solution such that the objective value was exact up to six decimal places for* $N = 35000$ *and saved the calculated solution as the optimal solution* $\overline{x}$ *to* (P_T), *where the corresponding objective value was* $v(P_T) = 3.099896$. *Then we ran the log-exponential algorithm a second time and noticed the number of iterations, the time needed to generate a solution which is within the maximum bound of* 10^{-6} *from the optimal solution* $\overline{x}$ *and the associated objective value. Note that, if we reduce the number of iterations in the log-exponential smoothing algorithm then the speed to generate a solution such that the objective value is close to the optimal objective value increases, but the accuracy decreases, i.e. then the algorithm fails to calculate a solution such that the objective value is exact up to six decimal places. For the scenarios in the Tables 6.15 - 6.17 we proceed in the same manner, where the corresponding values for* $N, v(P_T)$ *and CPU time are also presented.*

For the dual method we set in all situations N *to* 100000 *and saved the determined solutions for the second run, where then the number of iterations and the CPU time needed to get a solution which is within the maximum bound of* 10^{-6} *were noticed. The corresponding objective values of the dual problem were also recorded.*

As you may see in the Tables 6.14 - 6.17 the dual method performs again very well regarding speed and also precision in all four situations, which makes it a good candidate not only for problems where precision is of great importance but also for problems in high dimensions, that appear for instance in machine learning. Take also note, that if one has an optimal solution to the dual problem, then the optimal solution to the primal one can be reconstructed by using the formulae given in Remark 4.26.

Table 6.14: Performance evaluation for 10 points in $\mathbb{R}^{10}$

	dual ($\nu = 0.1$)	log-exp ($N = 35000, v(P_\mathcal{T}) = 3.099896$)
CPUtime	0.2889	55.4856
NumIt	1167	32265
ObjFunVal	3.099896	3.099896

Table 6.15: Performance evaluation for 50 points in $\mathbb{R}^{50}$

	dual ($\nu = 0.07$)	log-exp ($N = 45000, v(P_\mathcal{T}) = 6.066194$)
CPUtime	7.6268	70.3653
NumIt	1956	44173
ObjFunVal	6.066194	6.066195

Table 6.16: Performance evaluation for 100 points in $\mathbb{R}^{100}$

	dual ($\nu = 0.07$)	log-exp ($N = 70000, v(P_\mathcal{T}) = 9.700309$)
CPUtime	104.5634	145.2422
NumIt	3003	69163
ObjFunVal	9.700309	9.700310

Table 6.17: Performance evaluation for 100 points in $\mathbb{R}^{1000}$

	dual ($\nu = 0.01$)	log-exp ($N = 700000, v(P_\mathcal{T}) = 29.528790$)
CPUtime	5328.3671	7026.1593
NumIt	4017	691412
ObjFunVal	29.528790	29.528791

Remark 6.1. *The examples investigated in this section reveal that the origin seems to be a good starting point for running the proposed splitting proximal point method on the dual problem of a given nonlinear minmax location problem. This is actually not very surprising when one analyzes the constraints of the dual problems that do not allow (all) the components of the feasible dual solutions to wander far away from the origin.*

The next two remarks gives some hints how the classical multifacility minmax location problem (P^M) and its associated dual $(\widetilde{D}^M)$, both discussed in Section 4.5, can be solved numerically.

Remark 6.2. *By using the techniques described above one can also reformulate the classical multifacility minmax location problem (P^M) considered in Section 4.5 into an unconstrained optimization problem where the objective function is a sum of proper, lower semicontinuous functions as follows (note that $S \subseteq \mathcal{H}^m$)*

$$(P^M) \quad \inf_{(x_1,\ldots,x_m)\in S} \max\left\{ \left(\gamma_{C_{jk}}(x_j - x_k)\right)_{jk\in V}, \ \left(\gamma_{\widetilde{C}_{ji}}(x_j - p_i)\right)_{ji\in\widetilde{V}} \right\}$$

$$= \inf_{\substack{t\in\mathbb{R},\ x\in S,\ \gamma_{C_{jk}}(A_{jk}x)\leq t,\ jk\in V, \\ \gamma_{\widetilde{C}_{ji}}(B_{ji}x - p_i)\leq t,\ ji\in\widetilde{V}}} t = \inf_{\substack{t\in\mathbb{R},\ x\in S,\ (x,t)\in\mathrm{epi}\left(\gamma_{C_{jk}}(A_{jk}\cdot)\right),\ jk\in V, \\ (x,t)\in\mathrm{epi}\left(\gamma_{\widetilde{C}_{ji}}(B_{ji}\cdot - p_i)\right),\ ji\in\widetilde{V}}} t$$

$$= \inf_{(x,t)\in\mathcal{H}^m\times\mathbb{R}} \left\{ t + \delta_S(x) + \sum_{jk\in V} \delta_{\mathrm{epi}\left(\gamma_{C_{jk}}(A_{jk}\cdot)\right)}(x,t) + \sum_{ji\in\widetilde{V}} \delta_{\mathrm{epi}\left(\gamma_{\widetilde{C}_{ji}}(B_{ji}\cdot - p_i)\right)}(x,t) \right\}.$$

Now we can use the parallel splitting algorithm from Proposition 6.1 to solve (P^M) numerically. As the proximal operator of an indicator function collapses into a projection operator, one can apply here the formulae given in Theorem 5.2 and Remark 5.7 for the projection onto the epigraph of the gauge function composed with a linear operator.

While Remark 6.2 presented a method to solve the location problem (P^M) numerically, the next remark discusses a solving technique for the corresponding dual one $(\widetilde{D}^M)$.

Remark 6.3. *Set $w_{jk} \geq 0$, $\gamma_{C_{jk}}(\cdot) = w_{jk}\|\cdot\|_{\mathcal{H}}$, $jk \in J$, and $\widetilde{w}_{ji} \geq 0$, $w_{ji}\gamma_{\widetilde{C}_{ji}}(\cdot) = \|\cdot\|_{\mathcal{H}}$, $ji \in \widetilde{J}$. Additionally, we define $V = \{jk \in J : 1 \leq j < k \leq m,\ w_{jk} > 0\}$ (note that $|V| \leq (m/2)(m-1)$) and $\widetilde{V} = \{ji \in J : w_{ji} > 0\}$, then $\gamma_{C^0_{jk}}(\cdot) = (1/w_{jk})\|\cdot\|_{\mathcal{H}}$, $jk \in V$, $\gamma_{C^0_{ji}}(\cdot) = (1/\widetilde{w}_{ji})\|\cdot\|_{\mathcal{H}}$, $ji \in \widetilde{V}$, and the dual problem $(\widetilde{D}^M)$ transforms to (according to Remark 4.43, we omit the index sets I and $\widetilde{I}$ in the following formulation)*

$$(\widetilde{D}^M) \quad \max_{\substack{(z^*,\widetilde{z}^*)\in\mathcal{H}^{|V|}\times\mathcal{H}^{|\widetilde{V}|},\ \sum_{jk\in V}\frac{1}{w_{jk}}\|z^*_{jk}\|_{\mathcal{H}}+\sum_{ji\in\widetilde{V}}\frac{1}{\widetilde{w}_{ji}}\|\widetilde{z}^*_{ji}\|_{\mathcal{H}}\leq 1, \\ \sum_{jk\in V}A^*_{jk}z^*_{jk}+\sum_{ji\in\widetilde{V}}B^*_{ji}\widetilde{z}^*_{ji}=0_{\mathcal{H}\times\ldots\times\mathcal{H}}}} \left\{ -\sum_{ji\in\widetilde{V}} \langle\widetilde{z}^*_{ji}, p_i\rangle \right\}.$$

Further, by rewriting $(\widetilde{D}^M)$ into an unconstrained optimization problem of the form

$$(\widetilde{D}^M) \quad - \min_{(z^*,\widetilde{z}^*)\in\mathcal{H}^{|V|}\times\mathcal{H}^{|\widetilde{V}|}} \left\{ \sum_{ji\in\widetilde{V}} \langle\widetilde{z}^*_{ji}, p_i\rangle + \delta_{D_1}(z^*,\widetilde{z}^*) + \delta_{D_2}(z^*,\widetilde{z}^*) \right\},$$

$$(6.5)$$

where

$$D_1 = \left\{ (z^*,\widetilde{z}^*) \in \mathcal{H}^{|V|} \times \mathcal{H}^{|\widetilde{V}|} : \sum_{jk\in V} A^*_{jk}z^*_{jk} + \sum_{ji\in\widetilde{V}} B^*_{ji}\widetilde{z}^*_{ji} = 0_{\mathcal{H}\times\ldots\times\mathcal{H}} \right\}$$

and

$$D_2 = \left\{ (z^*, \widetilde{z}^*) \in \mathcal{H}^{|V|} \times \mathcal{H}^{|\widetilde{V}|} : \sum_{jk \in V} \frac{1}{w_{jk}} \|z_{jk}^*\|_{\mathcal{H}} + \sum_{ji \in \widetilde{V}} \frac{1}{\widetilde{w}_{ji}} \|\widetilde{z}_{ji}^*\|_{\mathcal{H}} \leq 1 \right\},$$

the minimization problem in (6.5) can also be solved numerically by the parallel splitting algorithm from Proposition 6.1. Here, one can apply the formulae given in Lemma 5.1 for the projection onto a unit ball generated by the weighted sum of norms, i.e. D_2. To get the projection operator onto D_1 one can use the formulas presented in [3, Example 28.14], while the proximal operators of the linear functions involved in the objective function in (6.5) can simply be determined by [3, Example 28.16] and the well-known Moreau's decomposition formula given in [3, Theorem 14.3(ii)].
The procedure for reconstruction of the optimal solution to (P^M) by using an optimal solution to $(\widetilde{D}^M)$ is described in Example 4.5.

Bibliography

[1] L. Altangerel, G. Wanka, O. Wilfer: *An Oriented Distance Function Application to Gap Functions for Vector Variational Inequalities*. Optimization, Simulation, and Control 76, 17-34, 2013.

[2] N.T. An, D. Giles, N.M. Nam, R.B. Rector: *The Log-Exponential Smoothing Technique and Nesterov's Accelerated Gradient Method for Generalized Sylvester Problems*. Journal of Optimization Theory and Applications 168, 559-583, 2016.

[3] H.H. Bauschke, P.L. Combettes: *Convex Analysis and Monotone Operator Theory in Hilbert Spaces*. CMS Books in Mathematics, Springer, New York, 2011.

[4] H.H. Bauschke, M.N. Buiy, X. Wang: *On the Sum of Projectors onto Convex Sets*. arXiv 1802.02287, 2018.

[5] C.R. Bector, S. Chandra, M.K. Bector: *Generalized Fractional Programming Duality: A Parametric Approach*. Journal of Optimization Theory and Applications 60, 243-260, 1989.

[6] A. Ben-Tal, A. Ben-Israel, M. Teboulle: *Certainty Equivalents and Information Measures: Duality and Extremal Principles*. Journal of Mathematical Analysis and Applications 157, 211-236, 1991.

[7] A. Berger, A. Grigoriev, A. Panin, A. Winokurow: *Location, Pricing and the Problem of Apollonius*. Optimization Letters 11, 1797-1805, 2017.

[8] J.M. Borwein, A.S. Lewis: *Partially Finite Convex Programming, Part I: Quasi Relative Interiors and Duality Theory*. Mathematical Programming 57, 15-48, 1992.

[9] R. I. Boţ, E. R. Csetnek: *On the Convergence Rate of a Forward-Backward Type Primal-Dual Splitting Algorithm for Convex Optimization Problems*. Optimization 64, 5-23, 2015.

[10] R. I. Boţ, E. R. Csetnek, A. Heinrich: *A Primal-Dual Splitting Algorithm for Finding Zeros of Sums of Maximal Monotone Operators*. SIAM Journal on Optimization 23, 2011-2036, 2013.

[11] R. I. Boţ, E. R. Csetnek, C. Hendrich, A. Heinrich: *On the Convergence Rate Improvement of a Primal-Dual Splitting Algorithm for Solving Monotone Inclusion Problems*. Mathematical Programming 150, 251-279, 2015.

[12] R. I. Boţ, E. R. Csetnek, C. Hendrich: *Recent Developments on Primal-Dual Splitting Methods with Applications to Convex Minimization*. in: P.M. Pardalos, T.M. Rassias (Eds.), "Mathematics Without Boundaries: Surveys in Interdisciplinary Research", Springer-Verlag, New York, 57-99, 2014.

© Springer Fachmedien Wiesbaden GmbH, part of Springer Nature 2020
O. Wilfer, *Multi-Composed Programming with Applications to Facility Location*,
Mathematische Optimierung und Wirtschaftsmathematik | Mathematical
Optimization and Economathematics, https://doi.org/10.1007/978-3-658-30580-2

[13] R.I. Boţ, S.-M. Grad, G. Wanka, *Duality for Optimization Problems with Entropy-Like Objective Functions*. Journal of Information and Optimization Sciences 26, 415-441, 2005.

[14] R. I. Boţ, S.-M. Grad, G. Wanka: *Duality in Vector Optimization*. Springer-Verlag, Berlin Heidelberg, 2009.

[15] R. I. Boţ, S.-M. Grad, G. Wanka: *New Constraint Qualification and Conjugate Duality for Composed Convex Optimization Problems*. Journal of Optimization Theory and Applications 135, 241-255, 2007.

[16] R. I. Boţ, A. Heinrich: *Regression Tasks in Machine Learning via Fenchel Duality*. Annals of Operations Research 222, 197-211, 2014.

[17] R. I. Boţ, C. Hendrich: *A Douglas-Rachford Type Primal-Dual Method for Solving Inclusions with Mixtures of Composite and Parallel-Sum Type Monotone Operators*. SIAM Journal on Optimization 23, 2541-2565, 2013.

[18] R. I. Boţ, C. Hendrich: *A Double Smoothing Technique for Solving Unconstrained Nondifferentiable Convex Optimization Problems*. Computational Optimization and Applications 54, 239-262, 2013.

[19] R. I. Boţ, C. Hendrich: *Convergence Analysis for a Primal-Dual Monotone + Skew Splitting Algorithm with Applications to Total Variation Minimization*. Journal of Mathematical Imaging and Vision 49, 551-568, 2014.

[20] R. I. Boţ, I. B. Hodrea, G. Wanka: *ε-Optimality Conditions for Composed Convex Optimization Problems*. Journal of Approximation Theory 153, 108-121, 2008.

[21] R. I. Boţ, N. Lorenz: *Optimization Problems in Statistical Learning: Duality and Optimality Conditions*. European Journal of Operational Research 213, 395-404, 2011.

[22] R. I. Boţ, N. Lorenz, G. Wanka: *Optimality Conditions for Portfolio Optimization Problems with Convex Deviation Measures as Objective Functions*. Taiwanese Journal of Mathematics 13, 515-533, 2009.

[23] R.I. Boţ, E. Vargyas, G. Wanka: *Duality for Location Problems with Unbounded Unit Balls*. European Journal of Operational Research 179, 1252-1265, 2007.

[24] R.I. Boţ, E. Vargyas, G. Wanka: *On the Relations Between Different Duals Assigned to Composed Optimization Problems*. Mathematical Methods of Operations Research (ZOR) 66, 47-68, 2007.

[25] R. I. Boţ, G. Wanka: *An Alternative Formulation for a New Closed Cone Constraint Qualification*. Nonlinear Analysis: Theory, Methods and Applications 64, 1367-1381, 2006.

[26] R. I. Boţ, G. Wanka: *Duality for Composed Convex Functions with Applications in Location Theory*. Multi-Criteria- und Fuzzy-Systeme in Theorie und Praxis, Deutscher Universitäts-Verlag, Wiesbaden, 2003.

[27] R.I. Boţ, G. Wanka: *The Conjugate of the Pointwise Maximum of Two Convex Functions Revisited*. Journal of Global Optimization 41, 625-632, 2008.

[28] M. Bounkhel: *Subdifferential Properties of Minimal Time Functions Associated with Set-Valued Mappings with Closed Convex Graphs in Hausdorff Topological Vector Spaces.* Journal of Function Spaces and Applications, Article ID 707603, 13 pages, 2013.

[29] S. Boyd, L. Vandenberghe: *Convex Optimization.* Cambridge University Press, 2004.

[30] Y. Censor, A.R. de Pierro, A.N. Iusem: *Optimization of Burg's Entropy over Linear Constraints.*, Applied Numerical Mathematics 7, 151-165, 1991.

[31] E.C. Chi, K. Lange: *Splitting Methods for Convex Clustering.* Journal of Computational and Graphical Statistics 24, 994-1013, 2015.

[32] M. Chiang: *Geometric Programming for Communication Systems.* Foundations and Trends in Communications and Information Theory 2, 1-154, 2005.

[33] G. Chierchia, N. Pustelnik, J.-C. Pesquet, B. Pesquet-Popescu: *Epigraphical Projection and Proximal Tools for Solving Constrained Convex Optimization Problems: Part I.* CoRR abs/1210.5844, 2012.

[34] P. L. Combettes, J.-C. Pesquet: *A Proximal Decomposition Method for Solving Convex Variational Inverse Problems.* Inverse Problems 24, 2008.

[35] O. Cornejo, C. Michelot: *A Proximal Solution for a Class of Extended Minimax Location Problem.* ICCSA, 712-721, 2005.

[36] E. R. Csetnek: *Overcoming the Failure of the Classical Generalized Interior-Point Regularity Conditions in Convex Optimization. Applications of the Duality Theory to Enlargements of Maximal Monotone Operators.* Logos Verlag Berlin, 2010.

[37] I. Csiszár: *On Generalized Entropy.* Studia Scientiarum Mathematicarum Hungarica 4, 404-419, 1969.

[38] P. M. Dearing: *Minimax Location Problems with Nonlinear Costs.* Journal of Research of the National Bureau of Standards 82, 65-72, 1977.

[39] P. M. Dearing: *On Some Minimax Location Problems Using Rectilinear Distances.* Ph.D. Dissertation, University of Florida, Gainesville, Florida, 1972.

[40] Z. Drezner: *The Weighted Minimax Location Problem with Set-up Costs and Extensions.* RAIRO-Operations Research. 25, 55-64, 1991.

[41] R. Durier, C. Michelot: *Geometrical Properties of the Fermat-Weber Problem.* European Journal of Operational Research 20, 332-343, 1985.

[42] I. Ekeland, R. Temam, *Convex Analysis and Variational Problems.* North-Holland Publishing Company, Amsterdam, 1976.

[43] J. Elzinga, D. W. Hearn: *Geometrical Solutions for Some Minimax Location Problems.* Transportation Science 6, 379-394, 1972.

[44] J. Elzinga, D. Hearn, W. Randolph: *Minimax Multifacility Location with Euclidean Distances.* Transportation Science 10, 321-336, 1976.

[45] J. Elzinga, D. W. Hearn: *The Minimum Covering Sphere Problem*. Management science 19, 96-104, 1972.

[46] D. H. Fang, C. Li, G. López, M.A. López: Stable and Total Fenchel Duality for Convex Optimization Problems in Locally Convex Spaces. SIAM Journal on Optimization 20, 1032-1051, 2009.

[47] D. H. Fang, C. Li, K.F. Ng: *Constraint Qualifications for Optimality Conditions and Total Lagrange Dualities in Convex Infinite Programming*. Nonlinear Analysis: Theory, Methods and Applications 73, 1143-1159, 2010.

[48] J. Fliege: *Coincidence Conditions in Multifacility Location Problems with Positive and Negative Weights*. European Journal of Operational Research 104, 310-320, 1998.

[49] R. L. Francis: *A Note on a Nonlinear Minimax Location Problem in Tree Networks*. Journal of Research of the National Bureau of Standards 82, 73-80, 1977.

[50] R. L. Francis: *Some Aspects of a Minimax Location Problem*. Operations Research 15, 1163-1169, 1967.

[51] C.R. Glassey: *Explicit Duality for Convex Homogeneous Programs*. Mathematical Programming 10, 176-191, 1976.

[52] A. Göpfert, H. Riahi, C. Tammer, C. Zălinescu: *Variational Methods in Partially Ordered Spaces*. Springer-Verlag, 2003.

[53] S.-M. Grad: *On Gauge Functions for Convex Cones with Possibly Empty Interiors*. Journal of Convex Analysis 24, 519-524, 2017.

[54] S.-M. Grad: *Vector Optimization and Monotone Operators via Convex Duality*. Springer-Verlag, Cham, 2015.

[55] S.-M. Grad, G. Wanka, O. Wilfer: *Duality and ε-Optimality Conditions for Multi-composed Optimization Problems with Applications to Fractional and Entropy Optimization*. Pure and Applied Functional Analysis 2, 43-63, 2017.

[56] S.-M. Grad, O. Wilfer: *A Proximal Method for Solving Nonlinear Minmax Location Problems with Perturbed Minimal Time Functions via Conjugate Duality*. Journal of Global Optimization 74, 121-160, 2019.

[57] Y. Hinojosa, J. Puerto: *Single Facility Location Problems with Unbounded Unit Balls*. Mathematical Methods of Operations Research 58, 87-104, 2003.

[58] J.-B. Hiriart-Urruty: *Conjugating the Inverse of a Concave Function*. Journal of Convex Analysis 12, 71-80, 2005.

[59] J.-B. Hiriart-Urruty, C. Lemaréchal: *Convex Analysis and Minimization Algorithms I*. Springer-Verlag, Berlin, 1993.

[60] J.-B. Hiriart-Urruty, C. Lemaréchal: *Convex Analysis and Minimization Algorithms II*. Springer-Verlag, Berlin, 1993.

[61] J.-B. Hiriart-Urruty, C. Lemaréchal: *Fundamentals of Convex Analysis.* Springer, Heidelberg, 2001.

[62] J.-B. Hiriart-Urruty, J.-E. Martínez-Legaz: *New Formulas for the Legendre-Fenchel Transform.* Journal of Mathematical Analysis and Applications 288, 544-555, 2003.

[63] I. B. Hodrea: *Farkas-Type Results for Convex and Non-Convex Inequality Systems.* Dissertation, 2007.

[64] H. Idrissi, O. Lefebvre, C. Michelot: *Duality Constrained Multifacility Location Problems with Mixed Norms and Applications.* Annals of Operations Research 18, 71-92, 1989.

[65] H. Jarchow: *Locally Convex Spaces.* B.G. Teubner, Stuttgart, 1981.

[66] T. R. Jefferson, S. Jorjani, C. H. Scott: *Quality Locations for the Constrained Minimax Location Model.* International Journal of Systems Science 24, 1009-1016, 1993.

[67] T. R. Jefferson, C. H. Scott: *Duality for Minmax Programs.* Journal of Mathematical Analysis and Applications 100, 385-392, 1984.

[68] H. Juel, R. F. Love: *Duality in Constrained Location Problems.* Operations Research Letters 6, 281-284, 1987.

[69] H. Juel, R. F. Love: *On the Dual of the Linearly Constrained Multi-Facility Location Problem with Arbitrary Norms.* Transportation Science 15, 329-337, 1981.

[70] H. Juel, R. F. Love: *The Dual of a Generalized Minimax Location Problem.* Annals of Operations Research 40, 261-264, 1992.

[71] H. W. Kuhn: *A Note on Fermat's Problem.* Mathematical Programming 4, 98-107, 1973.

[72] A.G. Kusraev, S.S. Kutateladze: *Subdifferentials: Theory and Applications.* Mathematics and Its Applications 323, Kluwer, Dordrecht, 1995.

[73] P. J. Laurent: *Approximation et Optimisation.* Hermann, Paris, 1972.

[74] D. T. Luc: *Theory of Vector Optimization.* Springer-Verlag, Berlin Heidelberg, 1989.

[75] T. Matutomi, H. Ishii: *Minimax Location Problem with A-Distance.* Journal of the Operations Research Society of Japan 41, 181-185, 1998.

[76] L. Meng, L. Peng, C.-F. Wen: *Fréchet and Proximal Subdifferentials of Perturbed Minimal Time Functions.* Journal of Nonlinear and Convex Analysis 17, 1729-1737, 2016.

[77] C. Michelot, F. Plastria: *An Extended Multifacility Minimax Location Problem Revisited.* Annals of Operations Research 111, 167-179, 2002.

[78] B.S. Mordukhovich, N.M. Nam: *An Easy Path to Convex Analysis and Applications.* Morgan and Claypool Publishers, Williston, 2014.

[79] B.S. Mordukhovich, N.M. Nam: *Limiting Subgradients of Minimal Time Functions in Banach spaces.* Journal of Global Optimization 46, 615-633, 2010.

[80] B.S. Mordukhovich, N.M. Nam: *Subgradients of Minimal Time Functions under Minimal Requirements*. Journal of Convex Analysis 18, 915-947, 2011.

[81] B. A. Murtagh, C. H. Scott, E. Sirri: *Solution of Constrained Minmax Location with Euclidean Distances via Conjugate Duality*. New Zealand Operational Research 13, 61-67, 1985.

[82] N.M. Nam, N.T. An, J. Salinas: *Applications of Convex Analysis to the Smallest Intersecting Ball Problem*. Journal of Convex Analysis 19, 497-518, 2012.

[83] N.M. Nam, N. Hoang, N.T. An: *Constructions of Solutions to Generalized Sylvester and Fermat-Torricelli Problems for Euclidean Balls*. Journal of Optimization Theory and Applications 160, 483-509, 2014

[84] N.M. Nam: *Subdifferential Formulas for a Class of Non-Convex Infimal Convolutions*. Optimization 64, 2213-2222, 2015.

[85] N.M. Nam, C. Zălinescu: *Variational Analysis of Directional Minimal Time Functions and Applications to Location Problems*. Set-Valued and Variational Analysis 21, 405-430, 2013.

[86] Y. Nesterov: *Introductory Lectures on Convex Optimization: A Basic Course* Applied optimization 87, Kluwer, Dordrecht, 2004´.

[87] F. Nielsen, R. Nock: *Approximating Smallest Enclosing Balls with Applications to Machine Learning*, International Journal of Computational Geometry & Applications 19:389-414, 2009.

[88] N. Parikh and S. Boyd: *Proximal Algorithms*. Foundations and Trends in Optimization, 1, 127-239, 2014.

[89] R.T. Rockafellar, *Convex Analysis*. Princeton University Press, Princeton, 1970.

[90] R.T. Rockafellar: *Level Sets and Continuity of Conjugate Convex Functions*. Transactions of the American Mathematical Society 123, 46-63, 1966.

[91] S. M. Robinson: *A Short Derivation of the Conjugate of a Supremum Function*. Journal of Convex Analysis 19, 569-574, 2012.

[92] C. H. Scott, T. R. Jefferson: *Conjugate Duals for Power Functions*. International Journal of Systems Science 20, 2153-2160, 1989.

[93] C. H. Scott, T. R. Jefferson: *Duality for a Sum of Convex Ratios*. Optimization 40, 303-312, 1997.

[94] C. H. Scott, T. R. Jefferson: *Duality for Minmax Programs*. Journal of Mathematical Analysis and Applications 100, 385-392, 1984.

[95] C. H. Scott, T. R. Jefferson, S. Jorani: *Conjugate Duality in Facility Location*. Facility Location: A Survey of Applications and Methods Edited by Z. Drezner, Springer-Verlag, Heidelberg, 89-101, 1995.

[96] J. Songsiri: *Projection onto an l1-Norm Ball with Application to Identification of Sparse Autoregressive Models.* ASAC, 2011.

[97] J. E. Spingarn: *Partial Inverse of a Monotone Operator.* Applied Mathematics and Optimization 10, 247-265, 1983.

[98] J.H. Wang, C. Li, H.K. Xu: *Subdifferentials of Perturbed Distance Function in Banach Spaces.* Journal of Global Optimization 46, 489-501, 2010.

[99] G. Wanka, U. Krallert: *Duality for Optimal Control-Approximation Problems with Gauges.* Journal for Analysis and its Applications 18, 491-504, 1999.

[100] G. Wanka, O. Wilfer: *A Lagrange Duality Approach for Multi-Composed Optimization Problems.* TOP 25, 288-313, 2017.

[101] G. Wanka, O. Wilfer: *Duality Results for Extended Multifacility Location Problems.* Optimization 67, 1095-1119, 2018.

[102] G. Wanka, O. Wilfer: *Duality Results for Nonlinear Single Minimax Location Problems via Multi-Composed Optimization.* Mathematical Methods of Operations Research 86, 401-439, 2017.

[103] G. Wanka, O. Wilfer: *Formulae of Epigraphical Projection for Solving Minimax Location Problems.* to appear in Pacific Journal of Optimization, 2019.

[104] G. Wanka, O. Wilfer: *Multifacility Minimax Location Problems via Multi-Composed Optimization.* Minimax Theory and its Applications 4(2), 355-386, 2019.

[105] O. Wilfer: *Duality Investigations for Multi-Composed Optimization Problems with Applications in Location Theory.* Dissertation, TU Chemnitz, 2017.

[106] C. Zălinescu: *Convex Analysis in General Vector Spaces.* World Scientific, Singapore, 2002.

[107] E. Zeidler: *Applied Functional Analysis.* Springer Verlag, Heidelberg, 1995.

[108] Y. Zhang, Y. He, Y. Jiang: *Subdifferentials of a Perturbed Minimal Time Function in Normed Spaces.* Optimization Letters 8, 1921-1930, 2014.

Index

© Springer Fachmedien Wiesbaden GmbH, part of Springer Nature 2020
O. Wilfer, *Multi-Composed Programming with Applications to Facility Location*,
Mathematische Optimierung und Wirtschaftsmathematik | Mathematical
Optimization and Economathematics, https://doi.org/10.1007/978-3-658-30580-2

Printed in the United States
By Bookmasters